Springer Textbooks in Earth Sciences, Geography and Environment

The Springer Textbooks series publishes a broad portfolio of textbooks on Earth Sciences, Geography and Environmental Science. Springer textbooks provide comprehensive introductions as well as in-depth knowledge for advanced studies. A clear, reader-friendly layout and features such as end-of-chapter summaries, work examples, exercises, and glossaries help the reader to access the subject. Springer textbooks are essential for students, researchers and applied scientists.

More information about this series at https://link.springer.com/bookseries/15201

Joaquim Sanz • Oriol Tomasa •
Abigail Jimenez-Franco •
Nor Sidki-Rius

Elements and Mineral Resources

 Springer

Joaquim Sanz
Department of Mining, Industrial and ICT
Universitat Politecnica de Catalunya (UPC)
Manresa, Spain

Oriol Tomasa
Department of Mining, Industrial and ICT
Universitat Politecnica de Catalunya (UPC)
Manresa, Spain

Abigail Jimenez-Franco
Universitat de Barcelona (UB)
Barcelona, Spain

Nor Sidki-Rius
Department of Mining, Industrial and ICT
Universitat Politecnica de Catalunya (UPC)
Manresa, Spain

ISSN 2510-1307   ISSN 2510-1315   (electronic)
Springer Textbooks in Earth Sciences, Geography and Environment
ISBN 978-3-030-85891-9   ISBN 978-3-030-85889-6   (eBook)
https://doi.org/10.1007/978-3-030-85889-6

Foreword

You are holding in your hands a special textbook on minerals. This is because books on mineralogy and geology tell us about minerals and rocks' properties, whereas ones on chemistry reveal the properties of elements and compounds. Consequently, most textbooks include neither reliable and understandable information on elements' origins nor up-to-date facts on their final application and recycling. My experience in the world of material science, particularly in the field of research into new or advanced materials, is that a researcher does not even consider whether it is possible to implement a new substance on a global scale on the basis of the requirements of its components whose raw materials are of questionable availability.

Besides, in general terms, people do not perceive material science issues as a whole picture. For example, who has not heard that tramways cause no pollution? Or that the electric car will virtually eliminate this problem? But have you ever stopped to consider that just a few centimeters of copper on a train's overhead line require blasting more than a ton of rock? To be specific, it involves crushing and grinding this ton of rock to the texture of flour, melting it at 1250 °C, converting the copper, and, finally, refining this metal electrolytically. We should also mention the waste rock and the sulfur gases emitted during the smelting process.

Many applications of elements or minerals can be surprising, and the speed of change in the applications of many materials has been dizzying in recent years, so much so that the phenomenon already has a name: the 'competition of materials'—which has the greatest application.

I should like to congratulate the authors for both the idea and the job done. I can only recommend the book to each of you interested in mineralogy, geology, chemistry, material sciences, or environmental sciences. Rigorous and straightforward, it will be useful to anyone interested in the origin of things, traveling routes, current applications, and recycling possibilities. It is simple and, at the same time, so very necessary.

Joan Viñals i Olià (Deceased)
Former Professor of Materials Science
and Metallurgical Engineering
University of Barcelona
Barcelona, Spain

The original version of the book was revised: The correction to the book is available at https://doi.org/10.1007/978-3-030-85889-6_102.

Acknowledgments

First of all, our gratitude is to our Editor **Alexis Vizcaino**; without him, this book would not have been possible.

We also sincerely thank the dedicated and patient collaboration of the Mining Engineer **Andrea Artacho Pradas** for her generous support in preparing the infographic images.

We also thank the people, companies, and institutions that have collaborated directly with the work and provided us either materials or photographs:

Ares Boyer Margalef. *Manresa (Catalonia)*
Jordi Casado Garriga. *Manresa (Catalonia)*
Daniel Calvo Torralba. *Manresa (Catalonia)*
Joan Carles Lacruz. *Barcelona (Catalonia)*
Carlos Domínguez. *Granada (Spain)*
Roger Gaona Boixader. *Sant Fruitós de Bages (Catalonia)*
Jose Lazuen (Roskill). *London (UK)*
Raul Osorio. *Manresa (Catalonia)*
Sílvia Palacios Ubach. *Lleida (Catalonia)*
Albert Prat. *Manresa (Catalonia)*
Rubèn Sanz Arcas. *Callús (Catalonia)*
Oleguer Serra. *Monistrol de Montserrat (Catalonia)*
Blancafort OM. *Collbató (Catalonia)*
Candela Medical España. Núria Vila. *Barcelona (Catalonia)*
Helion Tools. *Manresa (Catalonia)*
ICL. Rosa Vilajosana. *Súria (Catalonia)*
INTAN, SL. *Barcelona (Catalonia)*
Magneti Marelli, SA. *Santpedor (Catalonia)*
Òptica Soler. *Manresa (Catalonia)*
Regió 7. Mireia Arso. *Manresa (Catalonia)*
Regió 7. Jordi Morros. *Manresa (Catalonia)*
Regió 7. Salvador Redó. *Manresa (Catalonia)*
Stora Enso. *Castellbisbal (Catalonia)*
Tallers Ballús, Sl (Toyota). *Manresa (Catalonia)*
VHF-Technologies, SA. *(Switzerland)*
Vins Tomasa. *Manresa (Catalonia)*
Bombers de la Generalitat. *Manresa (Catalonia)*
Althaia—Hospital Sant Joan de Déu. Dr. Genís, Dr. Malet. *Manresa (Catalonia)*
Centre de Diagnosi per Imatge del Bages. Dr. J. A. Vila. *Manresa (Catalonia)*
Hospital de la Santa Creu i Sant Pau. Dr. Agustí Ruiz Martínez (Radiophysics and Radioprotection Unit) *Barcelona (Catalonia)*
Institut Mèdic per la Imatge (IMI). Dr. Manel Murillo. *Manresa (Catalonia)*
UAB-ICMAB-CSIC—Gerard Tobias Rossell. *Barcelona (Catalonia)*
UAB-CSIC—Francesca Campabadal Segura. *Barcelona (Catalonia)*

UAB—Javier Castelo Torras. *Miami Platja (Catalonia)*

UB—Joaquin Antonio Proenza Fernández. *Barcelona (Catalonia)*

UPC—Manel Romera. *Manresa (Catalonia)*

UPC—Francesca Sala. *Manresa (Catalonia)*

UPC—Fede Sanchez. *Manresa (Catalonia)*

UPC-INTE—Maria Amor Duch. *Barcelona (Catalonia)*

UPC-INTE—Arturo Vargas Drechsler. *Barcelona (Catalonia)*

Introduction

Thousands of scholars, professors, teachers, and members of the public have visited the Geological Museum Valentí Masachs, associated with the Universitat Politècnica de Catalunya (UPC) (Polytechnic University of Catalonia) in Manresa (Barcelona, Catalonia), since it opened in June 1980.

The Geological Museum Valentí Masachs is an unconventional one, very different from the traditional type. The exhibition's main aim is to visualize the connection between human activity and planet Earth (Fig. 1).

As a result of this distinctive pedagogical approach, distinguished members of the educational community demanded a much-needed book to deepen enquiry into the subject of minerals and elements, from their origin to their application and recycling (Fig. 2).

And here we have the book, *Elements and Mineral Resources*. We divided it into three parts: *Elements and Minerals*; *Rare Earth Elements*; and *Industrial Minerals*. Each includes the elements in that category. For each element, we have defined its most common features using a card-type format to show the essential data in a manageable way, with useful links, further reading, and a bibliography to satisfy your curiosity when needed.

Fig. 1 Museum room

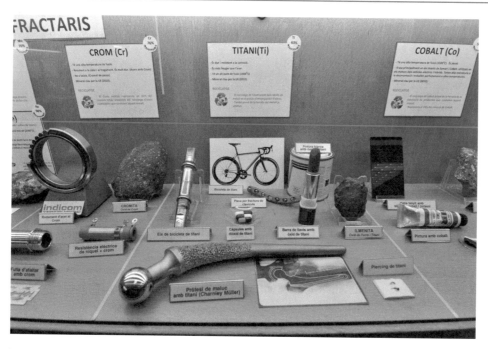

Fig. 2 Mineral applications showcase

This book aims to be a concise guide to the origin and usage of each element at a glance. However, we are aware that society evolves rapidly and that new needs, new applications, and new recycling opportunities arise constantly. Some applications included here might become obsolete or irrelevant, and some elements might become unprofitable to recycle in future. However, we hope that the book will be useful for teaching earth sciences, mining, chemistry, and engineering for university, high school, and personal interest. We are confident that it will promote awareness about how we should care for the Earth. There is no doubt we must use geological resources; however, it is imperative to use them sustainably to avoid exhaustion and preserve remaining material for future generations.

Keep in mind the three Rs: Recycling, Reusing, and Reducing. Remember that if the so-called circular economy of 7Rs (rethink, reduce, reuse, repair, refurbish, recover, recycle) is implemented with regard to the various objects and products obtained from the planet's natural resources, these resources will last much longer.

On the museum's Web site, http://geomuseu.upc.edu/index.php/en/inici-english/ (on the: *Slides: uses of minerals*), you will find 41 representative summary cards, presented as slides to be projected in class, to help you.

Bear in mind that we are not the last inhabitants of Earth, a great yet depletable planet!

Joaquim Sanz
Oriol Tomasa
Abigail Jimenez-Franco
Nor Sidki-Rius

Behind the Exploitation of Minerals, There are Always People…

The Geological Museum Valentí Masachs at the Polytechnic University of Catalonia (http://geomuseu.upc.edu/index.php/en/inici-english/) has always worked on two pedagogical fronts:

(1) To train people to discover the geological resources of our planet, which ultimately must be beneficial for our entire society, with no exclusions; and
(2) To promote critical thinking regarding the exploitation of mineral resources, because we always find people in mining. The miner's circumstances can be rugged yet have healthy working conditions and a salary, but many others suffer in extremely unfair and immoral conditions (child labor, to mention just one) (Fig. 3).

Fig. 3 Grinding rock by 'human force…' for the extraction of gold. Misky Mines (Camaná) (Perú) (*Image courtesy of* Silvia Palacios)

The extraction of certain strategic minerals in certain countries, regrettably, has caused and still causes wars, deaths, famine, and the displacement of people and animals from their own land.

The authors would like to encourage the readers and users of this book to take action. In other words, we invite you to make personal decisions while bearing both our planet and our society in mind. Use the Earth's resources sustainably, in the realization that we will not be the last inhabitants of our planet: Further generations have the right to geological resources with which to make their lives.

Contents

About the Authors

Joaquim Sanz completed Mining Engineering (Polytechnic University of Catalonia) (UPC) and is Gemmologist and Diamond Specialist (University of Barcelona) (UB) and (Gemmological Association of Great Britain).

Joaquim Sanz is Professor of Mineralogy (Polytechnic University of Catalonia) (UPC) and Director of Geological Museum Valentí Masachs (UPC), now retired.

Oriol Tomasa completed Mining Engineering and Master's Degree in Natural Resources Engineering from the Polytechnic University of Catalonia (UPC), is Lecturer in part time of Mineralogy (Polytechnic University of Catalonia) (UPC), and he is doing his Ph.D. in Natural Resources and Environmental Engineering.

Abigail Jimenez-Franco is Geologist (National Autonomous University of Mexico (UNAM), completed Ph.D. by the Polytechnic University of Catalonia (UPC) and is Gemmologist (University of Barcelona) (UB) and Co-editor of the Springer's Geology series. Currently, she is a Full-time Postdoctoral Researcher (University of Barcelona).

Nor Sidki-Rius completed Mining Engineering and Master's Degree in Mining Engineering (Polytechnic University of Catalonia) (UPC). Currently, she is working on her Ph.D. thesis titled 'Some approaches to improve knowledge about geomechanical behaviour in underground Potash mining' (Polytechnic University of Catalonia) (UPC).

Abbreviations

MGVM	Geological Museum Valentí Masachs. http://geomuseu.upc.edu/index.php/en/inici-english/
Mt	Million tons
t	Tons

Introduction

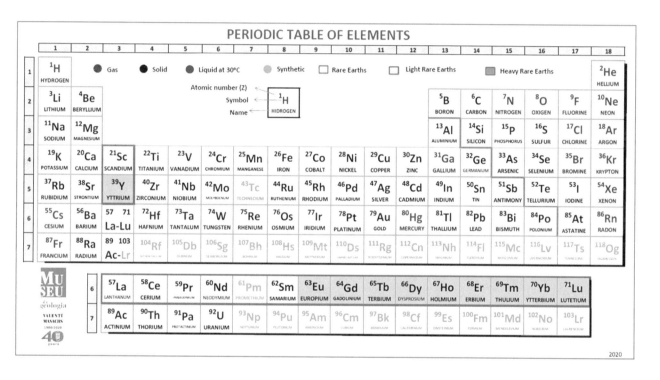

The great majority of elements in the periodic table are extracted from minerals. These are called ores. The elements are ordered alphabetically, showing their chemical symbol and atomic number (Z).

On each card in this book you will find a description of the element's main characteristics and a photograph of its mineral (ore). In addition, the position of the element in the periodic table is shown.

For some of these elements, you will see the phrase, 'Assigned the status of a strategic (metal or mineral) by the EU in 2017)'. This refers to the European Union's list of critical raw materials for industry in that year; a further seven were added in the 2020 update (see: critical-raw-materials-for-the-EU-2017 or EU-2020).

A list of the most important and current applications of each element or mineral follows, illustrated by photographs (Figs. 1.1, 1.2, 1.3, 1.4, 1.5 and 1.6).

A section on recycling follows, since it is vital in our times to reuse to the maximum the wealth hidden within each object or material that is manufactured using these elements. This is because of the savings achieved by reuse, in terms of both the economics and the consumption of mineral resources.

In these applications and recycling sections, all the information is included that we have been able to obtain from the companies involved in the transformation and recycling processes.

To finish are links (contrasted and updated) giving further information on the relevant element and/or mineral.

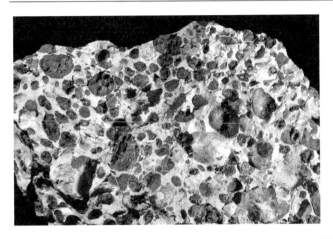

Fig. 1.1 Bauxite; and aluminum drinks cans. (*Photos* Joaquim Sanz. MGVM)

Fig. 1.2 Chalcopyrite; and electric vehicle with copper. (*Photos* Joaquim Sanz. MGVM)

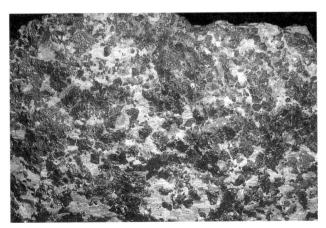

Fig. 1.3 Chromite; and a chrome-plated faucet. (*Photos* Joaquim Sanz. MGVM)

Fig. 1.4 Celestine; and fireworks containing strontium salts. (*Image courtesy of* Mireia Arso. Regió 7)

Fig. 1.5 Hematite; and an iron railway bridge. (*Photos* Joaquim Sanz. MGVM)

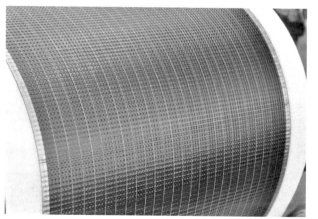

Fig. 1.6 Sphalerite with indium; and thin-film solar cells (CIGS). (*Photo* Joaquim Sanz. MGVM)

1.1 Changing Prices of Strategic Metals, 2009 to 2021

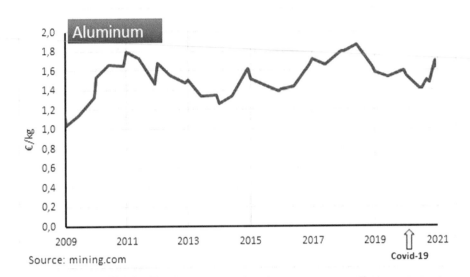

Source: mining.com

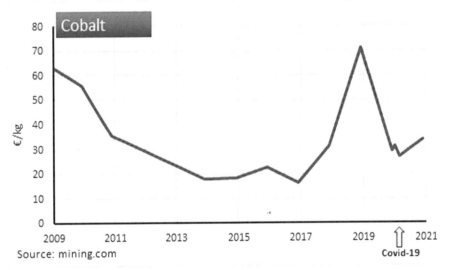

Source: mining.com

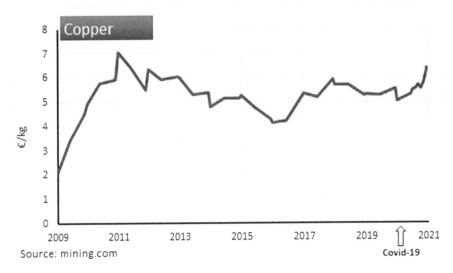

Source: mining.com

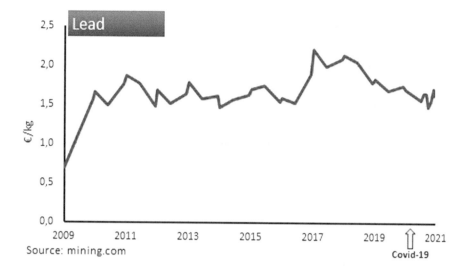

Source: mining.com

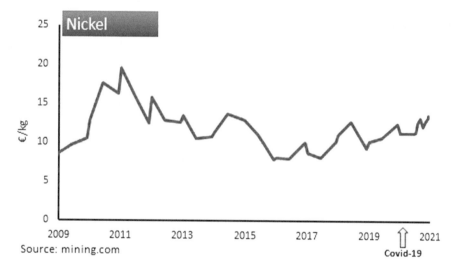

Source: mining.com

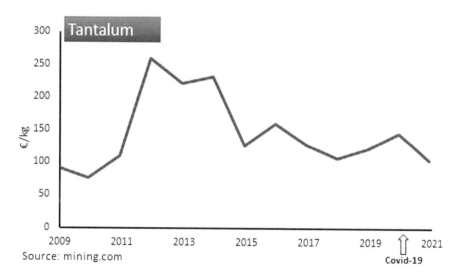

Source: mining.com

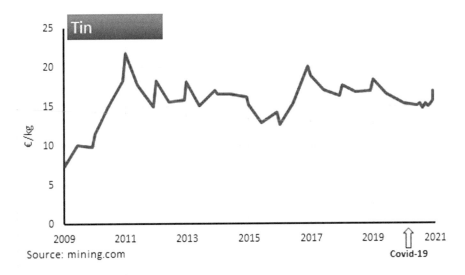

Source: mining.com

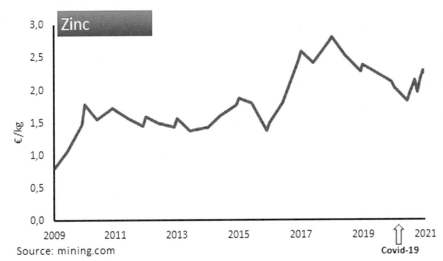

Source: mining.com

1.2   Changing Prices of Noble Metals
from 2009 to 2021

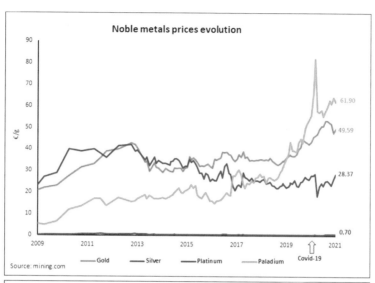

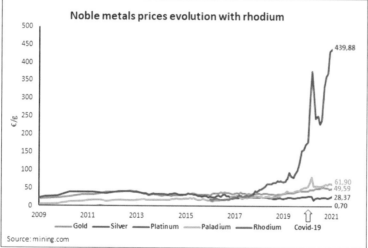

Aluminum (Al) [Z = 13]

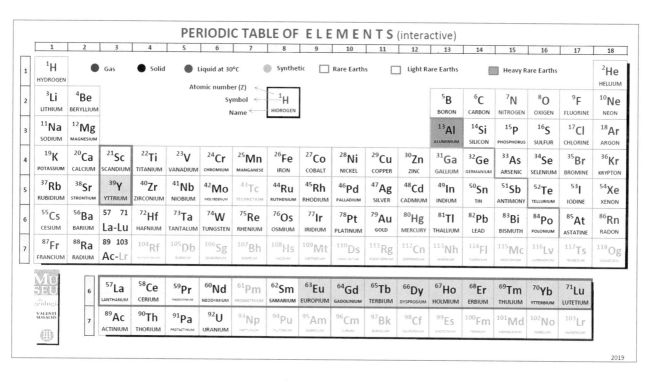

- The most abundant metallic element in the Earth's crust
- Despite its abundance, it was considered a rare and expensive metal until 1886, when Charles M. Hall of the United States and Paul L. T. Héroult of France independently discovered an inexpensive system for obtaining pure aluminum by electrolysis of its oxide (Al_2O_3)
- In 1888, the Bayer company developed a system for obtaining aluminum oxide from bauxite
- Good electrical and heat conductor, malleable, ductile, soft, and light
- Provides a metal barrier impervious to light, UV rays, corrosion, water vapor, oils and greases, oxygen, and microorganisms
- Obtained from bauxite rock (Fig. 2.1)
- Assigned the status of a strategic element by the EU in September 2020.

2.1 Geology

Bauxite is a type of rock containing one or more aluminum hydroxide minerals, most notably gibbsite ($Al(OH)_3$), boehmite (γ-AlO(OH)), and diaspore (α-AlO(OH)). Diaspore has the same chemical composition as boehmite but is denser and harder. A typical bauxite rock contains a mixture of goethite (FeO(OH)), hematite (Fe_2O_3), the clay mineral kaolinite, and a minor amount of anatase (TiO_2). Moreover,

J. Sanz et al., *Elements and Mineral Resources*, Springer Textbooks in Earth Sciences, Geography and Environment, https://doi.org/10.1007/978-3-030-85889-6_2

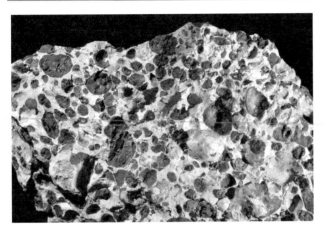

Fig. 2.1 Bauxite (rock formed by gibbsite, diaspore, and böhmite). *Miralles* (*Catalonia*) (*Photo* Joaquim Sanz. MGVM)

some bauxite has an appreciable RRE (Rare Earth Elements) content.

The formation of bauxites involves intense weathering in tropical or subtropical climates. Bauxites typically appear within thick profiles of laterites, which are formed by intense subaerial weathering and leaching of aluminosilicate rocks. These bauxites are referred to as lateritic bauxites.

Alternatively, bauxites may occur in paleokarst depressions as accumulations of clayey material within carbonate (limestone) sequences. These are referred to as karst bauxites, and they tend to be black to gray in color with mixed organic matter and, occasionally, chemically reduced minerals such as pyrite. Approximately 88% of global bauxite deposits are lateritic, and the remaining 12% belong to the karst bauxite group.

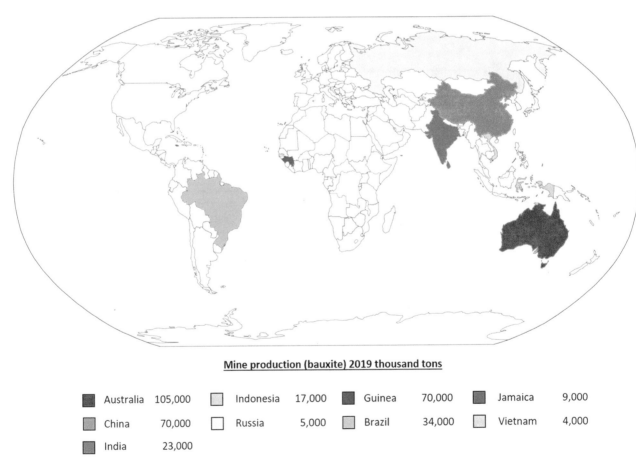

Mine production (bauxite) 2019 thousand tons

Australia	105,000	Indonesia	17,000	Guinea	70,000	Jamaica	9,000
China	70,000	Russia	5,000	Brazil	34,000	Vietnam	4,000
India	23,000						

Fig. 2.2 List of producing countries based on the US Geological Survey, Mineral Commodity Summaries

2.2 Producing Countries

The main producers of bauxite are Australia, China, Guinea, Brazil, and India, representing almost 92% of world production (Fig. 2.2).

World bauxite reserves are estimated at between 55 and 75 billion tons in Africa (32%), Oceania (23%), South America and the Caribbean (21%), and Asia (18%) .

2.3 Applications

Metallurgical industry

Aluminum is used in the manufacture of alloys such as Simagal (silica, magnesium, and aluminum), to which it confers a high mechanical resistance, and duralumin (aluminum, copper, manganese, and iron), enhancing the mechanical properties.

Electrical industry

Because it is a very good conductor, it is used in the manufacture of electrical cables and electronic components. The electrical industry is replacing copper with aluminum in high-, medium-, and low-voltage goods, because it carries electricity more cheaply even if the cables have to have more sections, and also it cuts out the risk of theft of copper cables.

Transport and construction

Its lightness and mechanical strength make aluminum the optimum material in the manufacture of aircraft (40% of an Airbus 380, the world's largest passenger transport aircraft, is of aluminum–lithium and carbon fiber-reinforced polymer (Fig. 2.3), cars, electric vehicles, trains, buses, and buildings (window profiles (Fig. 2.4), doors, and protection plates).

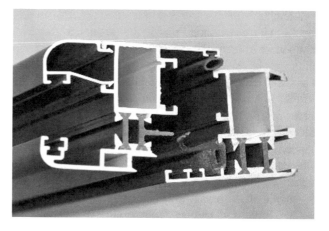

Fig. 2.4 Aluminum profile. (*Photo* Joaquim Sanz. MGVM)

Agbar Tower in Barcelona, standing at 142 m and with 34 floors, is completely covered in lacquered aluminum sheets in a variety of colors.

Other applications

In the food industry, aluminum is used in the manufacture of soft drink cans, containers, and Tetrabriks, because it is a hygienic material that does not affect the taste of the product. In thinly rolled form it constitutes aluminum foil, widely used in the kitchen and the food sector (Fig. 2.5).

Aluminum also forms part of CDs and computer hard disks.

In the cosmetics industry, aluminum salts are used in deodorants, because they stop sweat and prevent the growth of bacteria, which causes odors.

Bauxite is the main source of alumina, a high-temperature-resistant product widely used in the manufacture of steels, cements, non-ferrous metals, glass, and ceramics. At the same time, it is a good abrasive for polishing metals, ceramic supports, brake pads, etc.

Fig. 2.3 Airbus 380. (*Image courtesy of* Joan Carles Lacruz)

Fig. 2.5 Soft drink cans. (*Photo* Joaquim Sanz. MGVM)

Aluminum powder is a highly reactive metal when reinforced with magnesium powder, representing the main component of fireworks and giving a silvery white color of an intense brightness. The reactivity of aluminum powder is the basis of so-called alum-thermal welding. When the powder is mixed with iron oxide and external heat added, there is an exothermic reaction that reaches 1930 °C, allowing the ends of two steel beams or two train rails to be perfectly joined without the need for electric welding. The alumina formed remains as a protective slag over the weld.

Aluminum sulfates with elements such as potassium, sodium, iron, and chrome (among others) constitute the so-called alums, widely used in dyeing as a mordant in leather tanning and other industries.

Aluminum oxide (corundum), when transparent and red, constitutes the precious stone known as a ruby. The color is due to the presence of chromium impurities. Likewise, when it is blue (due to an iron and titanium content), it forms the precious stone known as a sapphire.

2.4   Recycling

Almost 100% of the aluminum consumed can be reused, although its reuse is not infinite.

Producing aluminum from cans of recycled soft drinks saves 95% of the electrical energy that would have been used if the metal had to be extracted from bauxite, and entails vast environmental savings as it also avoids the sludge normally generated during extraction (red mud). Contained in this red mud are important elements such as scandium, yttrium, and lanthanum, and it is also used to make bricks and cements.

In 2019, 36% of the aluminum used in Europe came from recycling, and it is expected that by 2050 this value will have risen to 50%, with a corresponding decrease in the emission of CO_2 into the atmosphere (European Aluminium, 2021). 75% of the aluminum produced is still in use, as it is a highly durable metal.

Further Readings

Almatis Alumina (2020) http://www.almatis.com/ (last accessed May 2021)

Aluminum Association (2021) https://www.aluminum.org/industries/production/bauxite (last accessed May 2021)

Bray EL (2021) Aluminium annual publication. US Geological Survey, Mineral Commodity Summaries 2021, pp 21–22. https://doi.org/10.3133/mcs2021

European Aluminium (2021) https://www.european-aluminium.eu/ (last accessed May 2021)

Gray T, Mann N (2009) The elements. Black Dog & Leventhal Publishers Inc., New York

Hsu PH (1989) Aluminum hydroxides and oxyhydroxides. Minerals in Soil Environments 1:331–378

Mata JM, Sanz J (2007) Guia d'identificació de Minerals. Manresa, Catalonia. Edicions UPC/Parcir (Catalan 2nd paper edition). 262p. ISBN: 9788483019023. http://hdl.handle.net/2117/90445

Quadbeck-Seeger H-J (2007) World of the elements: elements of the world. Wiley-VCH Verlag GmbH & Co, Germany

Red Mud Project (2019) https://redmud.org/ (last accessed May 2021)

Sanz J, Tomasa O (2017) Elements i Recursos minerals: Aplicacions i reciclatge. Manresa, Catalonia. Zenobita Edicions/Iniciativa Digital Politècnica (Catalan 3rd digital edition). http://hdl.handle.net/2117/105113

Sanz J, Tomasa O (2018) Elementos y Recursos minerales: Aplicaciones y reciclaje. Manresa, Catalonia. Zenobita Edicions/Iniciativa Digital Politècnica. (Spanish 1st digital edition). http://hdl.handle.net/2117/123674

Schulte RF, Foley NK (2013) Compilation of Gallium resource data for bauxite deposits. USGS. https://doi.org/10.3133/ofr20131272

Stwertka A (2018) A guide to the elements, 4th edn. Oxford University Press, England

USGS (2021) Commodity Statistics and Information: Aluminium. Available at: https://www.usgs.gov/centers/nmic/aluminum-statistics-and-information (last accessed May 2021)

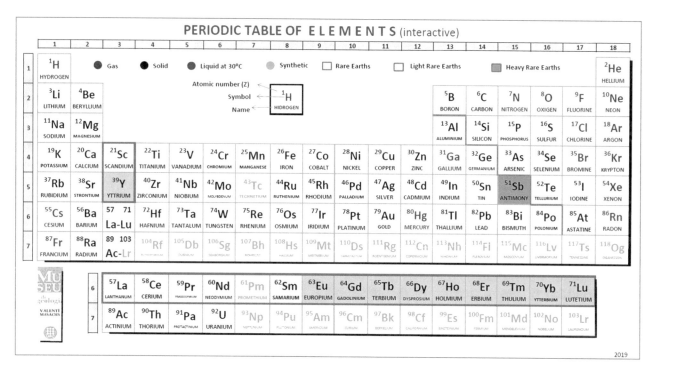

- A soft, brittle, semi-metal
- Low thermal and electrical conductivity
- Melts at low temperatures (630 °C)
- Gutenberg, the inventor of the printing press (1450), created his mobile type from an antimony and lead alloy, giving greater hardness to the lead and allowing it to be used multiple times in linotypes. At the same time, it permitted rapid melting in case of deterioration or the need for new fonts
- Assigned the status of a strategic metal by the EU in 2017

- Obtained from stibnite (Fig. 3.1) and lead with antimony recovered from electric vehicle batteries.

3.1 Geology

The primary mineral source of antimony (Sb) is stibnite (Sb_2S_3). Antimony's mineralization is commonly associated with igneous intrusion and hydrothermal-related ore systems. The classic antimony deposit is associated with epithermal polymetallic veins.

© The Author(s), under exclusive license to Springer Nature Switzerland AG 2022
J. Sanz et al., *Elements and Mineral Resources*, Springer Textbooks in Earth Sciences,
Geography and Environment, https://doi.org/10.1007/978-3-030-85889-6_3

Fig. 3.1 Stibnite (antimony sulfide). *Abella* (*Catalonia*) (*Photo* Joaquim Sanz. MGVM)

Fig. 3.3 Stage curtains. (*Photo* Joaquim Sanz. MGVM)

Stibnite can also be found in several types of deposits, such as intrusion-related, anorogenic, skarns, and porphyry systems. Antimony is recovered as a secondary product commonly related to silver epithermal vein deposits. Epithermal systems are temporally and genetically related to volcano-plutonic activity affinity at convergent tectonic plate margins. Many epithermal systems are transitional to porphyry environments.

3.2 Producing Countries

The most remarkable world reserves of antimony are in China (480,000 t), Russia (350,000 t), Bolivia (310,000 t), Australia (140,000 t), Turkey (100,000 t), United States (60,000 t), Tajikistan (50,000 t), Pakistan (26,000 t) and Mexico (18,000 t) (Fig. 3.2).

3.3 Applications

Chemical industry

The main use of antimony is in the production of flame retardants for paint, glues, plastics (for computer casings, televisions, and electrical cables, among others), and some fabrics (e.g. stage curtains) (Fig. 3.3).

Antimony also acts as a catalyst in the manufacture of PET plastics for water and beverage bottles.

Battery industry

Antimony in the classic alloy of lead plus antimony is used to manufacture lead-acid batteries (Fig. 3.4).Lead + calcium + tin batteries are appearing on the market with no need for maintenance, regulated by valves (VRLA). However, lead + antimony batteries and flame retardants together currently constitute 80% of market demand for antimony (Roskill 2020).

Electronics industry

Very pure antimony is used in the manufacture of semiconductors.

Metallurgical industry

Tin-alloyed antimony is used in welding and, alloyed with zinc, it adds hardness to the alloy.

Other fields

Ceramics, plastics, glass (especially solar panels), and industrial rubber industries consume appreciable amounts of antimony salts in their manufacture.

3.4 Recycling

Recycling is carried out on used lead-acid batteries from vehicles and industry.

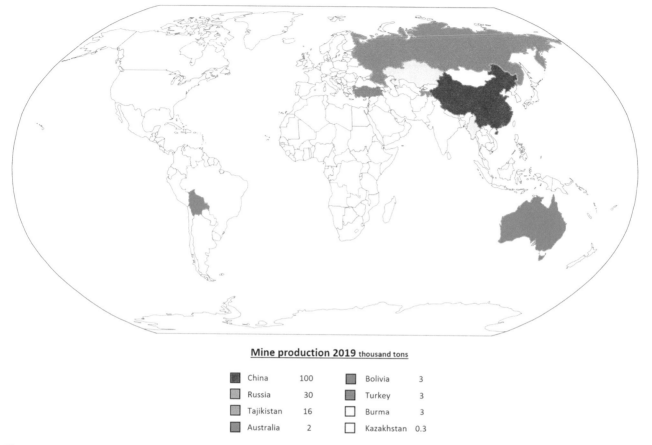

Mine production 2019 thousand tons

China	100		Bolivia	3
Russia	30		Turkey	3
Tajikistan	16		Burma	3
Australia	2		Kazakhstan	0.3

Fig. 3.2 List of producing countries based on the US Geological Survey, Mineral Commodity Summaries

Fig. 3.4 Car battery. (*Photo* Joaquim Sanz. MGVM)

According to Roskill (2020), by the mid-2020s the secondary supply of antimonial lead from recycling will be sufficient to meet metallurgical demand. Therefore, little primary antimony will be needed to meet metallurgical needs. On the other hand, non-metallurgical demand (flame retardants, plastics, and glass for solar panels) will increase to exceed the supply.

Further Reading

Gray T, Mann N (2009) The elements. Black Dog & Leventhal Publishers Inc., New York

Mata JM, Sanz J (2007) Guia d'identificació de minerals. Manresa, Catalonia. Edicions UPC/Parcir. (Catalan 2nd paper edition). 262p. ISBN: 9788483019023. http://hdl.handle.net/2117/90445

Pirajno F (2009) Hydrothermal processes and mineral systems. Hydrothermal Processes and Mineral Systems. https://doi.org/10.1007/978-1-4020-8613-7

Quadbeck-Seeger H-J (2007) World of the elements: elements of the world. Wiley-VCH Verlag GmbH & Co, Germany

Roskill (2020). Market Reports. Antimony. Available at: https://roskill.com/market-report/antimony/ (last accessed May 2021)

Sanz J, Tomasa O (2017) Elements i Recursos minerals: Aplicacions i reciclatge. Manresa, Catalonia. Zenobita Edicions /Iniciativa Digital

Politècnica. (Catalan 3rd digital edition). http://hdl.handle.net/2117/105113

Sanz J, Tomasa O (2018) Elementos y Recursos minerales: Aplicaciones y reciclaje. Manresa, Catalonia. Zenobita Edicions/Iniciativa Digital Politècnica (Spanish 1st digital edition). http://hdl.handle.net/2117/123674

Stwertka A (2018) A guide to the elements, 4th edn. Oxford University Press, England

USGS (2021) Commodity Statistics and Information. Antimony. Available at: https://www.usgs.gov/centers/nmic/antimony-statistics-and-information (last accessed May 2021)

Arsenic (As) [Z = 33]

PERIODIC TABLE OF E L E M E N T S (interactive)

- A brittle semi-metal with a metallic sheen
- A good heat conductor and a poor electrical conductor
- Highly toxic
- Obtained from arsenopyrite (iron arsenic sulfide) (Fig. 4.1), from nickel, cobalt, copper, and lead sulfides, as well as from realgar and orpiment, and from copper sulfides with gold.

4.1 Geology

The primary mineral source of arsenic (As) is arsenopyrite (FeAsS). Arsenic's mineralization is commonly associated with igneous intrusions and related ore systems. However, arsenopyrite can be found in many deposit types, such as vein systems, anorogenic, ring complexes, and breccia pipes. In some cases, there may be spatial, temporal, and genetic links with classic intrusive ore deposits such as porphyry and epithermal systems.

Fig. 4.1 Arsenopyrite (iron arsenic sulfide). *Queralbs* (*Catalonia*) (*Photo* Joaquim Sanz. MGVM)

Arsenic is recovered as a secondary product commonly related to epithermal vein deposits. Epithermal systems are temporally and genetically related to volcano-plutonic activity affinity at convergent tectonic plate margins. Many epithermal systems are transitional to porphyry environments.

4.2   Producing Countries

World reserves are very large, because arsenic is a by-product of obtaining copper, lead, gold, and silver from sulfides and arsenides. There are also notable reserves of orpiments and realgars, such as arsenic sulfides (Fig. 4.2).

4.3   Applications

Medicine
Arsenic is combined with transretinoic acid (ATRA) to form arsenic trioxide (ATO), a drug to fight acute promyelocytic leukemia (APL) as a last resort. Before, most patients who suffered from this disease died but nowadays, with this treatment, a higher survival rate is achieved.

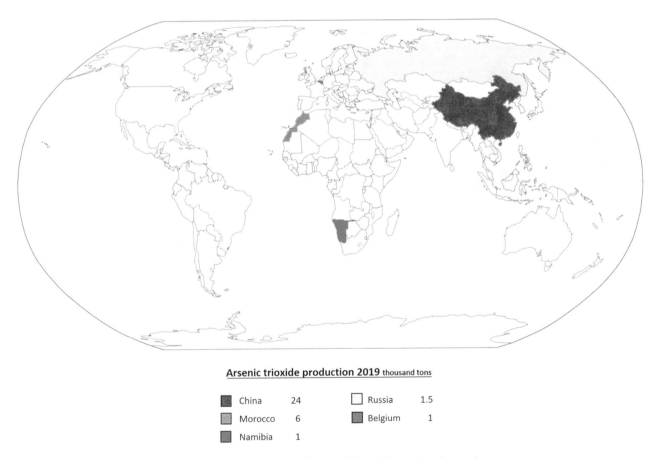

Arsenic trioxide production 2019 thousand tons

■ China	24	☐ Russia	1.5
▨ Morocco	6	▨ Belgium	1
■ Namibia	1		

Fig. 4.2 List of producing countries based on the US Geological Survey, Mineral Commodity Summaries

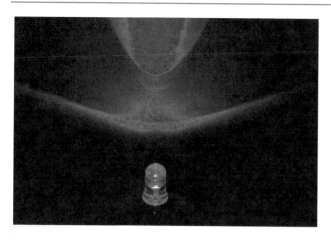

Fig. 4.3 Red-garnet LED. (*Photo* Joaquim Sanz. MGVM)

In China, arsenic has been used in traditional medicine for over two thousand years.

Metallurgical industry
Arsenic is used as an additive in lead alloys.

Electronics industry
Very high-purity gallium arsenide (99.9999%) is an important semiconductor used in solar cells, integrated circuits, laser diodes, and red-garnet LEDs (Fig. 4.3).

Past Applications in Europe (European Directive 2003/2/EC)

Copper chromium arsenate (CCA) was previously used as a wood preservative, and in certain countries, such as the United States, it is still used. CCA has been replaced by copper salt treatments that involve no arsenic or chrome.

Lead arsenate was used as an insecticide, and sodium arsenite as a herbicide.

Arsenic, being extremely toxic, has become a major environmental concern. Soil and water can be easily contaminated by arsenic, endangering people's health.

4.4 Recycling

Currently, the arsenic that is recycled comes is from the manufacture of gallium-arsenic semiconductors and electronic devices.

In countries where CCA (copper chromium arsenate) is permitted, arsenic is recovered from the water involved in wood treatment.

Further Reading

Gray T, Mann N (2009) The elements. Black Dog & Leventhal Publishers Inc., New York

Mata JM, Sanz J (2007) Guia d'identificació de minerals. Manresa, Catalonia. Edicions UPC/Parcir. (Catalan 2nd paper edition). 262 p. ISBN: 9788483019023. http://hdl.handle.net/2117/90445

Pirajno F (2009) Hydrothermal processes and mineral systems. https://doi.org/10.1007/978-1-4020-8613-7

Quadbeck-Seeger H-J (2007) World of the elements: elements of the world. Wiley-VCH Verlag GmbH & Co, Germany

Sanz J, Tomasa O (2017) Elements i Recursos minerals: Aplicacions i reciclatge. Manresa, Catalonia. Zenobita Edicions /Iniciativa Digital Politècnica. (Catalan 3rd digital edition). http://hdl.handle.net/2117/105113

Sanz J, Tomasa O (2018) Elementos y Recursos minerales: Aplicaciones y reciclaje. Manresa, Catalonia. Zenobita Edicions/Iniciativa Digital Politècnica (Spanish 1st digital edition). http://hdl.handle.net/2117/123674

Stwertka A (2018) A guide to the elements, 4th edn. Oxford University Press, England

USGS (2021) Commodity Statistics and Information. Arsenic. Available at: https://www.usgs.gov/centers/nmic/arsenic-statistics-and-information (last accessed May 2021)

PERIODIC TABLE OF E L E M E N T S (interactive)

(periodic table figure)

2019

- An alkaline earth metal, soft and heavy
- Reactive to air and water
- Absorbs x-rays
- Obtained from barite (Fig. 5.1)
- Barite (barium sulfate) was assigned strategic mineral status by the EU in 2017.

5.1 Geology

The primary mineral source of barium (Ba) is barite ($BaSO_4$). Barite deposits commonly consist of a strata-bound mantos type of an epigenetic character, typical of a VMS deposit type. They are found mostly in limestone. These ore deposits are mostly associated with magmatic or volcanic events. The host rocks are commonly associated

Fig. 5.1 Barite (barium sulfate). *Espinelves (Catalonia)* (*Photo* Joaquim Sanz. MGVM)

with dolomite formations. Ore mineralogy is nearly pure barite and the gangue minerals are mainly patches of coarse calcite with trace amounts of celestine, silica, Fe-(oxy) hydroxides, and Mn-oxides. It is also common to find

brecciated limestone cemented by barite. The largest in-situ deposits are in China, India, and the United States.

5.2   Producing Countries

The world's most notable reserves of barite are located in Kazakhstan (85 Mt), India (51 Mt), China (36 Mt), Turkey (35 Mt), Pakistan (26 Mt), Iran (24 Mt), and Russia (12 Mt) (Fig. 5.2).

5.3   Applications

Medicine

In medical procedures involving x-rays of the intestinal tract to examine the upper or lower part of the digestive system, patients are given an aqueous solution containing barium sulfate to coat the inside layer of their digestive organs. During the x-ray, the barium-sulfate coating absorbs much of the x-rays to highlight any areas of injury.

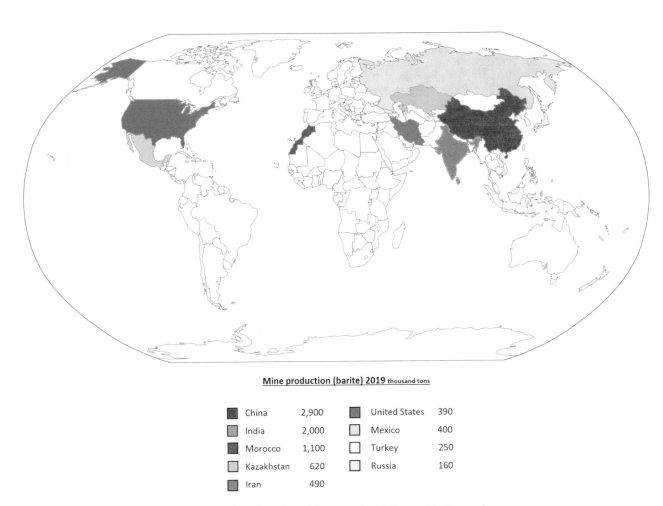

Mine production (barite) 2019 thousand tons

■	China	2,900	■	United States	390
■	India	2,000	■	Mexico	400
■	Morocco	1,100	□	Turkey	250
■	Kazakhstan	620	□	Russia	160
■	Iran	490			

Fig. 5.2 List of producing countries based on the US Geological Survey, Mineral Commodity Summaries

Glass and ceramics industry

Barium carbonate is used in the manufacture of glass, as it increases its refractive index and brilliance.

Various industries

Barium sulfate is used as a bulking agent in the manufacture of rubber (gloves, toys, tires, and erasers) to increase its elasticity, also in the manufacture of plastics.

Barium sulfate is one of the components of lithopone (zinc white), a white pigment used in paints and enamels to obtain coatings that do not darken in contact with sulfides. In car primer, it protects vehicles' metal parts.

Barium sulfate is used in the production of paper and cardboard because it provides both strength and density. It also gives a green color in fireworks and flares (Fig. 5.3).

Other fields

Barite is used in the manufacture of drilling muds to lubricate the drilling tools and improve the sustainability of the side walls in oil, natural gas, and water extraction wells (Fig. 5.4).

During 2019, 80% of barite consumption in the United States was destined for drilling oil and gas wells.

Barite is an additive in the preparation of barite concrete, which is used in the construction of walls and tunnels where linear accelerators, x-ray units, and nuclear plants are located, to prevent any leakage of ionizing radiation.

Iron oxide and barium oxide are involved in the manufacture of ferrite magnets.

The barite is used as a load element in the manufacture of vehicle brake pads.

Fig. 5.4 Drilling a water well. (*Image courtesy of* Raul Osorio)

5.4 Recycling

Barite is recycled from drilling muds from oil, natural gas, and water wells, as it is a strategic mineral.

Further Reading

Gray T, Mann N (2009) The elements. Black Dog & Leventhal Publishers Inc., New York

Mata JM, Sanz J (2007) Guia d'identificació de minerals. Manresa, Catalonia. Edicions UPC/Parcir (Catalan 2nd paper edition). 262 p. ISBN: 9788483019023. http://hdl.handle.net/2117/90445

Quadbeck-Seeger H-J (2007) World of the elements: elements of the world. Wiley-VCH Verlag GmbH & Co, Germany

Sanz J, Tomasa O(2017) Elements i Recursos minerals: Aplicacions i reciclatge. Manresa, Catalonia. Zenobita Edicions /Iniciativa Digital Politècnica (Catalan 3rd digital edition). http://hdl.handle.net/2117/105113

Fig. 5.3 Green fireworks. (*Image courtesy of* Mireia Arso)

Sanz J, Tomasa O (2018) Elementos y Recursos minerales: Aplicaciones y reciclaje. Manresa, Catalonia. Zenobita Edicions/Iniciativa Digital Politècnica (Spanish 1st digital edition). http://hdl.handle.net/2117/123674

Stwertka A (2018) A guide to the elements, 4th edn. Oxford University Press, England

USGS (2021) Commodity Statistics and Information. Barite https://www.usgs.gov/centers/nmic/barite-statistics-and-information (last accessed May 2021)

PERIODIC TABLE OF E L E M E N T S (interactive)

6.1 Geology

- An alkaline earth metal, light, and not very abundant
- Highly toxic
- Has a high melting point (1278 °C), a high heat capacity, and good thermal conductivity
- Assigned the status of a strategic metal status by the EU in 2017
- Obtained from bertrandite and beryl (Fig. 6.1).

The three most common beryllium-bearing minerals found in economic deposits are beryl ($Be_3Al_2Si_6O_{18}$), bertrandite ($Be_4Si_2O_7(OH)_2$), and phenakite (Be_2SiO_4). Beryl is the main beryllium mineral in igneous intrusions and metamorphic rocks formed at moderate to high temperatures. Deposits are found in some types of granites and pegmatite deposits. These contain beryl and phenakite: in high-grade pockets or seams and in the gravels formed by the weathering of such rocks; in high-temperature skarn and greisen related to

Fig. 6.1 Beryl (beryllium aluminosilicate). *Ferreira de Aves (Portugal)* (*Photo* Joaquim Sanz. MGVM)

specialized granitic intrusions; and in low-temperature epithermal deposits that mainly contain bertrandite and are related to volcanic rocks.

Beryl in rare-metal pegmatites can occur with other minerals that contain commodities of commercial importance, such as cesium, lithium, and tantalum, as well as clay minerals, feldspar, muscovite, and high-purity quartz. Beryl-bearing pegmatite districts occur in Brazil, Canada, China, Mozambique, Namibia, Portugal, and Zimbabwe. Non-pegmatite-related volcanogenic deposits form in a comparatively unusual geologic settings where bertrandite and other minerals replace carbonate clasts entrained as rock fragments in beryllium-rich tuff units. The Spor Mountain deposits of Utah (United States) are the only known economic example of this deposit type.

6.2   Producing Countries

Proven beryllium reserves in the United States, as the world's leading producer, are estimated at 21,000 tons (Fig. 6.2).

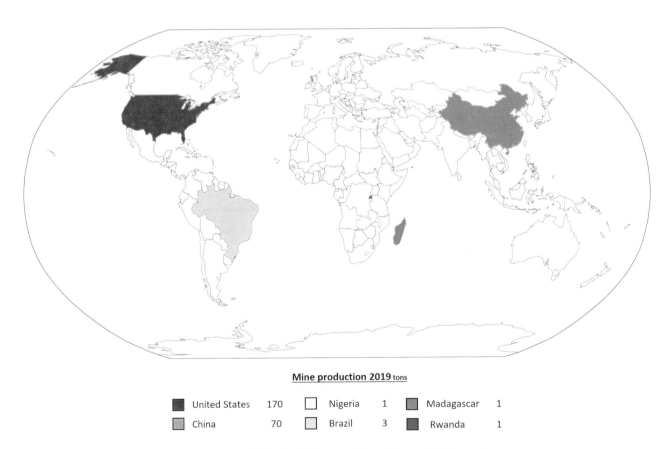

Mine production 2019 tons

| United States | 170 | Nigeria | 1 | Madagascar | 1 |
| China | 70 | Brazil | 3 | Rwanda | 1 |

Fig. 6.2 List of producing countries based on the US Geological Survey, Mineral Commodity Summaries

Fig. 6.3 Commercial aircraft. (*Photo* Joaquim Sanz. MGVM)

6.3 Applications

Metallurgical industry

Because of its lightness and strength, beryllium is used in the aerospace industry in the manufacture of aircraft and satellites (Fig. 6.3).

Beryllium is added to copper and aluminum alloys to increase strength and extend service life.

Beryllium and copper alloys are as strong as steel and do not produce sparks under friction. They are used in the manufacture of machinery that is in contact with flammable fluids or gases or in explosive environments.

Due to the toxicity of the metal dust and fumes generated in industries that use beryllium in the manufacture of various components, overall its consumption is reducing year on year. Control standards, both national and international, are increasingly stringent in these manufacturing industries.

Power generation

Beryllium is a component of the control rods in some nuclear power plants.

Electronics industry

By virtue of its lightness and resistance, beryllium is used in the manufacture of precision equipment such as gyroscopes, optical equipment supports, and computer equipment.

Electrical industry

Beryllium oxide is used as an electrical insulator and also as a heat sink, e.g. on high-power transistor insulating backplanes in telecommunications equipment.

Astronomy

The segments that make up the mirror of the astronomical telescope at the Roque de los Muchachos (La Palma) (Canary Islands) are built of beryllium, coated with a thin layer of nickel. The mirror segments of the James Webb Space Telescope are also made of beryllium, because it is light and strong and maintains its shape at cryogenic temperatures. Webb's mirrors are covered in a microscopically thin layer of gold, which optimizes them for reflecting infrared light, which is the primary wavelength of the light that this telescope will observe.

6.4 Recycling

Recycling of beryllium is carried out on offcuts generated in manufacturing products containing this metal. At the moment, recycled beryllium represents only 20–25% of the total use of this element.

Manufactured beryllium from recycling requires only 20% of the energy that it would consume if it were extracted from the ore.

Further Reading

Beryllium Science and Technology Association (2016) http://beryllium.eu/ (last accessed May 2021)

Eurometaux. Introducing Metals (2021) https://eurometaux.eu/about-our-industry/introducing-metals/ (last accessed May 2021)

Gray T, Mann N (2009) The elements. Black Dog & Leventhal Publishers Inc., New York

Instituto Astrofísica de las Canarias (2021) Beryllium. https://www.iac.es/es/observatorios-de-canarias/observatorio-del-roque-de-los-muchachos (last accessed May 2021)

James Webb Space Telescope (2021) Beryllium. https://www.jwst.nasa.gov/content/observatory/ote/mirrors/index.html (last accessed May 2021)

Mata JM, Sanz J (2007) Guia d'identificació de minerals. Manresa, Catalonia. Edicions UPC/Parcir (Catalan 2nd paper edition). 262p. ISBN: 9788483019023. http://hdl.handle.net/2117/90445

Quadbeck-Seeger H-J (2007) World of the elements: elements of the world. Wiley-VCH Verlag GmbH & Co, Germany

Sanz J, Tomasa O (2017) Elements i Recursos minerals: aplicacions i reciclatge. Manresa, Catalonia. Zenobita Edicions/Iniciativa Digital Politècnica (Catalan 3rd digital edition). http://hdl.handle.net/2117/105113

Sanz J, Tomasa O (2018) Elementos y Recursos minerales: Aplicaciones y reciclaje. Manresa, Catalonia. Zenobita Edicions/ Iniciativa Digital Politècnica (Spanish 1st digital edition). http://hdl.handle.net/2117/123674

Stwertka A (2018) A guide to the elements, 4th edn. Oxford University Press, England

USGS (2016) Beryllium—a critical mineral commodity—resources, production, and supply chain. https://pubs.usgs.gov/fs/2016/3081/fs20163081.pdf (last accessed May 2021)

USGS (2021) Commodity Statistics and Information. Beryllium https://www.usgs.gov/centers/nmic/beryllium-statistics-and-information (last accessed May 2021)

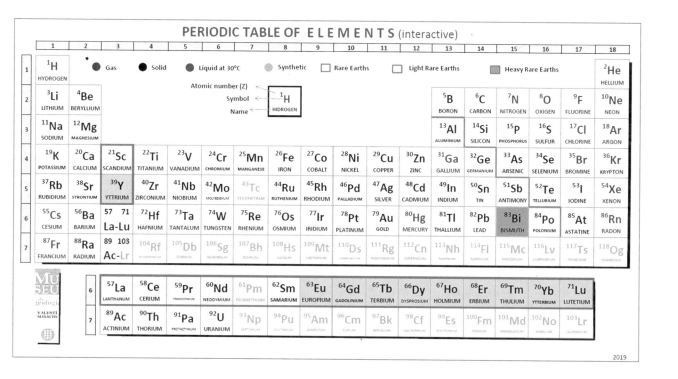

PERIODIC TABLE OF E L E M E N T S (interactive)

2019

- A silvery white metal with a pinkish hue
- Heavy and fragile
- Very low thermal conductivity
- Melts at 272 °C
- Obtained from bismuthinite and as a by-product of copper and lead metallurgy
- Also be found as native bismuth (Fig. 7.1)
- Assigned the status of a strategic metal by the EU in 2017
- Known in the early fifteenth century, but was confused with lead and tin. In 1713 Claude Geoffroy the Younger (a French nobleman) identified it definitively.

7.1 Geology

Bismuth (Bi) is mostly processed from lead ore. Bismuth's mineralization is commonly associated with igneous intrusion and related ore systems. The classic type of bismuth deposit is greisen. This is a kind of ore deposit characterized by a hydrothermally altered granitic rock. The mineralogical alteration association is mostly albite, quartz, and mica, very common in European deposits.

Bismuth can also be found as metallic inclusions in other minerals' cleavage or as bismuthinite (Bi_2S_3) in several kinds of deposit types, such as vein systems, anorogenic, ring complexes, and breccia pipes. In some cases there may be spatial, temporal, and genetic links with classical

© The Author(s), under exclusive license to Springer Nature Switzerland AG 2022
J. Sanz et al., *Elements and Mineral Resources*, Springer Textbooks in Earth Sciences,
Geography and Environment, https://doi.org/10.1007/978-3-030-85889-6_7

Fig. 7.1 Native bismuth. *Gualba* (*Catalonia*) (*Photo* Joaquim Sanz. MGVM)

intrusive ore deposits, as porphyry and epithermal systems. The main bismuth ore deposit is in Chenzhou (Hunan Province, China), a Co–Ni-Ag-Sn pegmatite deposit.

7.2 Producing Countries

There is not enough information to quantify the world's reserves, as most bismuth is obtained as a by-product of either lead or tungsten mining, depending on the country (Fig. 7.2).

7.3 Applications

Medicine

Bismuth subcitrate is used as an antidiarrheal and for the treatment of some gastrointestinal diseases, such as ulcers and gastritis (Fig. 7.3).

Metallurgical industry

Bismuth alloys expand as they cool, so they·are used in the manufacture of metal parts as they fill all the space in the mold that is used.

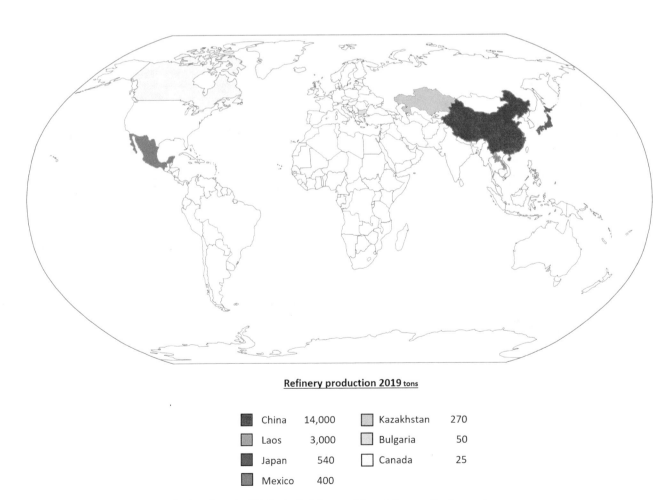

Refinery production 2019 tons

China	14,000	Kazakhstan	270
Laos	3,000	Bulgaria	50
Japan	540	Canada	25
Mexico	400		

Fig. 7.2 List of producing countries based on the US Geological Survey, Mineral Commodity Summaries

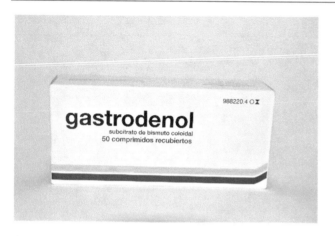

Fig. 7.3 Medicine. (*Photo* Joaquim Sanz. MGVM)

Tin–bismuth electroplating baths are used to coat electrotechnical and electronic components when the solderability of the finish is important.

Bismuth is relatively inert and harmless, so in model assembly, model-making, and so on it has been replacing lead, which is highly toxic, in special Sn-Bi solders that have a lower melting point (114–200 °C) than Sn–Pb.

Electronics industry

Bismuth forms part of Peltier plates, which produce cold or heat as a result of the passage of an electric current through the union of two metals, alloys, or semiconductors (the Peltier effect) (Fig. 7.4).

Fig. 7.4 Peltier plate (*Photo* Joaquim Sanz. MGVM)

Due to its high cost and low coefficient of energy performance, this is restricted to cooling certain computer CPUs, where the temperatures to be dissipated are not very high.

Other uses

Yellow bismuth oxide is a pigment, and it is used in cosmetics.

Other bismuth compounds are used in the manufacture of paint with a pearl-like finish and as a glaze for ceramics.

7.4 Recycling

Bismuth is recycled from materials made from alloys containing this metal. However, recycling is very limited due to the difficulty of separating bismuth from the other components of the alloys.

Further Reading

Eurometaux. (2021) Introducing Metals https://eurometaux.eu/about-our-industry/introducing-metals/ (last accesed May 2021)

Gray T, Mann N (2009) The elements. Black Dog & Leventhal Publishers Inc., New York

Mata JM, Sanz J (2007) Guia d'identificació de minerals. Manresa, Catalonia. Edicions UPC/Parcir (Catalan 2nd paper edition). 262 p. ISBN: 9788483019023. http://hdl.handle.net/2117/90445

Pirajno F (2009) Hydrothermal processes and mineral systems. Springer

Quadbeck-Seeger H-J (2007) World of the elements: elements of the world. Wiley-VCH Verlag GmbH & Co, Germany

Sanz J, Tomasa O (2017) Elements i Recursos minerals: Aplicacions i reciclatge. Manresa, Catalonia. Zenobita Edicions /Iniciativa Digital Politècnica (Catalan 3rd digital edition). http://hdl.handle.net/2117/105113

Sanz J, Tomasa O (2018) Elementos y Recursos minerales: Aplicaciones y reciclaje. Manresa, Catalonia. Zenobita Edicions/Iniciativa Digital Politècnica (Spanish 1st digital edition). http://hdl.handle.net/2117/123674

Stwertka A (2018) A guide to the elements, 4th edn. Oxford University Press, England

USGS (2021) Commodity Statistics and Information. Bismuth. https://www.usgs.gov/centers/nmic/bismuth-statistics-and-information (last accessed May 2021)

Boron (B) [Z = 5]

PERIODIC TABLE OF ELEMENTS (interactive)

- A semi-metal semiconductor
- Very hard
- Obtained from borax, colemanite, kernite, ulexite, and boron-rich brines (Fig. 8.1)
- Borates were assigned the status of strategic minerals by the EU in 2017.

8.1 Geology

Boron is widely distributed in the environment, including in soil, water, and animals, and it is essential to plant life. However, in nature it does not occur in an elemental state: it is always found in oxidized states as borate salts. Although borates comprise over 250 boron-bearing minerals (mostly salts of sodium, calcium, and magnesium), there are only four economically viable deposit types in the world: two

Borate deposits are formed in tectonically active areas in arid climates. The most are extracted in California and Turkey, and there are also large deposits in the Tethyan belt in western Asia and the Andean belt of South America.

8.2 Producing Countries

Of those countries with boron ore reserves, Turkey stands out, with 950,000 tons, followed by the United States, Russia, Chile, China, and Peru (Fig. 8.2).

8.3 Applications

Electric and electronic industry

Boron is an important dopant for silicon and germanium semiconductors.

Chemical industry

Sodium perborate is a source of active oxygen, with bleaching effects in detergents and other cleaning products.

Fig. 8.1 Brines with boron. Salar de Uyuni (Bolivia) (*Photo* Oriol Tomasa. MGVM)

sodium borates, borax ($Na_2B_4O_5(OH)_4 \cdot 8H_2O$) and kernite ($Na_2B_4O_6(OH)_2 \cdot 3H_2O$); a calcium borate, colemanite ($Ca_2B_6O_{11} \cdot 5H_2O$); and a sodium-calcium borate, ulexite ($NaCaB_5O_6(OH)_6 \cdot 5H_2O$).

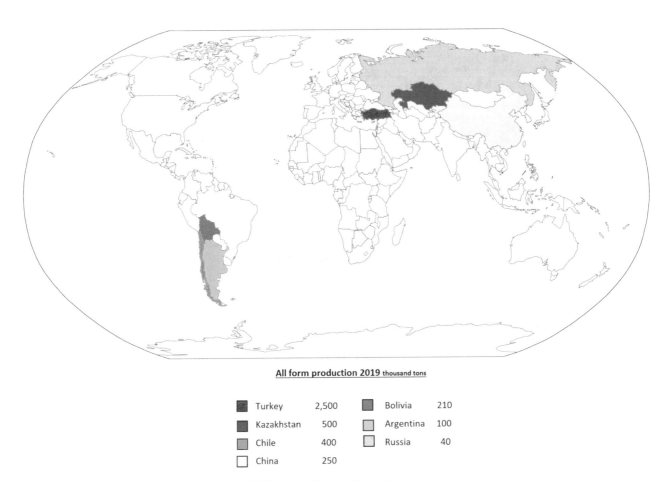

All form production 2019 thousand tons

Turkey	2,500	Bolivia	210
Kazakhstan	500	Argentina	100
Chile	400	Russia	40
China	250		

Fig. 8.2 List of producing countries based on the US Geological Survey, Mineral Commodity Summaries

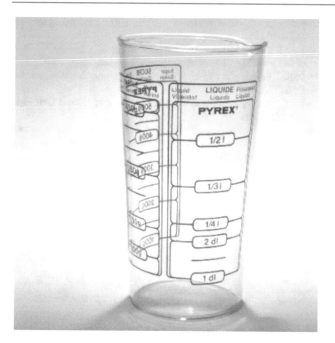

Fig. 8.3 Pyrex container. (*Photo* Joaquim Sanz. MGVM)

It is found in toothpaste as a tooth whitener. In addition, it is used in the production of fireworks to impart a green color.

Glass and ceramics industry

The main use of boron specifically in the form of aluminum sodium borosilicate is in the manufacture of high-strength glass, as it provides a low coefficient of thermal expansion. Pyrex is the brand name of this type of high-temperature resistant glass (Fig. 8.3). Borates are used in the ceramic industry, in the manufacture of glass fiber for insulation (IGFG), and in glazes such as for textile-grade glass fiber (TGFG).

A ceramic with high thermal conductivity and, at the same time, low thermal expansion is prepared with boron nitride, resulting in excellent resistance to thermal shock. The working temperature range for this ceramic is 850–1800 °C, depending on how it has been prepared.

Power generation

Boron carbide is used in the control rods of BWR-type (Boiling Water Reactor) nuclear power plants (GEE—Grupo Empresarial Electromédico). Boron is used as a neutron absorber in the pools of borated water where the spent fuel assemblies are cooled (Fig. 8.4).

Other fields

Boron carbide, a material with a hardness of 9.3 out of 10 on the Mohs scale, is used in the manufacture of tank armor and bulletproof guns, and as an abrasive. The energy released by

Fig. 8.4 Nuclear fuel cooling pool. (*Image courtesy of* Javier Castelo)

the rapid combustion of amorphous boron acts as a propellant in vehicle safety cushions (airbags).

In agriculture, boron salts are a vital micronutrient for many plants, especially palm oil crops.

Boron is a basic component of the manufacture of neodymium magnets, the most powerful permanent magnets on the market (see neodymium).

Boric acid is also involved in the manufacture of insecticides.

8.4 Recycling

The amount of boron that is recycled is very small.

Further Reading

Earth Magazine: Mineral Resource of the Month (2019) Boron. https://www.earthmagazine.org/article/mineral-resource-month-boron-0 (last accessed May 2021)

Encyclopedia of Geology (2020) 2nd edn. Elsevier, pp 489–504. https://www.elsevier.com/books/encyclopedia-of-geology/elias/978-0-08-102908-4

Gray T, Mann N (2009) The elements. Black Dog & Leventhal Publishers Inc., New York

Goodfellow (2021) Boron nitride. https://www.goodfellow-ceramics.com/sp/productos/ceramicas/nitrato-de-boro-bn/ (last accessed May 2021)

Mata JM, Sanz J (2007) Guia d'identificació de minerals. Manresa, Catalonia. Edicions UPC/Parcir (Catalan 2nd paper edition) 262p. ISBN: 9788483019023. http://hdl.handle.net/2117/90445

Quadbeck-Seeger H-J (2007) World of the elements: elements of the world. Wiley-VCH Verlag GmbH & Co, Germany

Roskill (2015) Market Reports. Boron. https://roskill.com/market-report/boron (last accessed May 2021)

Sanz J, Tomasa O (2017) Elements i Recursos minerals: Aplicacions i reciclatge. Manresa, Catalonia. Zenobita Edicions /Iniciativa Digital Politècnica (Catalan 3th digital edition). URL: http://hdl.handle.net/2117/105113

Sanz J, Tomasa O (2018) Elementos y Recursos minerales: Aplicaciones y reciclaje. Manresa, Catalonia. Zenobita Edicions/ Iniciativa Digital Politècnica (Spanish 1st digital edition). http://hdl.handle.net/2117/123674

Stwertka A (2018) A guide to the elements, 4th edn. Oxford University Press, England

USGS (2021) Commodity Statistics and Information. Boron https://www.usgs.gov/centers/nmic/boron-statistics-and-information (last accessed May 2021)

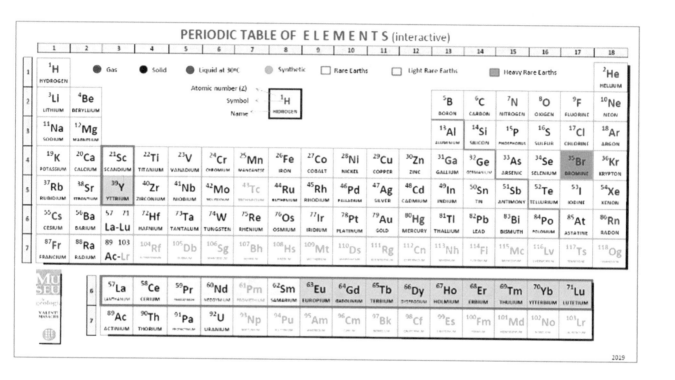

PERIODIC TABLE OF ELEMENTS (interactive)

9.1 Geology

- A non-metal
- Liquid at room temperature
- Slightly soluble in water
- Dense, highly aggressive, corrosive, and unpleasant-smelling
- Vapors are highly toxic when inhaled
- Obtained from the water of some seas, from salt lakes, and from the brine associated with some oil deposits (Fig. 9.1)

Bromine is abundant in nature in the form of bromide salts or as organobromine compounds, which are produced by many types of marine organisms. The most recoverable form of bromine is the soluble salts found in seawater, salt lakes, inland seas, and brine wells. Seawater contains bromine at about 65 ppm. Bromine at much higher concentrations (2500–10,000 ppm) is found in inland seas and brine wells, for instance the Dead Sea (approximately 5 gr/l), in some

Fig. 9.1 Bromine-rich seawater. *Dead Sea* (*Israel*) (*Image courtesy of* ICL)

thermal springs, and in rare insoluble silver bromide minerals (such as bromargyrite (AgBr), found in Mexico and Chile).

The major areas of bromine production are salt brines in the United States and China, the Dead Sea in Israel and Jordan, and oceanic water in Wales and Japan.

9.2 Producing Countries

The main world reserves of bromine are in Israel, Jordan, and the United States (Fig. 9.2).

9.3 Applications

Electricity generation

The combustion of coal to produce electricity in thermal power plants generates mercury emissions. Bromine additives in the coal convert the elemental mercury to its oxidized form, as this is more easily recovered by the emission-control system.

Chemical industry

Brominated compounds are used in the production of brominated flame retardants (BFR) and plastics, paint, and building materials, to make them less combustible. However, in Europe the use of certain BFRs is banned in

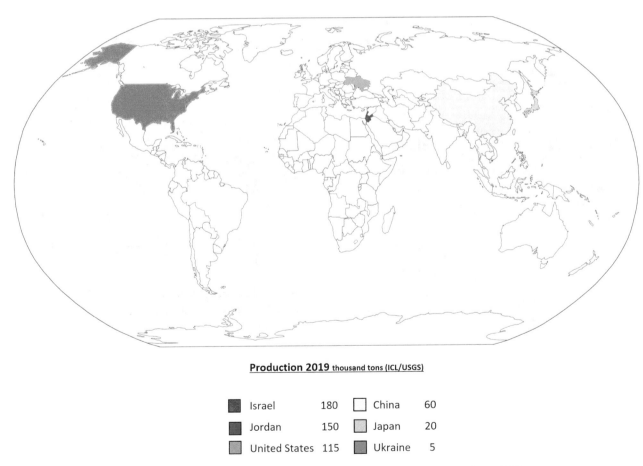

Production 2019 thousand tons (ICL/USGS)

■ Israel	180	□ China	60
■ Jordan	150	▨ Japan	20
▨ United States	115	■ Ukraine	5

Fig. 9.2 List of producing countries based on the US Geological Survey, Mineral Commodity Summaries and ICL

Fig. 9.3 Printed circuit board. (*Photo* Joaquim Sanz. MGVM)

electronic display plastics because, when they burn, they release environmentally toxic gases and prevent recycling (International Bromine Council BSEF 2021). They are added in the manufacture of polyesters and the epoxy resins in printed circuit boards (Fig. 9.3).

Oil industry

Bromine compounds have many applications in drilling fluids for oil and gas wells.

Agriculture

Methyl bromide is used as a pesticide for soil fumigation.

Food

Bromine is used in the production of brominated vegetable oil (BVO), used as an emulsifier in many citrus non-alcoholic beverages. The amount of this additive in soft drinks is regulated, as too much can be harmful to health. However, in certain brands of soft drinks BVOs are being gradually replaced by other products.

Other fields

Silver bromide constitutes the basic element of photographic film as it is photosensitive, and, despite digital photography, still finds a use among certain professionals and amateurs.

The addition of bromine to butyl rubber imparts considerable physical resistance to industrial tires and rubbers (conveyor belts, caps, and medical membranes), as well as low permeability and high resistance to shock, to weather (cold-heat) and to aging. The rubber blades of windscreen wipers benefit from this contribution, lasting longer and improving the driving visibility.

Bromine compounds are in the composition of certain pharmaceutical products and used in the treatment of industrial waters.

9.4 Recycling

Recycled bromine is from the treatment of chemical solutions containing this element, preventing the waste from causing environmental damage, as well as from the incineration of plastics containing flame retardants based on this element's derivatives.

Further Reading

AZOM (2021) Azo Materials. https://www.azom.com/article.aspx?ArticleID=3528 (last accessed May 2021)

BSEF (2021) International Bromine Council. www.bsef.com (last accessed May 2021)

Encyclopedia Britannica. Bromine. https://www.britannica.com/science/bromine

Gray T, Mann N (2009) The elements. Black Dog & Leventhal Publishers Inc., New York

ICL (2021) Bromine Retardants. https://www.icl-ip.com/flame-retardants/ (last accessed May 2021)

Mata JM, Sanz J (2007) Guia d'identificació de minerals. Manresa, Catalonia. Edicions UPC/Parcir (Catalan 2nd paper edition). 262p. http://hdl.handle.net/2117/90445

Quadbeck-Seeger H-J (2007) World of the elements: elements of the world. Wiley-VCH Verlag GmbH & Co, Germany

Sanz J, Tomasa O (2017) Elements i Recursos minerals: Aplicacions i reciclatge. Manresa, Catalonia. Zenobita Edicions/Iniciativa Digital Politècnica (Catalan 3rd digital edition). http://hdl.handle.net/2117/105113

Sanz J, Tomasa O (2018) Elementos y Recursos minerales: Aplicaciones y reciclaje. Manresa, Catalonia. Zenobita Edicions/Iniciativa Digital Politècnica. (Spanish 1st digital edition). http://hdl.handle.net/2117/123674

Stwertka A (2018) A guide to the elements, 4th edn. Oxford University Press, England

USGS (2021) Commodity Statistics and Information. Bromine. https://www.usgs.gov/centers/nmic/bromine-statistics-and-information (last accessed May 2021)

Cadmium (Cd) [Z = 48]

PERIODIC TABLE OF E L E M E N T S (interactive)

- A soft, malleable, and ductile metal
- Toxic
- In contact with the air, it forms an oxidation patina
- A by-product of the extraction of zinc from zinc sulfides such as sphalerite (Fig. 10.1)

10.1 Geology

Cadmium minerals are very scarce. Greenockite (CdS) is the main cadmium mineral, but cadmium is most often found in zinc ores, replacing zinc in the structure of sphalerite (ZnS). This is because cadmium shares certain chemical properties with zinc and often substitutes for it in the sphalerite crystal

Fig. 10.1 Sphalerite (zinc sulfide with cadmium). *Picos de Europa (Santander, Spain)* (*Photo* Joaquim Sanz. MGVM)

lattice. Greenockite is frequently associated with weathered sphalerite (ZnS) and wurtzite ((Zn,Fe)S).

Because of its association with zinc minerals, cadmium is usually sourced as a by-product of the recovery of primary zinc from zinc ores. It can be also obtained from some lead or complex copper-lead–zinc ores. However,

cadmium in lead and copper ores is associated with zinc sulfide rather than other minerals. China has the most extensive cadmium reserves. The United States, Mexico, Peru, Russia, and India are other countries with substantial deposits.

10.2 Producing Countries

Since cadmium is a by-product of zinc production, the world reserves of this element relate to the reserves of that metal (see: zinc). The cadmium content of zinc ores is of the order of 0.03% (Fig. 10.2).

10.3 Applications

Power generation

Cadmium is used in the control rods of many nuclear reactors because of its ability to absorb neutrons.

Cadmium telluride is used in the manufacture of the conductive thin film on photovoltaic solar cells.

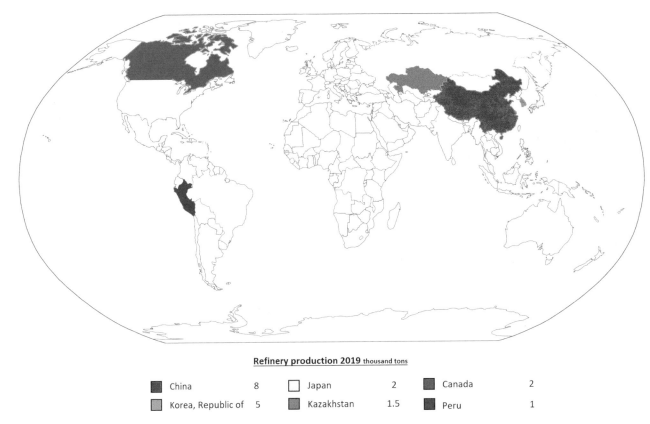

Refinery production 2019 thousand tons

■ China	8	□ Japan	2	■ Canada	2
■ Korea, Republic of	5	■ Kazakhstan	1.5	■ Peru	1

Fig. 10.2 List of producing countries based on the US Geological Survey, Mineral Commodity Summaries

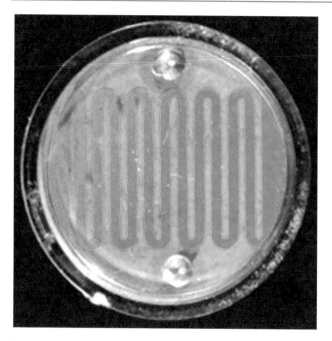

Fig. 10.3 Cadmium sulfide cell. (*Photo* Joaquim Sanz. MGVM)

Cadmium sulfide cells act as photoresistors, varying their resistance according to the intensity of light that they receive (Fig. 10.3).

Battery industry

In 2009, 89% of the cadmium consumed was for the production of rechargeable nickel–cadmium (Ni–Cd) batteries for use in portable electronic equipment and toys; however, these batteries were heavy and could not store as much energy as the Ni-MH (nickel-metal hydride) batteries (see: lanthanum) or Li-ion batteries (see: lithium) now replacing them.

From 31 December 2016, the EU has prohibited the use and trade of batteries containing more than 0.002% of cadmium by weight (Directive 2006/66/EC), and they may be used only in emergency systems and lighting, such as alarms and medical equipment, due to the toxicity of the material, with the recommendation of progressive replacement by Li-ion batteries.

Electroplating industry

Cadmium is used in electroplating (cadmium plating) of only certain materials, due to its toxicity. It is in certain special alloys for bearings because of its low coefficient of friction and high fatigue resistance (Fig. 10.4).

Fig. 10.4 Cadmium-plated steel. (*Photo* Joaquim Sanz. MGVM)

Chemical industry

Cadmium sulfide is a yellow pigment and cadmium selenide a red pigment, are both are used in paint.

10.4 Recycling

Cadmium is mainly recovered from spent industrial nickel–cadmium batteries and from the scrap generated during the production of metallic materials containing cadmium.

Further Reading

Gray T, Mann N (2009) The elements. Black Dog & Leventhal Publishers Inc., New York

Mata JM, Sanz J (2007) Guia d'identificació de minerals. Manresa, Catalonia. Edicions UPC/Parcir (Catalan 2nd paper edition). 262p. ISBN: 9788483019023. http://hdl.handle.net/2117/90445

Quadbeck-Seeger H-J (2007) World of the elements: elements of the world. Wiley-VCH Verlag GmbH & Co, Germany

Sanz J, Tomasa O (2017) Elements i Recursos minerals: Aplicacions i reciclatge. Manresa, Catalonia. Zenobita Edicions/Iniciativa Digital Politècnica (Catalan 3rd digital edition). http://hdl.handle.net/2117/105113

Sanz J, Tomasa O (2018) Elementos y Recursos minerales: Aplicaciones y reciclaje. Manresa, Catalonia. Zenobita Edicions/Iniciativa Digital Politècnica (Spanish 1st digital edition). http://hdl.handle.net/2117/123674

Stwertka A (2018) A guide to the elements, 4th edn. Oxford University Press, England

USGS (2021) Commodity Statistics and Information. Cadmium. https://www.usgs.gov/centers/nmic/cadmium-statistics-and-information (last accessed May 2021)

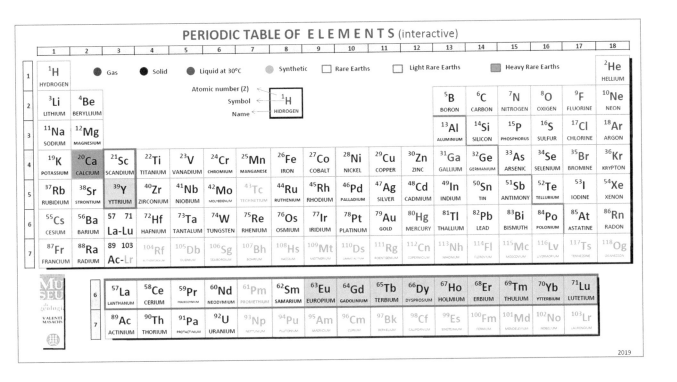

- A soft, alkaline earth metal
- The fifth most abundant element on Earth
- A basic element for all living things
- Bones of humans and animals contain much calcium, as do shells and skeletons of many marine creatures
- Fundamental component of calcite, limestone, chalk, marble, and gypsum (Fig. 11.1)
- Many waters, inland and marine, contain dissolved calcium as Ca^{2+} ions. The classification as hard or soft is on the basis of the amount of calcium carbonate per mg/l

- Stalactite and stalagmite formation in karst caves demonstrates the high of calcium carbonate content achieved over time by filtering water.

11.1 Geology

The main sources of calcium are rocks with a high percentage of $CaCO_3$, such as limestone and chalk.

Limestone deposits are extensive all around the world. Therefore, there is much variability in limestone deposits.

Fig. 11.1 Calcite (calcium carbonate). *Arties* (*Catalonia*) (*Photo* Joaquim Sanz. MGVM)

Typically, they are formed in two main environments. The first is shallow, calm, and warm-water seas, nowadays seen at latitudes between 30° N and 30° S, e.g. the Caribbean Sea and around the Pacific Ocean islands. In this setting, to form calcium carbonate shells and skeletons, organisms are capable of extracting the necessary ingredients from ocean water. After death, their skeletons and shells accumulate as sediment and are lithified. Limestones formed in this type of deposits are biogenic sedimentary rock (Geyssant 2001). After they were uplifted, such deposits became widespread worldwide. The other way that limestone can be formed is through evaporation of water with a high percentage of $CaCO_3$. When droplets of water seep from fractures or other pore spaces to a cave's roof, they can evaporate rather than fall to the cave floor. Any calcium carbonate dissolved in the droplet is deposited on the cave's roof, forming stalactites. If such droplets do fall, evaporation will contribute to the construction of a stalagmite. Stalactites and stalagmites are not used as a source of calcium: they have only ornamental and aesthetic value.

Chalk is a non-clastic carbonate sedimentary rock, notably white, soft, and fine-grained. On the whole, it was formed during the Cretaceous Period, 90 million years ago, in deep marine conditions from the gradual accumulation of ooze at the bottom of a great sea. Oozes are a mixture of sub-microscopic skeletons and protozoa shells that have fallen to the seafloor, then consolidated and compressed during diagenesis into chalk. The purest varieties can have up to 99% calcium carbonate as calcite mineral; however, others can have impurities, such as small amounts of silica or detrital grains of quartz and chert derived from sponge spicules, diatoms, and radiolarians. Some may contain small proportions of mineral impurities, e.g. clay minerals, among others (*Encyclopaedia Britannica*).

There is a wide variety of famous deposits of European chalk, such as the White Cliffs of Dover in England, the Champagne region of France, and some of the world's highest chalk cliffs at Jasmund National Park in Germany.

11.2 Producing Countries

The world's reserves of rocks with exploitable calcium are very large and at the same time variable, depending on the geology of each country (Fig. 11.2).

11.3 Applications

Depending on the application, crushed limestone and chalk may be used directly or after first extracting the calcium carbonate, calcium oxide (lime), etc. (see: calcite) by means of certain thermal and/or chemical treatments.

Various industries
Crushed quality limestone is essential to the manufacture of cement.

The world's largest consumption of calcium carbonate (both crushed (GCC) and precipitated (PCC)) is in the manufacture of paper (there is less consumption of vegetable pulp) (Fig. 11.3). PVC-type plastics, paint, and adhesives are used to give consistency (filler) to the final product.

Calcium carbonate is an excellent feed supplement for laying hens, to give consistency to their eggshells, and is a key ingredient of chewing gum (Fig. 11.4).

This compound, together with barium sulfate, is an additive in the production of rubber, to which it imparts greater elasticity.

Calcium carbide is the raw material for the manufacture of acetylene gas. Together with oxygen it forms a gaseous mixture that allows welding or cutting of many metals.

Calcium sulfate (gypsum) is used in wall lining in construction, cement, and surgical casts (see: gypsum).

Calcium hypochlorite is employed as a water disinfectant, a bleaching agent, and a deodorant.

Calcium phosphate is involved in the production of fertilizers and animal feed (see: phosphorus).

Calcium oxide (lime) is used in the manufacture of steel and glass, in sugar refinery and flue gas desulfurization, and to obtain precipitated calcium carbonate.

Medicine
Very pure calcium carbonate is used as an antacid to neutralize an excess of hydrochloric acid in the stomach. It is the most important and inexpensive calcium dietary supplement.

Hydroxyapatite, a calcium phosphate, is a drug used to strengthen the enamel in teeth and bones and to treat osteoporosis.

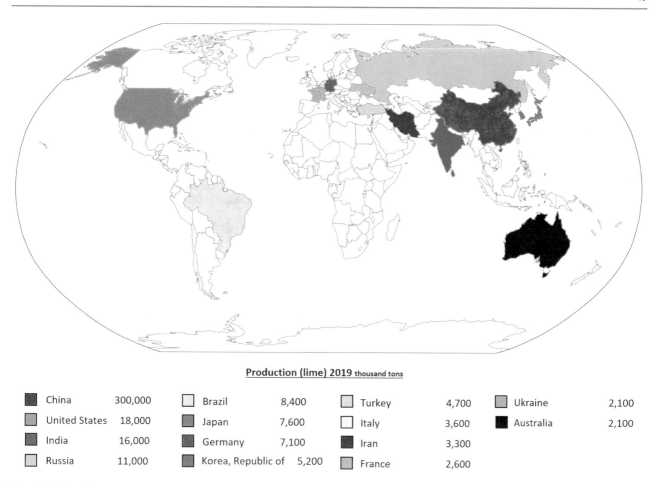

Production (lime) 2019 thousand tons

China	300,000	Brazil	8,400	Turkey	4,700	Ukraine	2,100
United States	18,000	Japan	7,600	Italy	3,600	Australia	2,100
India	16,000	Germany	7,100	Iran	3,300		
Russia	11,000	Korea, Republic of	5,200	France	2,600		

Fig. 11.2 List of producing countries based on the US Geological Survey, Mineral Commodity Summaries

Fig. 11.3 Paper manufacturing. (*Image courtesy of* STORA ENSO)

Fig. 11.4 Hen's egg. (*Photo* Joaquim Sanz. MGVM)

Calcium carbonate is an essential component of toothpaste.

Other fields

Calcium carbonate is still useful for sticks of chalk for the classic blackboards in those schools without whiteboards (these use marker pens) or smartboards.

11.4 Recycling

In the paper industry, as much calcium oxide as possible is recovered.

Further Reading

Britannica (2021) Calcite. https://www.britannica.com/science/calcite

Britannica (2021) Calcium. https://www.britannica.com/science/calcium

Britannica (2021) Chalk. https://www.britannica.com/science/chalk

Geyssant J (2001) Limestone deposits. In: Calcium Carbonate. Birkhäuser, Basel, pp 31–51. https://doi.org/10.1007/978-3-0348-8245-3_3

Gray T, Mann N (2009) The elements. Black Dog & Leventhal Publishers Inc., New York

Mata JM, Sanz J (2007) Guia d'identificació de minerals. Manresa, Catalonia. Edicions UPC/Parcir (Catalan 2nd paper edition). 262p. ISBN: 9788483019023. http://hdl.handle.net/2117/90445

Middleton GV, Church MJ, Coniglio M, Hardi LA, Longstaffe FJ (eds) (2003) Encyclopedia of sediments and sedimentary rocks. Encyclopedia of Earth Sciences Series. https://doi.org/10.1007/978-1-4020-3609-5

Quadbeck-Seeger H-J (2007) World of the elements: elements of the world. Wiley-VCH Verlag GmbH & Co, Germany

Sanz J, Tomasa O (2017) Elements i Recursos minerals: Aplicacions i reciclatge. Manresa, Catalonia. Zenobita Edicions /Iniciativa Digital Politècnica (Catalan 3rd digital edition). http://hdl.handle.net/2117/105113

Sanz J, Tomasa O (2018) Elementos y Recursos minerales: Aplicaciones y reciclaje. Manresa, Catalonia. Zenobita Edicions/Iniciativa Digital Politècnica. (Spanish 1st digital edition). http://hdl.handle.net/2117/123674

Stwertka A (2018) A guide to the elements, 4th edn. Oxford University Press, England

USGS (2021) Commodity Statistics and Information. Lime. https://www.usgs.gov/centers/nmic/lime-statistics-and-information?qt-science_support_page_related_con=0#qt-science_support_page_related_con (last accessed May 2021)

Warren JK (2016) Interpreting evaporite textures. In: Evaporites. Springer, Cham, pp 1–83. https://doi.org/10.1007/978-3-319-13512-0_1

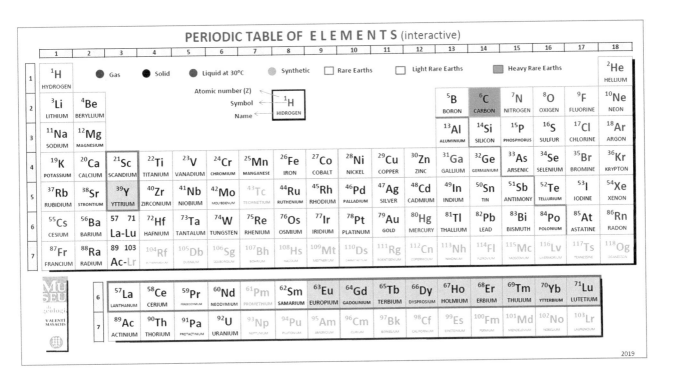

- The key non-metal in organic chemistry
- A vital component of all living beings
- The basis of life on Earth
- A component of CO_2 gas, a pollutant causing global warming, acid rain, and climate change. Our civilization is constantly emitting large amounts of CO_2 from the combustion of fossil fuels (oil, coal, natural gas) (Fig. 12.1), in addition to natural emissions from volcanoes and animal and plant metabolism

- Found also in nature in the form of diamond (the hardest) and graphite (very soft).

12.1 Applications

Energy generation

Oil, like natural gas and coal, has carbon as its basic component. Fossil fuels, they are all sources of heat to

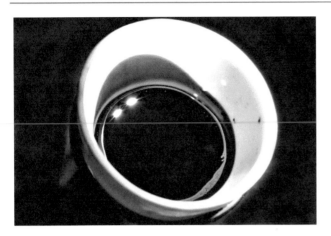

Fig. 12.1 Oil (crude). Tarragona (Catalonia) (*Photo* Joaquim Sanz. MGVM)

Fig. 12.3 Carbon steel pan (*Photo* Joaquim Sanz. MGVM)

create the steam to turn turbines and thus generate electrical energy.

Oil and petrochemical industry

The oil industry operates drilling wells for the extraction of crude oil, later distilled by petrochemical companies to obtain gases (butane, propane), gasoline, gas oil, kerosene, lubricating oils, and tars. In turn, from petroleum derivatives we obtain plastics, synthetic textile fibers, among many other products (Fig. 12.2).

Metallurgical industry

Carbon steel is an alloy of iron and carbon for the manufacture of metal structures, vehicles, utensils, etc. (see: iron) (Fig. 12.3).

Other fields

The radioactive carbon-14 isotope, with a half-life of 5730 years, is behind a system for dating organic remains up 50,000 years old to determine the approximate age of archaeological remains, as well as fossils.

Due to the strength and lightness properties of carbon fiber composites, they are used in the manufacture of sports materials (see: carbon-graphite).

12.2 Recycling

Carbon recycling is unknown.

Further Reading

Gray T, Mann N (2009) The elements. Black Dog & Leventhal Publishers Inc., New York

Mata JM, Sanz J (2007) Guia d'identificació de minerals. Manresa, Catalonia. Edicions UPC/Parcir (Catalan 2nd paper edition). 262p. ISBN: 9788483019023. http://hdl.handle.net/2117/90445

Quadbeck-Seeger H-J (2007) World of the elements: elements of the world. Wiley-VCH Verlag GmbH & Co, Germany

Sanz J, Tomasa O (2017) Elements i Recursos minerals: Aplicacions i reciclatge. Manresa, Catalonia. Zenobita Edicions/Iniciativa Digital

Fig. 12.2 Oil-drilling platform. (*Image courtesy of* Creative Commons Attribution 3.0 Brazil)

Politècnica (Catalan 3rd digital edition). http://hdl.handle.net/2117/105113

Sanz J, Tomasa O (2018) Elementos y Recursos minerales: Aplicaciones y reciclaje. Manresa, Catalonia. Zenobita Edicions/Iniciativa Digital Politècnica (Spanish 1st digital edition). http://hdl.handle.net/2117/123674

Stwertka A (2018) A guide to the elements, 4th edn. Oxford University Press, England

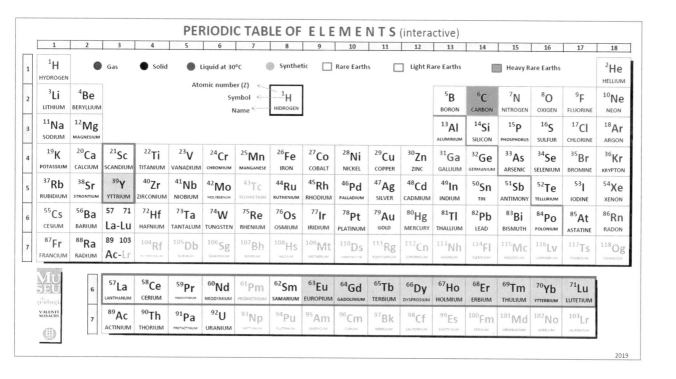

13.1 Geology

- An allotropic form of carbon
- The hardest natural material in existence (10 on the Mohs scale) (Fig. 13.1)
- High thermal conductivity and an excellent electrical insulator
- Blue diamonds (those with boron impurities) are natural semiconductors.

Diamond ore deposits are confined to a few geological settings. Diamonds are typically found in ancient regions of the Earth's crust, known as cratons. These are extensive parts that achieved long-term stability so have been poorly deformed over a long period.

Cratons are composed of crustal basement rocks, such as granites and gneisses. The lithospheric mantle roots beneath cratons can reach to depths of more than 250 km, and due to

Fig. 13.1 Diamond (carbon). Kimberley (South Africa). (*Photo* Joaquim Sanz. MGVM)

their relatively low temperatures and extreme thickness they are the main locations of diamond formation in the Earth's mantle (Tappert and Tappert 2011).

It is believed that diamonds crystallize within large chambers hosted by mantle rocks, reaching to the lower–upper mantle and transition zone (Litvin 2017). Chambers are thought to be formed of ultrabasic mantle rocks that suffered early carbonation metasomatism. The eventual dissolution and dissemination of primary and accessory minerals and the dissemination of carbon-hosted phases in carbonate melts within chambers make up a magma enriched with elements such silica-oxide-carbonatitic with dissolved carbon (Litvin 2017).

Depending on the mantle region (upper, transition, or lower), diamond-forming melts are associated with different types of minerals. In the upper mantle they relate to peridotite eclogite, in the transition zone to magnesium silicate minerals, and in the lower zone to magnesium, iron, calcium, and titanium oxides and silicates (Litvin 2017).

Magmas can reach the surface by ascending underground volcanic structures known as pipes and craters of a few thousand meters in diameter. The volcanic eruptions associated with these processes are believed to be small yet violent. When the magmas cool after the eruption, they can form a volcanic rock known as kimberlite, which is the main type of diamond-bearing volcanic rock. There are further diamond-bearing volcanic rocks, such as lamproites, lamprophyres, and komatiites, yet only lamproites are economically significant.

The best-known kimberlite deposits are near the town of Kimberley in South Africa, where the discovery of the diamond called 'Star of South Africa' in 1869 spawned a diamond rush and created the Big Hole, claimed to be the largest ever excavated by hand. However, nowadays the main production of kimberlitic diamonds is known to be in Russia,

Canada, and Botswana. In the case of the lamproite type, one of the best well-known ore deposits is Argyle mine in the north of Western Australia, in the East Kimberley region.

Kimberlitic deposits located on the surface can suffer erosion; however, diamonds are highly resistant to chemical and mechanical weathering, therefore can become widely dispersed over long distances. When diamonds from different kimberlite sources join, other heavy minerals can comprise economic placer deposits. Diamonds are also frequently concentrated in alluvial deposits. Consequently, they can be found in sandstones or conglomerates. Nowadays, the largest and economically significant placer mining district is along the west coast of southern Africa, where diamonds are recovered from beach and marine sediments, as well as from riverbeds (e.g. Namaqualand, Oranjemund, and Lüderitz).

13.2   Producing Countries

The world's most important reserves of natural diamonds are in South Africa, followed by Australia, Canada, and Russia (Fig. 13.2).

13.3   Applications

Jewelry

A natural gem-quality diamond is considered to be one of the most valuable precious stones for both jewelry and the investment world, unfortunately financing many conflicts in African countries. Diamond dust is used as an abrasive to cut and polish both diamonds and other very hard materials (Fig. 13.3).

Industry

Natural diamonds that are not of gem quality (bort) and synthetic diamonds are employed in the manufacture of bits, saws, disks, and so on, as cutting and polishing tools for rock and concrete, for wellheads, and for public works (Fig. 13.4).

The 8% of diamonds for industry are of synthetic origin, as their quality can be controlled and their physical characteristics customized to the specific requirements of the work to be performed.

Electronics Industry

The ability to obtain synthetic diamonds from methane gas (Akhan, 2021) and to create the semiconductor properties or high thermal conductivity is contributing to alternatives to silicon in the manufacture of wide bandgap (WBG) semiconductors.

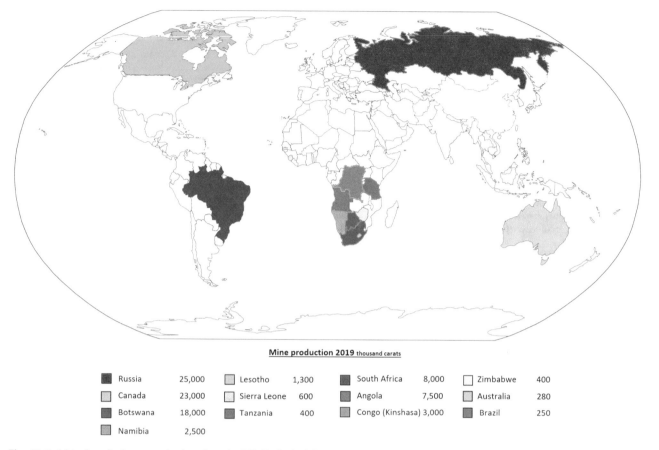

Mine production 2019 thousand carats

Russia	25,000	Lesotho	1,300	South Africa	8,000	Zimbabwe	400
Canada	23,000	Sierra Leone	600	Angola	7,500	Australia	280
Botswana	18,000	Tanzania	400	Congo (Kinshasa)	3,000	Brazil	250
Namibia	2,500						

Fig. 13.2 List of producing countries based on the US Geological Survey, Mineral Commodity Summaries

Fig. 13.3 Diamond jewel (brilliant cut). (*Photo* Joaquim Sanz. MGVM)

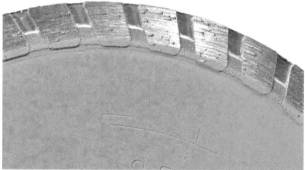

Fig. 13.4 Diamond saw blade. (*Photo* Joaquim Sanz. MGVM)

13.4   Recycling

All non-gem natural industrial diamond (bort), small crystals (grit) and diamond dust (dust) can be recovered, but the decrease in prices caused by the industrial production of synthetic diamonds has led to a reduction in their recycling.

References

Litvin YA (2017) Mantle-carbonatite conception of diamond and associated phases genesis. In: Genesis of diamonds and associated phases. Springer Mineralogy, Cham. https://doi.org/10.1007/978-3-319-54543-1_7

Tappert R, Tappert MC (2011) The origin of diamonds. In: Diamonds in nature. Springer, Berlin. https://doi.org/10.1007/978-3-642-12572-0_1

Further Reading

Akhan Semiconductor (2021) Home. https://www.akhansemi.com/home.html. Last accessed May 2021.

Gray T, Mann N (2009) The elements. Black Dog & Leventhal Publishers Inc., New York

Gurney JJ, Helmstaedt HH, Richardson SH, Shirey SB (2010) Diamonds through time. Econ Geol 105(3):689–712. https://doi.org/10.2113/gsecongeo.105.3.689

Mata JM, Sanz J (2007) Guia d'identificació de minerals. Manresa, Catalonia. Edicions UPC/Parcir. (Catalan 2nd paper edition). 262 p. ISBN: 9788483019023. http://hdl.handle.net/2117/90445

Quadbeck-Seeger H-J (2007) World of the elements: elements of the world. Wiley-VCH Verlag GmbH & Co, Germany

Revuelta MB (2018) Mineral deposits: types and geology. In: Mineral resources. Springer, Cham, pp. 49–119. https://doi.org/10.1007/978-3-319-58760-8_2

Sanz J, Tomasa O (2017) Elements i Recursos minerals: Aplicacions i reciclatge. Manresa, Catalonia. Zenobita Edicions /Iniciativa Digital Politècnica (Catalan 3rd digital edition). http://hdl.handle.net/2117/105113

Sanz J, Tomasa O (2018) Elementos y Recursos minerales: Aplicaciones y reciclaje. Manresa, Catalonia. Zenobita Edicions/ Iniciativa Digital Politècnica (Spanish 1st digital edition). http://hdl.handle.net/2117/123674

Smith BHS, Smith SC (2009) The economic implications of kimberlite emplacement. Lithos 112:10–22. https://doi.org/10.1016/j.lithos.2009.04.041

Stwertka A (2018) A guide to the elements, 4th edn. Oxford University Press, England

USGS (2021a) Commodity statistics and information. Diamond. https://pubs.usgs.gov/periodicals/mcs2021/mcs2021-diamond-industrial.pdf. Last accessed May 2021

USGS (2021b). Commodity statistics and information. Gemstones. https://pubs.usgs.gov/periodicals/mcs2021/mcs2021-gemstones.pdf. Last accessed May 2021

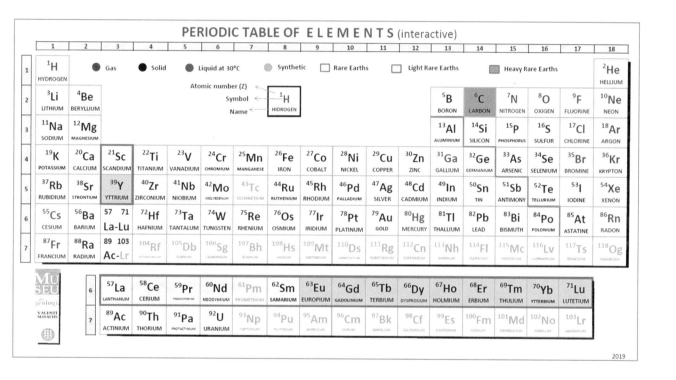

14.1 Geology

- An allotropic form of carbon
- A soft mineral (1–2 on the Mohs scale), in contrast to the hardness of diamond (Fig. 14.1)
- A good solid lubricant
- Low electrical conductivity
- Refractory, resisting high temperatures very well
- Assigned the status of a strategic mineral by the EU in 2017.

Graphite ore deposits fall into three main categories: microcrystalline graphite deposits, recently named 'graphitic carbon' (Beyssac and Rumble 2014); disseminated flake-graphite; and chip graphite (Robinson et al. 2017).

Microcrystalline graphite deposits are formed in coal seams by contact or regional metamorphism. Microcrystalline graphite deposits are mainly stratiform or lens shaped. They generally range from a few to tens of meters thick and from

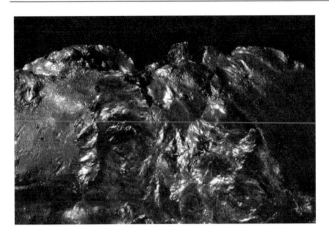

Fig. 14.1 Graphite (carbon). *Huelva (Jaén) Spain* (*Photo* Joaquim Sanz. MGVM)

pegmatite, and syenite. Commonly, flake-graphite deposits occur as strata-bound lenses. There are several deposits of great importance, for instance those in Canada, China, Brazil, and Mozambique (Simandl et al. 1995).

Chip graphite deposits are structured in well-defined veins, formed as a result of deposition from carbon-bearing fluids. Generally, vein graphite deposits are related to igneous rocks or a few metamorphic rocks, such as granulite. The largest deposits are in Sri Lanka and India (Luque et al. 2014).

In addition, synthetic graphite can be produced from hydrocarbon sources using a technology based on a high-temperature heat treatment; however, synthetic graphite is more expensive to produce than natural graphite. The United States is a major producer (Robinson et al. 2017).

hundreds of meters to a kilometer in length. Usually they lie within metasedimentary rocks, for instance the metamorphosed coal deposits of Sonora in Mexico (Taylor 2006).

Graphite flake-deposits can be found in a diversity of host rocks, such as marble, paragneiss, iron formation, quartzite,

14.2 Producing Countries

The world's graphite reserves are abundant and are in Turkey (90 Mt), China (73 Mt), Brazil (72 Mt), Mozambique (25 Mt), Canada, the United States, Tanzania (18 Mt), India

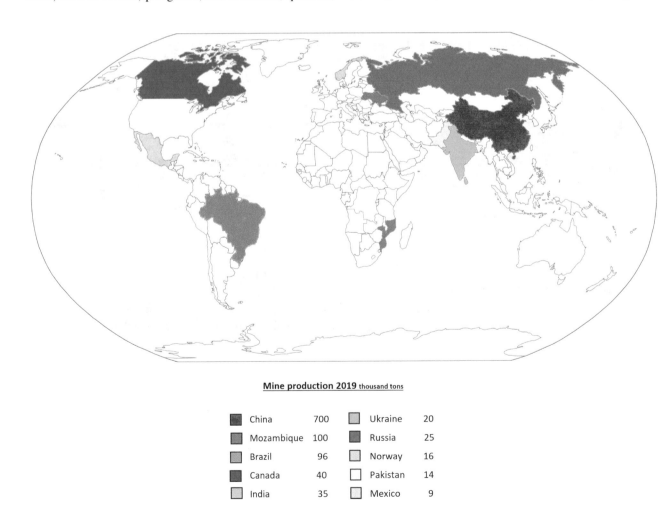

Mine production 2019 thousand tons

China	700		Ukraine	20
Mozambique	100		Russia	25
Brazil	96		Norway	16
Canada	40		Pakistan	14
India	35		Mexico	9

Fig. 14.2 List of producing countries based on the US Geological Survey, Mineral Commodity Summaries

(8 Mt), Vietnam (7.6 Mt), Mexico (3 Mt), Madagascar (1.6 Mt), and other countries (Fig. 14.2).

14.3 Applications

Refractory Applications

Graphite is mainly used as a refractory in linings, crucibles, and molds in the manufacture of steels and vehicle brake linings, which need to withstand high temperatures. It is a good natural lubricant that works well at high temperatures (Fig. 14.3).

Expanded graphite (made from flakes of natural graphite in a chromic acid bath, then in sulfuric acid) is used as a firewall in the manufacture of fire doors, because it resists heat well.

Electronics Industry

Graphene is a flat, lamellar structure of graphite that is just one carbon atom thick. It has high thermal and electrical conductivity. It also has a high elasticity and great hardness (it is two hundred times harder than steel). These characteristics make graphene an ideal material for components for integrated circuits and transparent touch screens.

Power Generation

Flake-graphite is used as an anode in the manufacture of Li-ion batteries (Fig. 14.4). According to Roskill, significant growth is expected in the use of this element in the coming years, both in mobile batteries and electric vehicles (EV).

Other Fields

Graphite is an important element in the manufacture of vehicle brake pads, and its applications as a natural lubricant are numerous.

Fig. 14.4 Li-ion battery. (*Photo* Joaquim Sanz. MGVM)

Increasingly, carbon fiber-reinforced polymers (PRF) are being employed. Aircraft such as the A350 use them, achieving a fuel economy of 5% (Fig. 14.5).

Carbon nanotubes are made of concentrically rolled graphite layers, forming tubes of nano-metric diameter. They are used as reinforcement in sports materials due to their high strength and low density.

Graphite combined with clay comprises the 'lead' in a writing pencil.

14.4 Recycling

Due to the abundance of natural graphite, the reuse of graphite is growing only slightly. However, the graphite from refractory bricks and molds for electric furnaces and metal melting processes is recovered, as well as that from vehicle brakes.

Fig. 14.3 Electric steel furnace. (*Image courtesy of* Roskill)

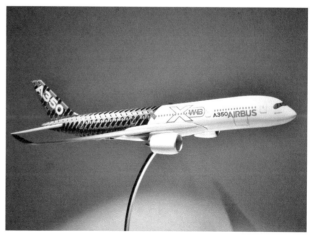

Fig. 14.5 Airbus A350. (*Photo* Joaquim Sanz. MGVM)

References

Beyssac O; Rumble D (2014) Graphitic carbon: a ubiquitous, diverse, and useful geomaterial. Elements 10(6):415–420. https://doi.org/10.2113/gselements.10.6.415

Luque FJ, Huizenga JM, Crespo-Feo E, Wada H, Ortega L, Barrenechea JF (2014) Vein graphite deposits: geological settings, origin, and economic significance. Mineralium Deposita, 49(2): 261–277. https://doi.org/10.1007/s00126-013-0489-9

Robinson Jr GR, Hammarstrom JM, Olson DW (2017) Graphite (No. 1802-J). US Geological Survey. https://pubs.er.usgs.gov/publication/pp1802J

Simandl GJ, Paradis S, Valiquette G, Jacob HL (1995) Crystalline graphite deposits, classification and economic potential. In: Proceedings of 28th forum on the geology of industrial minerals. Marinsburg, West Virginia, pp. 168–174, May 1992

Taylor HA (2006) Graphite. In: Kogel JE, Trivedi NC, Barker JM, Krukowski ST (eds) Industrial Minerals and Rocks, 7th edn. Society for Mining, Metallurgy, and Exploration Inc., Littleton, CO, pp 507–518

Further Readings

Gray T, Mann N (2009) The elements. Black Dog & Leventhal Publishers Inc., New York

Graphenea (2021) https://www.graphenea.com/. Last accessed May 2021

Mata JM, Sanz J, (2007) Guia d'identificació de minerals. Manresa, Catalonia. Edicions UPC/Parcir (Catalan 2nd paper edition). 262 p. ISBN: 9788483019023. http://hdl.handle.net/2117/90445

Olsen D (2020) Graphite annual publication. US Geological Survey, Mineral Commodity Summaries. URL: https://pubs.usgs.gov/periodicals/mcs2020/mcs2020-graphite.pdf

Quadbeck-Seeger H-J (2007) World of the elements: elements of the world. Wiley-VCH Verlag GmbH & Co, Germany

Roskill (2020) Market reports. Natural & synthetic graphite. Available at: https://roskill.com/market-report/natural-synthetic-graphite/. Last accessed May 2021

Sanz J, Tomasa O (2017) Elements i Recursos minerals: Aplicacions i reciclatge. Manresa, Catalonia. Zenobita Edicions /Iniciativa Digital Politècnica (Catalan 3rd digital edition). URL: http://hdl.handle.net/2117/105113

Sanz J, Tomasa O (2018) Elementos y Recursos minerales: Aplicaciones y reciclaje. Manresa, Catalonia. Zenobita Edicions/Iniciativa Digital Politècnica (Spanish 1st digital edition). URL: http://hdl.handle.net/2117/123674

Simandl GJ, Paradis S, Akam C (2015) Graphite deposit types, their origin, and economic significance. British Columbia Ministry of Energy and Mines & British Columbia Geol Surv 3:163–171

Stwertka A (2018) A guide to the elements, 4th edn. Oxford University Press, England

Touzain P, Balasooriya N, Bandaranayake K, Descolas-Gros C (2010) Vein graphite from the Bogala and Kahatagaha–Kolongaha mines, Sri Lanka: a possible origin. Canadian Mineralogist 48(6):1373–1384. https://doi.org/10.3749/canmin.48.5.1373

USGS (2021) Commodity Statistics and Information. Graphite. Available at: https://www.usgs.gov/centers/nmic/graphite-statistics-and-information. Last accessed May 2021

PERIODIC TABLE OF E L E M E N T S (interactive)

- An alkaline metal, not very abundant
- Lightweight and ductile
- Liquid at room temperature
- Very low melting point (28.4 °C)
- Very reactive in contact with water and oxygen
- Found in pollucite and lepidolite (Fig. 15.1).

15.1 Geology

The main source of cesium is pollucite, a primary lithium-cesium-rubidium mineral. It is also found in association with lithium-rich, lepidolite-bearing, and petalite-bearing zoned granite pegmatites.

LCT pegmatites (lithium–cesium–tantalum) form in compressional settings, in subduction zones, and in the thickened crust of high mountain ranges formed by the

Fig. 15.1 Pollucite (cesium aluminosilicate). *Mesquitela (Portugal)* (*Photo* Joaquim Sanz. MGVM)

collision of two continents. Giant deposits of these pegmatites are known at Tanco in Canada, Greenbushes in Australia, and Bikita in Zimbabwe.

15.2 Producing Countries

Cesium reserves relate to pollucite deposits, which in turn are associated with lithium-lepidolite mines (Australia, Namibia, United States, Zimbabwe, Canada). Cesium is also found in certain brines in Chile and China.

Currently, the Namibia and Zimbabwe fields have known reserves of cesium oxide of 30,000 and 60,000 tons respectively, and the reserves of other countries (Australia, Canada, United States) are unknown (Fig. 15.2).

15.3 Applications

Oil and Gas Drilling

Cesium formate is used as a drilling fluid in oil and gas extraction wells; it lubricates the drilling tools, maintains a constant pressure on the ground, and helps to remove debris.

Mine production 2019 tons

(No official resources for production data)

■ Namibia □ Zimbabwe

Fig. 15.2 List of producing countries based on the US Geological Survey, Mineral Commodity Summaries

Fig. 15.3 Paper manufacturing. (*Image courtesy of* Stora Enso)

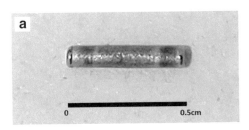

Fig. 15.4 **a**, **b** Cesium-137 'seeds' for brachytherapy (5 mm). (*Image courtesy of* Agustí Ruiz. Hospital Sant Pau (BCN))

Electrical Industry

Due to its attribute as a photosensitive element, cesium is used in photoelectric cells as it converts light into a flow of electrons (electricity).

Technical Measurements

The cesium-137 isotope, with a decay period of thirty years, is used in small quantities to calibrate radiation detection equipment. It is also used industrially to measure liquid flows and material thicknesses (such as paper) (Fig. 15.3).

Medicine

Cesium-137 is used in brachytherapy procedures for the treatment of cancer in small capsules known as 'seeds', which are inserted into the area where a tumor is present (Fig. 15.4).

Electronic Industry

Cesium is the chief component of the cesium atomic clock, the most accurate in the world with an error of one nanosecond per day, taking as its reference point the vibration frequency of cesium-133. Such clocks control the frequency of television and telephone transmitters, and global positioning systems (GPS).

Several cesium compounds are used in the manufacture of photoelectric cells, spectrophotometers, infrared detectors, and scintillometers. The last are used in SPECT photomultipliers (gamma-ray detectors) and hospital CT scanners.

15.4 Recycling

Cesium formate is recovered from the drilling fluids from oil and natural gas wells for reuse in new drillings.

Further Reading

Gray T, Mann N (2009) The elements. Black Dog & Leventhal Publishers Inc., New York

Mata JM, Sanz J (2007) Guia d'identificació de minerals. Manresa, Catalonia. Edicions UPC/Parcir (Catalan 2nd paper edition). 262 p. ISBN: 9788483019023. http://hdl.handle.net/2117/90445

Mindat (2021) Cesium https://www.mindat.org/min-5243.html. Last accessed May 2021

Quadbeck-Seeger H-J (2007) World of the elements: elements of the world. Wiley-VCH Verlag GmbH & Co, Germany

Sanz J, Tomasa O (2017) Elements i Recursos minerals: Aplicacions i reciclatge. Manresa, Catalonia. Zenobita Edicions/Iniciativa Digital

Politècnica (Catalan 3rd digital edition). http://hdl.handle.net/2117/105113

Sanz J, Tomasa O (2018) Elementos y Recursos minerales: Aplicaciones y reciclaje. Manresa, Catalonia. Zenobita Edicions/Iniciativa Digital Politècnica (Spanish 1st digital edition). http://hdl.handle.net/2117/123674

Stwertka A (2018) A guide to the elements, 4th edn. Oxford University Press, England

USGS (2021) Commodity Statistics and Information. Cesium. Available at: https://www.usgs.gov/centers/nmic/cesium-and-rubidium-statistics-and-information. Last accessed May 2021

Chlorine (Cl) [Z = 17]

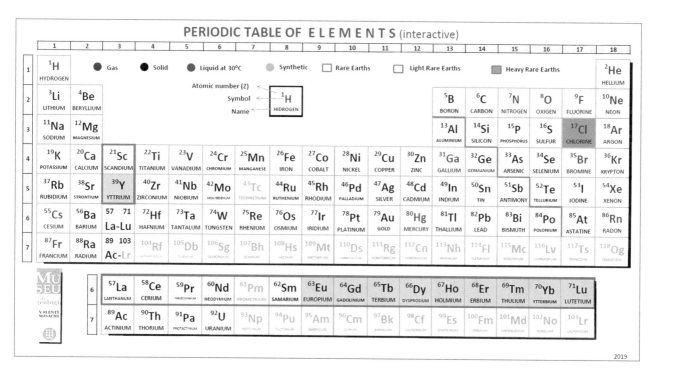

- A highly reactive and toxic gas
- An effective disinfectant, even in small amounts
- Obtained by electrolysis of sodium chloride obtained from halite and brines (Fig. 16.1).

16.1 Geology

The main source of chlorine is halite (NaCl) and brines associated with the evaporitic deposits. Brine is a water solution with a high content of halite (NaCl), greater than 5%. The main halite deposits are saline flats characteristic of

Fig. 16.1 Halite (sodium chloride). *Súria (Catalonia)* (*Photo* Joaquim Sanz. MGVM)

arid basins. They are shallow and dry apart from when storm flooding turns the saltpan and its surroundings into a temporary lake. Continental saline flats occupy the lowest areas of closed arid basins. The surrounding brine-soaked mudflats are permeated with evaporite minerals that have grown within the sediment. Saline flats occur in both continental

and marine margins, called sabkha. These deposits vary in size from one square kilometer to the largest, of thousands of square kilometers, such as Lake Uyuni in Bolivia.

Another type of halite deposit is salt domes. These are formed when salt deposits are buried in layers of sediment. Confined salt has the capacity to deform plastically under pressures and elevated temperatures, and it is forced by unequal pressures to flow upwards through weaker overlying strata. The resulting formation takes the characteristic shape of an elongated mushroom and it is composed of relatively pure halite.

16.2   Producing Countries

Most important salt producing countries in the world for the extraction of chlorine (Fig. 16.2).

16.3   Applications

Chemical Industry

Chlorine is mainly used in the purification of water (Fig. 16.3).

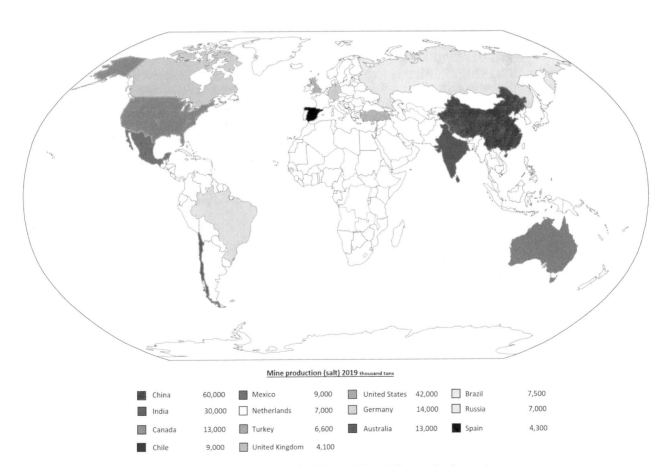

Mine production (salt) 2019 thousand tons

China	60,000	Mexico	9,000	United States	42,000	Brazil	7,500
India	30,000	Netherlands	7,000	Germany	14,000	Russia	7,000
Canada	13,000	Turkey	6,600	Australia	13,000	Spain	4,300
Chile	9,000	United Kingdom	4,100				

Fig. 16.2 List of producing countries based on the US Geological Survey, Mineral Commodity Summaries

Fig. 16.3 Water treatment plant. (*Photo* Joaquim Sanz. MGVM)

Fig. 16.5 PVC pipe. (*Photo* Joaquim Sanz. MGVM)

Sodium hypochlorite (bleach) is used to bleach pulp in paper manufacture, to bleach textiles, and as a disinfectant (Fig. 16.4).

One of the important utilities of chlorine is the manufacture of hydrochloric acid, with multiple industrial uses such as: the regeneration of ion-exchange resins; a metal deoxidizer in a previous step to galvanization and extrusion; and a descaling agent for calcium carbonate precipitates.

The main component of the acidity of our stomach is natural hydrochloric acid, facilitating our digestive processes.

Chlorine is used in the production of vinyl chloride, a compound used to make polyvinyl chloride (PVC) (Fig. 16.5).

Chlorine is involved in the preparation of epichlorodine (EPI), a resin derived from chlorine that bonds the layers of polyester on the blades of wind turbines.

Carbon tetrachloride is an excellent solvent of fats and oils, but its use is being reduced due to its toxicity.

Another chlorine compound that was previously widely used in surgery for its anesthetic power is chloroform, a highly volatile liquid that is now being replaced by more suitable anesthetics due to its toxic effects on the liver and kidneys. It is currently used as a solvent in certain industries.

Chlorine is also involved in the composition of certain refrigerant gases (CFCs and HCFCs—chlorofluorocarbons) which, due to their polluting effects on the ozone layer, have been replaced by HFOs (hydrofluoolefins); however, some countries still use CFCs and HCFCs.

16.4 Recycling

Chlorine recycling is unknown.

Further Readings

Eurochlor (2021) Association of Chlor-alkali plant operators in Europe. www.eurochlor.org. Last accessed February 2021

Gray T, Mann N (2009) The elements. Black Dog & Leventhal Publishers Inc., New York

Mata JM, Sanz J (2007) Guia d'identificació de minerals. Manresa, Catalonia. Edicions UPC/Parcir (Catalan 2nd paper edition). 262 p. ISBN: 9788483019023. http://hdl.handle.net/2117/90445

Pirajno F (2009) Hydrothermal processes and mineral systems. Springer

Quadbeck-Seeger H-J (2007) World of the elements: elements of the world. Wiley-VCH Verlag GmbH & Co, Germany

Sanz J, Tomasa O (2017) Elements i Recursos minerals: aplicacions i reciclatge. Manresa, Catalonia. Zenobita Edicions/Iniciativa Digital Politècnica (Catalan 3rd digital edition). http://hdl.handle.net/2117/105113

Sanz J, Tomasa O (2018) Elementos y Recursos minerales: aplicaciones y reciclaje. Manresa, Catalonia. Zenobita Edicions/Iniciativa Digital Politècnica (Spanish 1st digital edition). http://hdl.handle.net/2117/123674

Schwab FL (2003) Sedimentary petrology. In: Meyers RA (ed) Encyclopedia of physical science and technology. Academic Press, San Diego, pp. 495–529. ISBN: 9780122274107

Fig. 16.4 Paper manufacture. (*Image courtesy of* Stora Enso)

Stwertka A (2018) A guide to the elements, 4th edn. Oxford University Press, England

Warren JK (2016) Evaporites: a geological compendium. Springer, Cham. https://doi.org/10.1007/978-3-319-13512-0

USGS (2021) Commodity statistics and information. Salt. Available at: https://www.usgs.gov/centers/nmic/salt-statistics-and-information. Last accessed February 2021

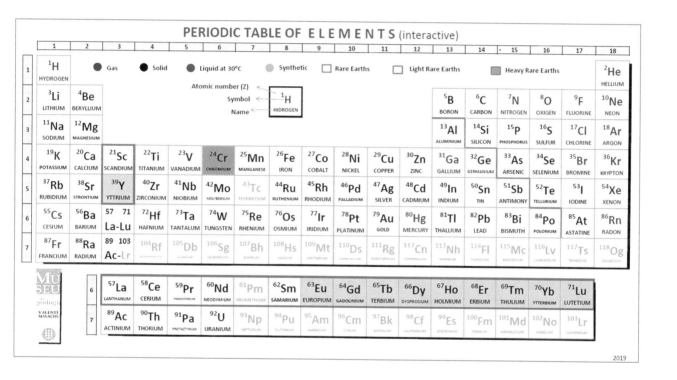

17.1 Geology

- Very hard metal
- High melting temperature
- Does not rust
- Resists heat and friction well
- Can be polished to a very high 'mirror' finish
- Obtained from chromite (Fig. 17.1).

The primary mineral source of chromium (Cr) is chromite (Fe, Mg) Cr_2O_4. Chromite deposits can be classified into four deposit types: stratiform chromite; podiform chromite; placer chromite; and laterite.

1. In stratiform chromite deposits, the chromite ore is located primarily in massive chromitite seams and, less

Fig. 17.1 Chromite (chrome oxide). *Turkey* (*Photo* Joaquim Sanz. MGVM)

abundantly, in disseminated chromite-bearing layers, both of which occur in the ultramafic section of large, layered mafic–ultramafic stratiform complexes. These mafic–ultramafic intrusions mainly formed in stable cratonic settings or during rift-related events during the Archean or early Proterozoic, although there are exceptions. The chromitite seams are cyclic in nature, as well as laterally contiguous throughout the entire intrusion. Gangue minerals include pyrite, chalcopyrite, pyrrhotite, pentlandite, bornite, and platinum-group metals (PGMs). Stratiform chromite deposits are primarily hosted by peridotites, harzburgites, dunites, pyroxenites, troctolites, and anorthosites. Although metamorphism may have altered the ultramafic regions of layered intrusions post-deposition, only igneous processes are responsible for this formation.

2. Podiform chromite deposits are small magmatic chromite bodies formed in the ultramafic section of an ophiolite complex in the oceanic crust. These have been found in mid-oceanic ridge, off-ridge, and suprasubduction tectonic settings. Most podiform chromite deposits are found in dunite or peridotite near the contact of tectonite zones in ophiolites.

3. Placer deposits are formed by gravity separation from podiform or stratiform chromites. Placer environments typically contain black sand, a conspicuous, shiny, black mixture of iron oxides, commonly magnetite, ilmenite, hematite, monazite, rutile, zircon, wolframite, or cassiterite.

4. In laterite deposits, the deep in-situ weathering of ultramafic rocks of varying composition has caused the formation of chromium-rich limonitic soil. The chromitiferous rocks consist of disseminations and minor segregations of chrome-spinels in dunite or harzburgite. The chrome-spinels are preserved in the in-situ deposit because of their high resistance to weathering. As a result

of the removal of Mg and Si from the ultramafics, the chromite is enriched in lateritic soils by a factor ranging from 2 to more than 10. In some areas, soils of different ultramafic rocks are mixed due to alluvial processes, which can cause further enrichment of chromite. These mixed soils often occur as transported soils above non-ultramafic rocks. The abundance of chromite and the content of magnetite, ilmenite, and rutile in the limonitic laterite vary considerably, depending on the occurrence of the accessory opaque minerals of the various types of ultramafics and on local weathering conditions.

The world's main resources are in stratiform chromite deposits, such as the Bushveld Complex in South Africa. Many of the major stratiform chromite deposits also contain PGMs like platinum, palladium, rhodium, osmium, iridium, and ruthenium.

17.2   Producing Countries

The world's largest chromite reserves are in South Africa (230,000 t), Kazakhstan (200,000 t) and India (100,000 t), followed at some distance by Turkey and Finland (Fig. 17.2).

17.3   Applications

Metallurgical Industry

Over 90% of chromite consumption is for metallurgical applications.

The main use of chrome (67%) is in the preparation of ferrochrome for stainless steels (Fig. 17.3).

Heat and corrosion-resistant chromium superalloys are used in the manufacture of aircraft engines and gas turbines.

Thanks to its high resistance to corrosion, this metal is used for the electrodeposition (chrome) coating of all types of parts (metal or plastic) to impart hardness and an unalterable bright finish (Fig. 17.4).

Due to its resistance to high temperatures, chrome is used as a refractory element in the manufacture of molds for brickmaking, for cement kilns, and for metal casting.

Glass and Ceramics Industry

Chromium (III) oxide gives glass its green color and is also used in paint. It is a dopant of corundum (aluminum oxide), giving this substance its red color; it is used in the manufacture of synthetic rubies.

Medicine

Chromium, alloyed with cobalt and molybdenum, among other uses is currently used to make coronary stents to

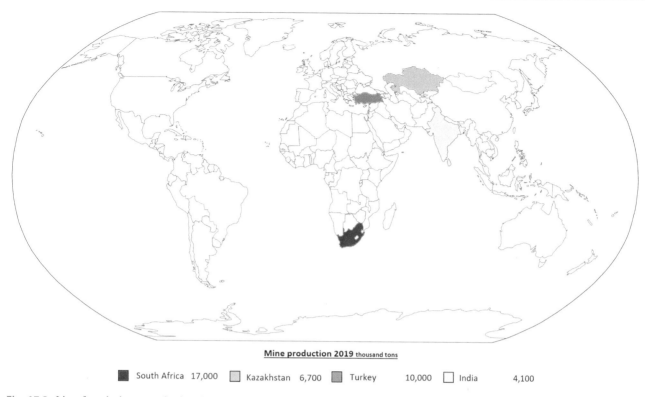

Mine production 2019 thousand tons

  ■ South Africa 17,000   ☐ Kazakhstan 6,700   ■ Turkey   10,000   ☐ India   4,100

Fig. 17.2 List of producing countries based on the US Geological Survey, Mineral Commodity Summaries

Fig. 17.3 Stainless-steel casserole dish. (*Photo* Joaquim Sanz. MGVM)

Fig. 17.4 Chrome plating on a faucet. (*Photo* Joaquim Sanz. MGVM)

reduce coronary artery stenosis, which could lead to heart attacks (Fig. 17.5).

Cobalt chromium alloy is widely used in dental prostheses. Although it has a high cost, it has long durability.

Elgiloy, an alloy of cobalt (40%), chrome (20%), nickel (15%), iron, and molybdenum, is used in certain dental braces (orthodontics).

Other Fields

Chromium (III) salts are used in leather tanning, although alternative products are being sought that are not as harmful.

Chromium is responsible for the red color of rubies, the green color of emeralds, and the dichroism of alexandrite (green in sunlight and red in incandescent light).

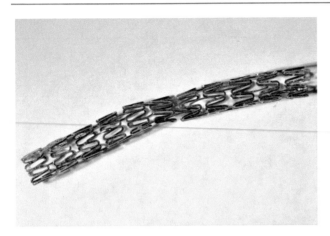

Fig. 17.5 Cardiovascular stent. (*Photo* Joaquim Sanz. MGVM)

17.4   Recycling

Chrome recycling is undertaken from stainless steels that contain it; however, the recovery rate does not exceed 30%.

Further Readings

Eurometaux (2021) Introducing the metals. https://eurometaux.eu/about-our-industry/introducing-metals/. Last accessed May 2021

Friedrich G (1982) Genesis of low-grade chromite ore deposits in lateritic soils from the Philippines. In: Amstutz GC et al (eds) Ore genesis. Special publication of the society for geology applied to mineral deposits, vol 2. Springer, Berlin. https://doi.org/10.1007/978-3-642-68344-2_25

Gray T, Mann N (2009) The elements. Black Dog & Leventhal Publishers Inc., New York

Mata JM, Sanz J (2007) Guia d'identificació de minerals. Manresa, Catalonia. Edicions UPC/Parcir (Catalan 2nd paper edition). 262 p. ISBN: 9788483019023. http://hdl.handle.net/2117/90445

Mosier DL, Singer DA, Moring BC, Galloway JP (2012) Podiform chromite deposits—database and grade and tonnage model. US Geological Survey Scientific Investigations Report 2012-5157. https://pubs.usgs.gov/sir/2012/5157/

Mineral Resource of the Month: Chromium. USGS Mineral Commodities Team. Available at: https://www.earthmagazine.org/article/mineral-resource-month-chromium/. Last accessed May 2021

Quadbeck-Seeger H-J (2007) World of the elements: elements of the world. Wiley-VCH Verlag GmbH & Co, Germany

Roskill (2020) Market reports. Chromium. Available at: https://roskill.com/market-report/chromium/. Last accessed May 2021

Sanz J, Tomasa O (2017) Elements i Recursos minerals: aplicacions i reciclatge. Manresa, Catalonia. Zenobita Edicions/Iniciativa Digital Politècnica (Catalan 3rd digital edition). http://hdl.handle.net/2117/105113

Sanz J, Tomasa O (2018) Elementos y Recursos minerales: aplicaciones y reciclaje. Manresa, Catalonia. Zenobita Edicions/Iniciativa Digital Politècnica (Spanish 1st digital edition). http://hdl.handle.net/2117/123674

Schulte RF, Taylor RD, Piatak NM, Seal RR (2012) Stratiform chromite deposit model, Chap. E of mineral deposit models for resource assessment. U.S. Geological Survey Scientific Investigations Report 2010–5070–E, 13

Stwertka A (2018) A guide to the elements, 4th edn. Oxford University Press, England

USGS (2021) Commodity Statistics and Information. Chromium. Available at: https://www.usgs.gov/centers/nmic/chromium-statistics-and-information. Last accessed May 2021

Cobalt (Co) [Z = 27]

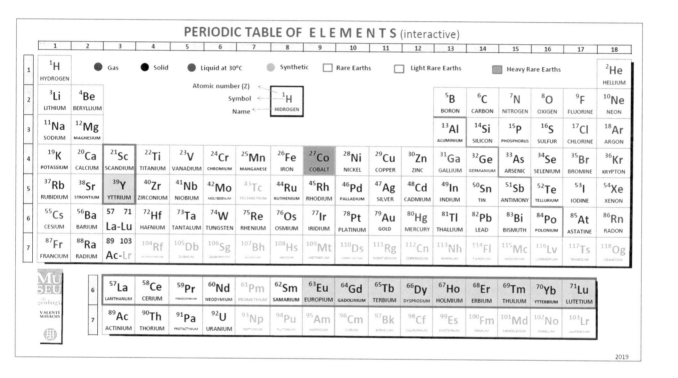

18.1 Geology

- A ferromagnetic metal
- High melting point (1500 °C)
- Heavy
- Assigned the status of a strategic metal by the EU in 2017
- Obtained as a by-product of nickel and copper ores, such as carrollite (Fig. 18.1) and skutterudite, and laterites with nickel and cobalt. Primary cobalt is also obtained from cobaltine.

Most cobalt deposits are associated with nickel deposits. These occur in two geological settings: magmatic sulfide deposits; and laterite deposits. Mines currently in operation exploit both kinds; however, laterite ore deposits comprise about 70% of known nickel–cobalt resources. Besides nickel and cobalt (frequently found in association), there are other important ore minerals of cobalt: as sulfides like cobaltite ($CoAsS$); arsenates like erythrite $Co_3(AsO_4)_2.8H_2O$; and arsenides like skutterudite ($CoAs_{2-3}$).

J. Sanz et al., *Elements and Mineral Resources*, Springer Textbooks in Earth Sciences, Geography and Environment, https://doi.org/10.1007/978-3-030-85889-6_18

Fig. 18.1 Carrollite (copper, cobalt, and nickel sulfide). *Kambobwe (Democratic Republic of Congo) (Photo* Joaquim Sanz. MGVM)

Lateritic Ni-Co deposits arose from chemical weathering in tropical and subtropical conditions, with high temperatures and intense rain. They were formed in ultramafic rocks and a serpentine-saprolite horizon, normally covered by a thinner layer of limonite (a lateritic horizon, sensu stricto).

Such deposits are formed under tectonic uplift conditions with a restricted water column.

18.2 Producing Countries

The DR Congo is the world's largest cobalt producer, accounting for over 70% of cobalt mines' supply in 2020. It is important to mention that UNICEF estimates that 40,000 children are working in artisan- and small-scale mining (ASM) across the south of DR Congo, the source of 20% of the cobalt that is exported. Roskill estimates that over a million Congolese people are dependent on the revenues generated by this type of cobalt mining and its associated logistic and support businesses. Furthermore, in the DRC it is estimated that each worker supports an average of nine dependents (IPIS—International Peace Information Services).

The world's leading cobalt reserves are in the Congo (Kinshasa) (3.6 Mt), followed by Australia (1.2 Mt) and, some way below, Cuba, Canada, the Philippines, and Russia, among others (Fig. 18.2).

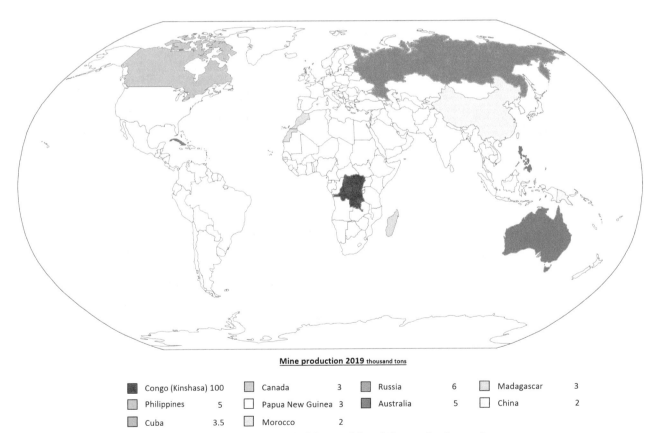

Mine production 2019 thousand tons

Congo (Kinshasa) 100		Canada	3	Russia	6	Madagascar	3
Philippines	5	Papua New Guinea	3	Australia	5	China	2
Cuba	3.5	Morocco	2				

Fig. 18.2 List of producing countries based on the US Geological Survey, Mineral Commodity Summaries

18.3   Applications

Metallurgical Industry

Cobalt can be combined with two or more elements (such as molybdenum and chromium (Stellite)) or other non-metallic elements, lending it combined physical–chemical attributes such as magnetism and resistance to wear, high temperatures, and corrosion.

Cobalt-containing alloys are divided into superalloys, magnetic alloys, prosthetic alloys, and wear-resistant alloys.

Generally, the hardest alloys containing cobalt are used for parts that have to be highly resistant to wear, such as metal-cutting tools (cemented carbides), while those that are less hard are used in high-temperature resistant parts, such as the blades of gas turbines and aircraft (Cobalt Institute 2021) (Fig. 18.3).

Magnet Manufacture

Cobalt is used in the preparation of alloys that are magnetic, strong, and resistant to high temperatures (Al–Ni–Co). Cobalt-samarium magnets have a high demagnetization resistance (SmCo), but are expensive. Neodymium-boron-dysprosium magnets are replacing these magnets, especially in hybrid and electric vehicles, as they are more powerful, less expensive, and more versatile (see: neodymium, dysprosium).

Some of these magnets are used in electric guitar pickups.

Battery Industry

Cobalt consumption is increasing in the cathode manufacture for rechargeable lithium-ion (Li-ion) batteries, due to the growing number of electric vehicles, and nickel-metal hydride (Ni-MH) batteries (see: nickel, lithium and lanthanum) (Fig. 18.4).

Fig. 18.4   Li-ion battery. (*Photo* Joaquim Sanz. MGVM)

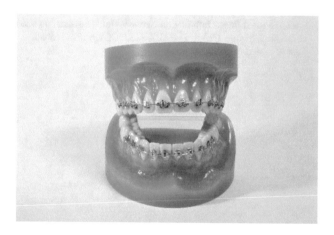

Fig. 18.5   Elgiloy dental braces. (*Photo* Joaquim Sanz. MGVM)

Medicine

Cobalt-60 isotope produces gamma rays (γ) and is used in sterilizing surgical instruments and in industrial controls.

Cobalt chromium alloy is widely used in dental prostheses. Although it is costly, it has good durability.

Elgiloy, an alloy of cobalt (40%), chrome (20%), nickel (15%), iron, and molybdenum, is used in certain dental braces (orthodontics) (Fig. 18.5).

Other Fields

Cobalt salts are an ingredient in the manufacture of paint pigments, blue glass, and ceramics.

Cobalt-molybdenum oxide (CoMOX) is used as a catalyst in the desulfurization of crude oil in order to convert the sulfur that it contains into hydrogen sulfide (H_2S), which can be reconverted into elemental sulfur or sulfuric acid (H_2SO_4).

Fig. 18.3   Aircraft turbine. (*Photo* Joaquim Sanz. MGVM)

18.4 Recycling

Cobalt recycling is carried out on offcuts generated by the manufacture of products containing this metal, from the cathodes of Li-ion batteries, and from catalysts.

The end-of-life recycling rate for cobalt is estimated by the UNEP (United Nations Environment Programme) at 68%.

Further Readings

Cobalt Institute (2021) https://www.cobaltinstitute.org/. Last accessed May 2021

Eurometaux (2021) Introducing the metals. https://eurometaux.eu/about-our-industry/introducing-metals/. Last accessed May 2021

Gray T, Mann N (2009) The elements. Black Dog & Leventhal Publishers Inc., New York

IPIS International Peace Information Services. https://es.search.yahoo.com/search?fr=mcafee&type=E210ES1377G0&p=IPIS+cobalt. Last accessed May 2021

Mata JM, Sanz J (2007) Guia d'identificació de minerals. Manresa, Catalonia. Edicions UPC/Parcir (Catalan 2nd paper edition). 262 p. ISBN: 9788483019023. http://hdl.handle.net/2117/90445

Quadbeck-Seeger H-J (2007) World of the elements: elements of the world. Wiley-VCH Verlag GmbH & Co, Germany

Roskill (2021a) Market reports. Cobalt. https://roskill.com/market-report/cobalt/. Last accessed May 2021

Roskill (2021b) Cobalt. Sustainability monitor. https://roskill.com/roskill-product/cobalt-sustainability-monitor/. Last accessed May 2021

Sanz J, Tomasa O (2017) Elements i Recursos minerals: aplicacions i reciclatge. Manresa, Catalonia. Zenobita Edicions/Iniciativa Digital Politècnica (Catalan 3rd digital edition). http://hdl.handle.net/2117/105113

Sanz J, Tomasa O (2018) Elementos y Recursos minerales: aplicaciones y reciclaje. Manresa, Catalonia. Zenobita Edicions/Iniciativa Digital Politècnica (Spanish 1st digital edition). http://hdl.handle.net/2117/123674

Stwertka A (2018) A guide to the elements, 4th edn. Oxford University Press, England

USGS (2021) Commodity statistics and information. Cobalt. https://www.usgs.gov/centers/nmic/cobalt-statistics-and-information. Last accessed May 2021

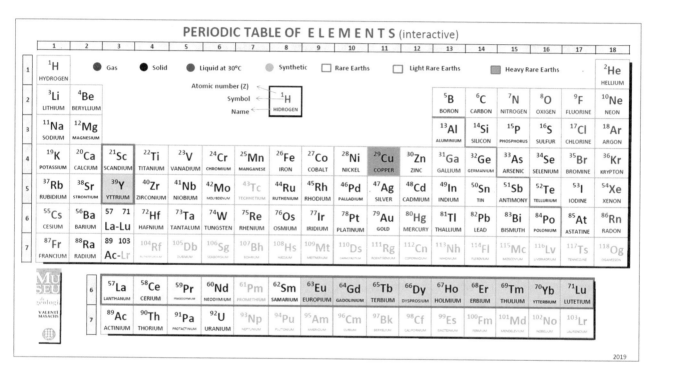

PERIODIC TABLE OF E L E M E N T S (interactive)

- Very good electrical and thermal conductor
- A malleable, ductile, and dense metal
- A micronutrient essential for the proper growth and well-being of plants and animals
- Resists oxidation by forming a protective greenish patina of copper carbonate that withstands the passage of time
- Antimicrobial properties
- One of the oldest metals used by humans. Jewelry dating back 9000 years has been found in Iraq
- From the Latin word *cuprum*, referring to Cyprus (in Greek, Kupros), from where the Romans sourced it

- Obtained from chalcopyrite (Fig. 19.1), bornite, chalcocite, and oxidized minerals such as brochantite and antlerite.

19.1 Geology

The main copper ores are chalcopyrite ($CuFeS_2$) and bornite (Cu_5FeS_4). Porphyry systems are the main source of copper's mineralization.

Fig. 19.1 Chalcopyrite (copper and iron sulfide). *El Brull (Catalonia)* (*Photo* Joaquim Sanz. MGVM)

exceptionally acidic environment in the above-ground water table, inducing the formation of the characteristic gossans.

19.2 Producing Countries

The world's most important known reserves are the porphyry copper deposits of Chile, with a volume of 200,000 t, followed by Australia with 87,000 t, Peru with 87,000 t, Russia with 61,000 tons, Mexico with 53,000 t, the United States with 51,000 t, and Indonesia, China, Congo (Kinshasa), and Zambia (Fig. 19.2).

19.3 Applications

Electrical Industry

In 2017, world consumption of copper for electrical applications was of the order of 75% of the total demand of more than 30 Mt for copper in all its forms and utilities.

The rapid growth in numbers of electric vehicles (EVs), hybrids (HEVs), plug-in hybrids (PHEVs) and their charging stations, together with the trend for ever-larger electric vehicles and batteries with greater capacity and increased range, is increasing copper consumption and will continue to do so (Fig. 19.3).

Copper is used in the manufacture of low- and medium-voltage electrical cables, because it is a very good conductor (the electrification of the high-speed Barcelona–Madrid train took 6590 t of cable). For the same reason, it is used in the manufacture of electrical components, thermal transfer devices such as radiators and refrigerators, and electronic components such as printed circuits, windings, and busbars.

In railway catenaries, copper is alloyed with silver to increase its recrystallization temperature to make it more resistant to the high temperatures caused by the friction of the pantograph.

Building Sector

Both the construction of new buildings and the renovation of older ones are increasing. Copper makes a fundamental contribution to ecological and healthy buildings due to its many applications (the utilities of water, gas, electricity, and air conditioning; building automation; solar panels; charging points for electric vehicles; and so on). Since it is a natural, durable, and resistant material with hardly any maintenance costs and is completely recyclable at the end of its useful life, copper represents a step toward constructing sustainable buildings (Copper alliance).

1. Porphyry ore systems are so named because of the porphyritic texture of the mineralized intrusions, originating from high-temperature magmatic-hydro-thermal fluids. Such deposits are characterized by large tonnage but low grade. Mineralization typically occurs as disseminations, stockworks, and veins of sulfides of Fe, Mo, Pb, Zn, W, Bi, and Sn, as well as native Au.

2. The Central African Copper Belt (CACB) has the world's largest and highest-grade sedimentary rock-hosted stratiform copper deposits (sedimentary copper), endowed with a unique metallogenic diversity. Sedimentary copper deposits generally occur within reduced facies above oxidized facies at the lowest redox boundary. Many deposits are in the vicinity of macro-structural features, primarily growth faults and large anticlines and synclines formed during Lufilian inversion of such faults.

3. There are some copper deposits in the Iberian Pyritic Belt, related to gossan deposits. Gossans are exceedingly ferruginous, the product of oxidation by weathering and leaching of sulfide mineralization. The colors depend significantly on the mineralogical composition of iron hydroxides and oxides phases, varying between red (hematite), yellow (jarosite), brown, and black (galena), with stains of azure blue, malachite green, and peacock blue (copper). The weathering of sulfide deposits usually gives a yellow–brown color with a coarse cellular boxwork and sponge structure. The chalcopyrite often changes to native copper, melaconite (CuO), azurite ($Cu_3(CO_3)_2(OH)_2$), and malachite ($CuCO_3Cu(OH)_3$). Dark shades of peacock blue, green, and black colors and triangular cellular structures are easily recognizable, associated with primary copper sulfide at depth. Massive sulfide deposits typically contain large quantities of iron sulfides and carbonates (pyrite, pyrrhotite, marcasite, siderite, and ankerite) that oxidize to produce an

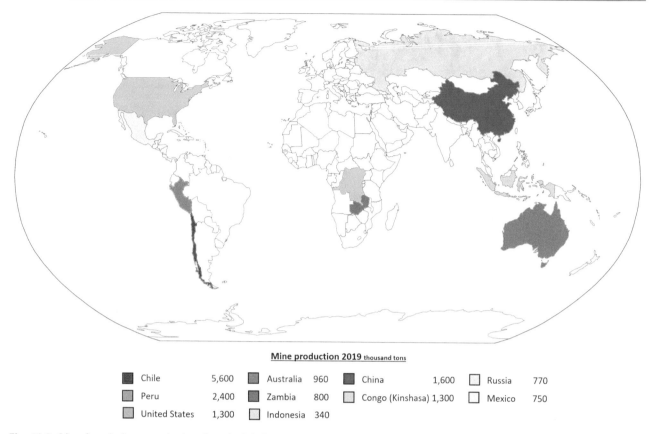

Mine production 2019 thousand tons

■ Chile	5,600	■ Australia	960	■ China	1,600	□ Russia	770
■ Peru	2,400	■ Zambia	800	□ Congo (Kinshasa)	1,300	□ Mexico	750
■ United States	1,300	■ Indonesia	340				

Fig. 19.2 List of producing countries based on the US Geological Survey, Mineral Commodity Summaries

Fig. 19.3 Electric vehicle being charged. (*Photo* Joaquim Sanz. MGVM)

Fig. 19.4 Coins. (*Photo* Joaquim Sanz. MGVM)

Other Uses

Due to its ductile and malleable characteristics, copper is widely used in the manufacture of plates, tubes and wires, as well as coins (Fig. 19.4).

Cupronickel alloys are used in pipes for seawater and in desalination plants, and also to protect offshore structures (Copper alliance).

Brass, a copper and zinc alloy, is used in the manufacture of wind instruments (Fig. 19.5), faucets, and so on. Bronze,

Fig. 19.5 The fiscorn is the Catalan bass-flugelhorn. (*Photo* Joaquim Sanz. MGVM)

a copper and tin alloy, goes into making propellers, statues, bells, bearings, and similar.

Copper sulfate is used as a fungicide and bactericide in agriculture.

19.4 Recycling

Copper is entirely recyclable without losing any of its characteristic properties. Recycling it saves natural resources, reduces SO_2 emissions by 86%, CO_2 emissions by 94%, and the generation of solid waste by 99%. It consumes 85% less energy than obtaining it *ex novo* and reduces water consumption by 98%.

More than 40% of the demand by EU member states for copper is met by recycling, and 48% of this material is recovered from end-of-life equipment. Increasingly, due to the growth in demand, the reuse of copper scrap is making sense, becoming a much-appreciated secondary source

(by-product) and leading to 'urban mining', or the recovery of substances contained in obsolete products, which are so present in our society.

Further Readings

Broughton DW (2014) Geology and ore deposits of the Central African Copperbelt. Doctoral thesis, Colorado School of Mines.

European Copper Institute (2021) Copper and the Circular Economy. Available at: https://copperalliance.eu/. Last accessed May 2021

Gossan—An overview. ScienceDirect topics. Available at: https://www.sciencedirect.com/topics/earth-and-planetary-sciences/gossans

Gray T, Mann N (2009) The elements. Black Dog & Leventhal Publishers Inc., New York

La Farga (2021) The Copper Museum. https://www.lafarga.es/en/the-group/the-copper-museum/introduction. Last accessed May 2021

Mata JM, Sanz J (2007) *Guia d'identificació de minerals*. Manresa, Catalonia. Edicions UPC/Parcir (Catalan 2nd paper edition). 262 p. ISBN: 9788483019023. http://hdl.handle.net/2117/90445

Nornickel (2021) Copper. URL: https://ar2019.nornickel.com/commodity-market-overview/copper. Last accessed May 2021

Pirajno F (2009) Hydrothermal processes and mineral systems. Springer

Quadbeck-Seeger H-J (2007) World of the elements: elements of the world. Wiley-VCH Verlag GmbH & Co, Germany

Roskill (2020) Market Reports. Copper. Available at: https://roskill.com/market-report/copper-demand-to-2035/. Last accessed May 2021

Sanz J, Tomasa O (2017) *Elements i Recursos minerals: aplicacions i reciclatge*. Manresa, Catalonia. Zenobita Edicions/Iniciativa Digital Politècnica (Catalan 3rd digital edition). http://hdl.handle.net/2117/105113

Sanz J, Tomasa O (2018) Elementos y Recursos minerales: aplicaciones y reciclaje. Manresa, Catalonia. Zenobita Edicions/Iniciativa Digital Politècnica (Spanish 1st digital edition). http://hdl.handle.net/2117/123674

Stwertka A (2018) A guide to the elements, 4th edn. Oxford University Press, England

USGS (2021) Commodity statistics and information. Copper. Available at: https://www.usgs.gov/centers/nmic/copper-statistics-and-information. Last accessed May 2021

PERIODIC TABLE OF E L E M E N T S (interactive)

- Highly reactive, corrosive, and toxic gas
- Reacts with humid air and water to give hydrofluoric acid
- The most electronegative element known
- Obtained from fluorite (Fig. 20.1)
- Phosphate rocks are an important source
- Assigned the status of a strategic mineral by the EU in 2017.

20.1   Geology of Fluorite

Fluorite occurs as an accessory mineral in granite, granite pegmatites, and syenites, and is also found in hydrothermal deposits. The following deposit types are in order of importance:

1. Subalkaline volcanic-related epithermal fluorspar deposits. Many fluorspar deposits worldwide occur in association with subalkaline volcanic rocks. The most plentiful

Fig. 20.1 Fluorite (calcium fluoride). *Sant Cugat del Vallès (Catalonia)* (*Photo* Joaquim Sanz. MGVM)

deposits are in China, specifically on the southeastern coast in Zhejiang Province.

2. Alkaline volcanic-related epithermal fluorspar deposits. Northern Mexico, Mongolia, and Kenya have fluorspar deposits and occurrences related to alkaline volcanic rocks.

3. Mississippi Valley-type (MVT) fluorspar deposits. Districts with this type, which contains significant amounts of fluorite, account for three of the world's top-ten sources of fluorspar: the Illinois Kentucky fluorspar district in the United States; the Northern Pennines and the Southern Pennines ore fields in the United Kingdom; and the Marico district in South Africa. A subclass of this type is salt-related carbonate-hosted fluorspar deposits, with the best-known example in the Asturias region of Spain.

4. Carbonatite-related fluorspar deposits. The most abundant are the Fission deposit in Canada and the Speewah deposit in Australia. Two further examples are the Okorusu deposit in Namibia and the Amba Dongar deposit in India.

5. Alkaline intrusion-related fluorspar deposits. The Yermakovskoye deposit in Buryatiya Republic was the first Russian alkaline intrusion-related deposit to be mined.

6. Fluorspar deposits related to granites. The most important are the Voznesenka and the Pogranichny deposits in Russia and the Lost River deposit in Alaska.
   The El Hammam deposit in Morocco, the Yinkuangchong deposit in China, and the Taskaynar deposit in Kazakhstan have a spatial association with granitic rocks. Fluorspar ores from these places contain tin, tungsten, bismuth, or molybdenum minerals.

7. Fluorspar deposits that appear within tuffaceous limy lacustrine sediments. Together with a single fluorite occurrence in southeastern Oregon, a number of deposits in west-central Italy make up this class of apparently conformable deposits within tuffaceous limy lacustrine sediments.

20.2 Producing Countries

World reserves of fluorite are led by Mexico (68,000 t), followed by China and South Africa (42,000 t and 41,000 t respectively), Mongolia (22,000 t), Spain (10,000 t), and smaller reserves are in Vietnam, the United Kingdom, the United States, Iran, and Thailand, among other countries (Fig. 20.2).

There are large reserves of fluorine in phosphate rocks, with worldwide distribution.

20.3 Applications

Metallurgical Industry

Fluorite is a metallurgical flux in smelting aluminum and steel since it makes the slag more manageable and more reactive, facilitating the removal of impurities and giving better performance in the melting process.

Chemical Industry

All fluorinated chemicals are derived from hydrofluoric acid (HF) obtained from fluorite. The chief application of HF in the chemical sector is in the production of fluorocarbonates.

Hydrofluoric acid (HF) is the only acid that dissolves glass, so it is used in engraving and making frosted glass.

High-purity HF is a key chemical in the electronics industry to manufacture semiconductors and printed circuit boards, as it can selectively attack silica (SiO_2).

HF is also used in the production of organofluorinated compounds, such as Teflon polymer (PTFE), which is used in coatings for aircraft, rockets, cables, capacitors, pans, etc. (Fig. 20.3), since it is a good insulator with great resistance to high temperatures. To keep the coating on the metal base of non-stick cookware, an adhesive is necessary. One such adhesive contains PFOA (perfluorooctanoic acid), a carcinogenic substance, so its use in Europe has been prohibited from 1 July 2020; however, for a long time many manufacturers have used an alternative that is not dangerous to health.

Teflon is used in the manufacture of medical prostheses (heart valves), since it does not react with any substance or tissue. It is also in Gore-Tex, a fabric that is breathable yet protects against air, water, and wind.

HF is a key ingredient in the manufacture of third-generation fluorinated refrigerant gases (HFCs), less aggressive to the environment than CFCs and HCFCs;

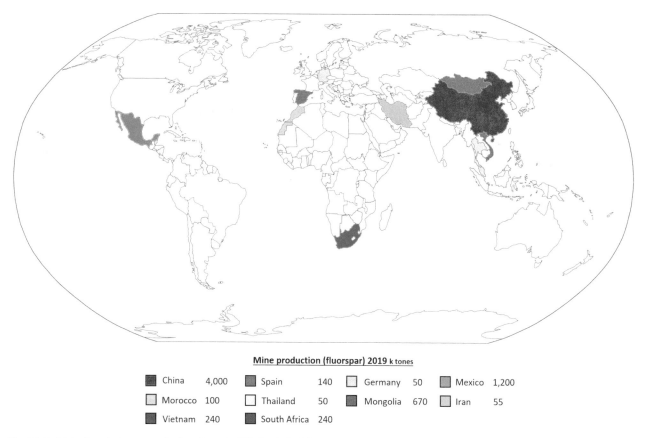

Mine production (fluorspar) 2019 k tones

■ China	4,000	■ Spain	140	□ Germany	50	■ Mexico	1,200
□ Morocco	100	□ Thailand	50	■ Mongolia	670	□ Iran	55
■ Vietnam	240	■ South Africa	240				

Fig. 20.2 List of producing countries based on the US Geological Survey, Mineral Commodity Summaries

Fig. 20.3 Frying pan with Teflon coating. (*Photo* Joaquim Sanz. MGVM)

however, their presence in the atmosphere does contribute to the greenhouse effect, so their use must be restricted to minimize the emissions.

HFO (hydrofluoroolefin) refrigerants are fourth-generation fluorine-based gases and have a lower global warming potential (GWP) than the HFCs that they replace.

Medicine

Hydrofluoric acid is used in the production of antidepressant drugs, such as fluoxetine (Prozac).

Fluorodeoxyglucose, with the radioactive isotope fluorine-18 (half-life 109.8 min), is a radiopharmaceutical in positron emission medical imaging (PET) (see lutetium) to assess glucose metabolism in the heart, lungs, and brain.

Power Generation

Uranium hexafluoride (UF_6) is used in uranium enrichment to produce the uranium-235 fissile fuel to generate electricity in nuclear power plants.

Synthetic fluorite is obtained as a by-product of the uranium enrichment process.

Other Fields

Fluoride compounds are used in the preparation of toothpaste (Fig. 20.4).

Fluorosilicic acid (FSA) is used in the fluoridation of drinking waters (see: phosphorus) to prevent tooth decay. Not all countries agree on the use of fluoride in drinking water.

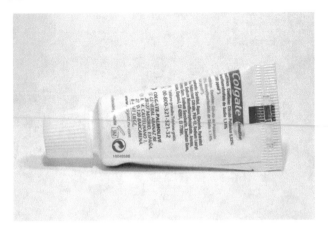

Fig. 20.4 Fluoride-containing toothpaste. (*Photo* Joaquim Sanz. MGVM)

The cement industry uses fluorite to reduce the clinkerization temperature.

Fluorite is also used in the glass industry in the manufacturing process.

20.4 Recycling

In aluminum smelting, part of the hydrofluoric acid generated is recycled by using fluorosilicic acid to extract aluminum from the bauxite.

Further Readings

Azo Mining (2013) https://www.azomining.com/Article.aspx? ArticleID=289. Last accessed March 2021

Gray T, Mann N (2009) The elements. Black Dog & Leventhal Publishers Inc., New York

Mata JM, Sanz J (2007) Guia d'identificació de minerals. Manresa, Catalonia. Edicions UPC/Parcir (Catalan 2nd paper edition). 262 p. ISBN: 9788483019023. http://hdl.handle.net/2117/90445

Minersa Group (2021) Fluorite. Available at: http://www.minersa.com/eng/. Last accessed May 2021

New World Encyclopedia (2017) Online. https://www.newworldencyclopedia.org/entry/fluorite. Last accessed March 2021

Orgiva Mining (2013) Uses of Fluorite. Available at: http://www.mineradeorgiva.com/index.php/en/fluorite.html. Last accessed May 2021

Pirajno F (2009) Hydrothermal Processes and Mineral Systems. Springer

Quadbeck-Seeger H-J (2007) World of the elements: elements of the world. Wiley-VCH Verlag GmbH & Co, Germany

Roskill (2021) Market Reports. Fluorspar. Available at: https://roskill.com/market-report/fluorspar/. Last accessed May 2021

Sanz J, Tomasa O (2017) Elements i Recursos minerals: aplicacions i reciclatge. Manresa, Catalonia. Zenobita Edicions/Iniciativa Digital Politècnica (Catalan 3rd digital edition). http://hdl.handle.net/2117/105113

Sanz J, Tomasa O (2018) Elementos y Recursos minerales: aplicaciones y reciclaje. Manresa, Catalonia. Zenobita Edicions/Iniciativa Digital Politècnica (Spanish 1st digital edition). http://hdl.handle.net/2117/123674

Stwertka A (2018) A guide to the elements, 4th edn. Oxford University Press, England

USGS (2017) Critical Mineral Resources of the United States—Economic and Environmental Geology and Prospects for Future Supply. https://pubs.usgs.gov/pp/1802/g/pp1802g.pdf. Last accessed March 2021

USGS (2021) Commodity Statistics and Information. Fluorspar. Available at: https://www.usgs.gov/centers/nmic/fluorspar-statistics-and-information. Last accessed May 2021

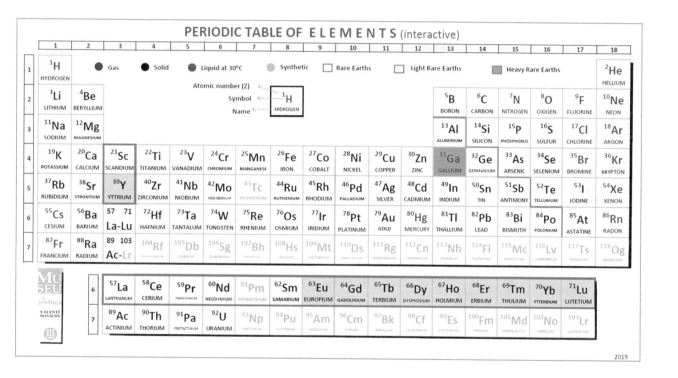

PERIODIC TABLE OF E L E M E N T S (interactive)

- Ductile and malleable metal
- Liquid at a temperature of 30 °C
- Rare semiconductor metal
- Assigned the status of a strategic metal by the EU in 2017
- Global supply from processing bauxite into alumina (Fig. 21.1) and, to a lesser extent, as a by-product of sphalerite (zinc ore).

21.1 Geology

According to USGS, most of the world's gallium is from bauxite mining and from sphalerite. Bauxite deposits are traditionally regarded as the economic source of aluminum (Al); however, they are also an important source of gallium as a by-product, because its close geochemical affinity with Al enables gallium to substitute easily in rock-forming aluminosilicates such as feldspar. Gallium also shows an

Fig. 21.1 Bauxite. *Fontespatlla (Catalonia)* (*Photo* Joaquim Sanz. MGVM)

germanium (Ge), silicon (Si), indium (In), cadmium (Cd), and tin (Sn).

Bauxites consist of one or more aluminum hydroxide minerals, most notably gibbsite ($Al(OH)_3$), boehmite (γ-AlO (OH)), and diaspore (α-AlO(OH)). Depending on the host rock, they are classified as either laterite or karst. Laterite bauxites result from intense subaerial weathering of aluminosilicate rocks such as granite, gneiss, basalt, syenite, and shale. Karst bauxites form in paleokarst depressions within carbonate sequences such as limestone or dolomite. The typical concentration of gallium in bauxite deposits is 10–160 ppm, and there is no substantial difference in concentration between karst- and laterite-type bauxites.

affinity with iron (Fe) and zinc (Zn), enabling it to substitute for these elements in common rock-forming minerals. Gallium can also be found in geochemical association with

21.2   Producing Countries

World reserves of gallium are directly related to bauxite and zinc sulfide deposits (see: aluminum and zinc) (Fig. 21.2).

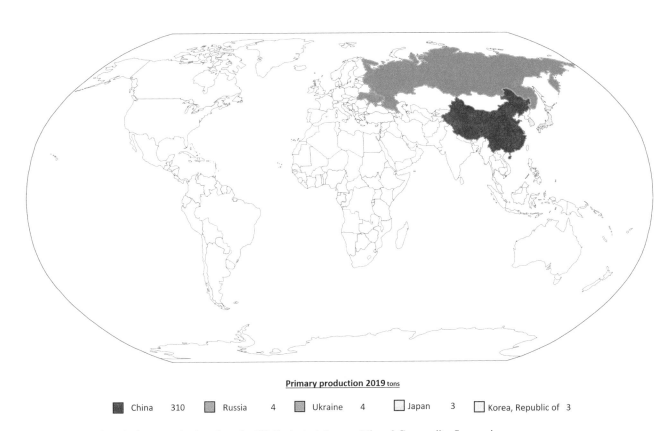

Primary production 2019 tons

| | China | 310 | | Russia | 4 | | Ukraine | 4 | | Japan | 3 | | Korea, Republic of | 3 |

Fig. 21.2 List of producing countries based on the US Geological Survey, Mineral Commodity Summaries

Fig. 21.3 The 'Thin-film' solar cells. (IGSC) (*Photo* Joaquim Sanz. MGVM)

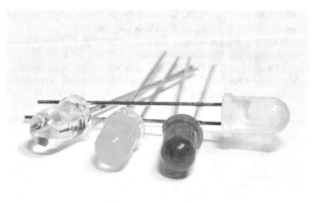

Fig. 21.4 LEDs with nitride and gallium arsenide. (*Photo* Joaquim Sanz. MGVM)

21.3 Applications

Power Generation

Copper indium gallium selenide (CIGS) is a semiconductor used in the manufacture of the conductive 'thin layer' solar cells; it is an effective alternative to crystalline silicon, with 20.4% efficiency (Fig. 21.3).

Gallium-aluminum alloy provides a source of hydrogen by reacting with water. This can power engines and fuel cells with hydrogen.

Electronics Industry

In the form of wafers (thin sheets of semiconductor material on which microcircuits are incorporated using doping, chemical etching, or deposition of materials), gallium arsenide (GaAs), gallium nitride, and gallium phosphoride (GaP) represent almost 80% of total gallium consumption.

Gallium arsenide (GaAs) is a semiconductor used in manufacturing integrated circuits (ICs) and LEDs, and as a laser diode in CD and DVD disk readers and infrared emitting diodes (Fig. 21.4).

Gallium nitride (GaN) is used as a semiconductor in the manufacture of LED electroluminescent diodes for lighting vehicles, buses, trains, airplanes, traffic lights, buildings, and homes across entire cities, with significant savings in energy consumption without sacrificing the quality and intensity of incandescent or fluorescent lamps.

Gallium is an important compound in the manufacture of Blu-Ray DVD disk laser readers, as well as optoelectronic equipment for aerospace, medical, communications, and research applications.

The transition from 4 to 5G technology in smartphones and tablets will progressively increase the consumption of gallium nitride (GaN) and gallium arsenide (GaAs) (Roskill).

Medicine

The alloy of gallium, indium, and tin (galinstan) is used as a substitute for mercury in clinical thermometers.

Gallium nitrate is used as an intravenous drug to treat hypercalcemia, a disease associated with bone cancer, although its effectiveness in repeated use and long-term treatment has been questioned.

The radio isotopes gallium-67 and gallium-68 are used to assess inflammatory processes in general, and especially in the lungs and bones, using SPECT (see: technetium). However, gallium-67 has been superseded by fluorine-18 (see: fluorine).

21.4 Recycling

Offcuts from the manufacture of electronic components containing gallium arsenide are recycled.

Further Readings

Gray T, Mann N (2009) The elements. Black Dog & Leventhal Publishers Inc., New York

Indium Corporation (2021) https://www.indium.com/gallium/. Last accessed May 2021

Mata JM, Sanz J (2007) Guia d'identificació de minerals. Manresa, Catalonia. Edicions UPC/Parcir. (Catalan 2nd paper edition) 262 p. ISBN: 9788483019023. http://hdl.handle.net/2117/90445

Quadbeck-Seeger H-J (2007) World of the elements: elements of the world. Wiley-VCH Verlag GmbH & Co, Germany

Roskill (2020) Market reports. Gallium. Available at: https://roskill.com/market-report/gallium/. Last accessed May 2021

Sanz J, Tomasa O (2017) Elements i Recursos minerals: aplicacions i reciclatge. Manresa, Catalonia. Zenobita Edicions /Iniciativa Digital Politècnica (Catalan 3rd digital edition). http://hdl.handle.net/2117/105113

Sanz J, Tomasa O (2018) Elementos y Recursos minerales: aplicaciones y reciclaje. Manresa, Catalonia. Zenobita Edicions/Iniciativa Digital Politècnica (Spanish 1st digital edition). http://hdl.handle.net/2117/123674

Stwertka A (2018) A guide to the elements, 4th edn. Oxford University Press, England

USGS (2013) Compilation of Gallium resource data for Bauxite deposits. https://pubs.usgs.gov/of/2013/1272/pdf/ofr2013–1272.pdf

USGS (2021) Commodity statistics and information. Gallium. Available at: https://www.usgs.gov/centers/nmic/gallium-statistics-and-information. Last accessed May 2021

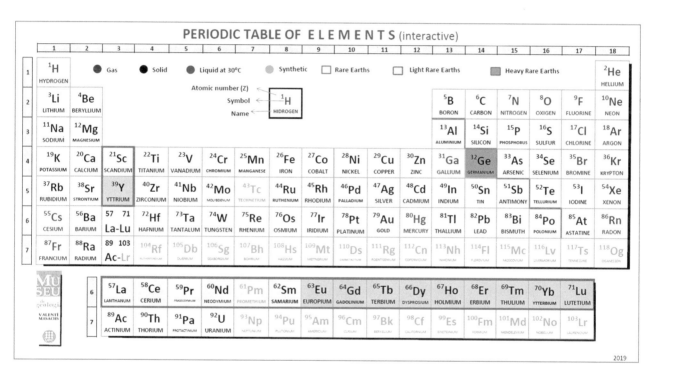

As early as 1871, Mendeleyev predicted that a new element would be discovered very similar to silicon, and between it and tin

- Discovered by German chemist Clemens Winkler in 1886, who named it after the Latin Germania (Germany)
- Excellent but not very abundant semiconductor semi-metal
- Hard and fragile
- Has a high refractive index
- Oxidizes slowly in contact with air
- Assigned the status of a strategic metal by the EU in 2017

- Found in sphalerite (zinc) (Fig. 22.1) as a by-product and in the ashes of certain coals for thermal power plants.

22.1 Geology

According to USGS (2021), germanium does not form specific deposits but occurs rather as a by-product in a variety of deposit types that contain copper, gold, lead, silver, and zinc. Germanium concentrations in sphalerite from these deposits are typically a few hundred parts per million.

J. Sanz et al., *Elements and Mineral Resources*, Springer Textbooks in Earth Sciences, Geography and Environment, https://doi.org/10.1007/978-3-030-85889-6_22

Fig. 22.1 Sphalerite (zinc sulfide with germanium). *Santander (Spain)* (*Photo* Joaquim Sanz. MGVM)

Because it is a by-product of metallurgical operations commonly fed by concentrates from any number of different deposits, it is difficult to track germanium production back to a specific location. The examples discussed below, however, are known to be significant contributors to major germanium-producing facilities.

Types of deposits that contain significant germanium include volcanogenic massive sulfide (VMS) deposits, sedimentary exhalative (SEDEX) deposits, Mississippi Valley-type (MVT) lead–zinc deposits (including Irish-type lead–zinc deposits), and Kipushi-type zinc-lead-copper replacement bodies in carbonate rocks. Germanium is most enriched in the Kipushi-type deposits, but worldwide production is mostly from low-temperature stratiform sphalerite deposits (where mineralization follows stratigraphic layering) and strata-bound sphalerite deposits (where mineralization may cross-cut strata but is restricted to a particular stratigraphic unit).

Another important source of germanium is coal and lignite deposits. China is a major producer of germanium from coal. The Lincang lignite mine close to Lincang City in Yunnan Province is a significant source.

Germanium is initially recovered from the leaching of zinc residues or coal ash, followed by precipitation of a germanium concentrate. The extraction of germanium from its ores involves two stages: the production of a germanium concentrate by retorting, roasting, or pyrometallurgy; and deposition of germanium sulfide or oxide. Concentrates are chlorinated to germanium tetrachloride ($GeCl_4$) and subsequently purified by hydrolysis to germanium dioxide (GeO_2), reduced pyrolytically with hydrogen gas (H_2) to germanium metal powder, then melted into bars.

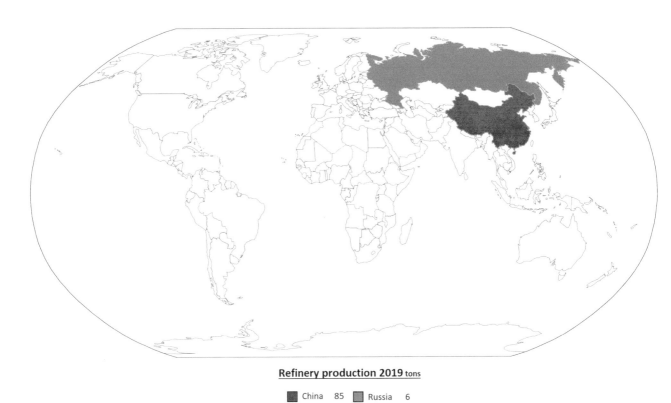

Refinery production 2019 tons

■ China   85   ■ Russia   6

Fig. 22.2 List of producing countries based on the US Geological Survey, Mineral Commodity Summaries

Fig. 22.3 Thermal camera. (*Photo* Joaquim Sanz. MGVM)

22.2 Producing Countries

Germanium reserves are related to zinc sulfide deposits with an approximate 3% germanium content (see: zinc) (Fig. 22.2).

Significant amounts of germanium are found in the ashes of certain coals used in power plants.

22.3 Applications

Optical Industry

A third of the consumption of germanium is destined for the manufacture of fiberglass.

Germanium oxide is obtained from germanium tetrachloride, which is added to silicon dioxide to impart a high refractive index and low optical dispersion, characteristics that make it useful in the manufacture of optical fibers and lasers.

Pure metallic germanium, in the form of disks placed inside the optics, is used in night-vision equipment in electrical installations as it is transparent to infrared rays and is useful in thermal cameras (Fig. 22.3) to detect hot-spots in forests.

Electronics and Lighting Industry

Most applications of germanium relate to solar cells (terrestrial and for satellites), fiber-optic systems, and infrared rays, among others.

Germanium antimony telluride is used in the production of rewritable Blu-Ray DVD recording layers.

Germanium is a semiconductor material used together with silicon in high-speed integrated circuits, and is replacing gallium arsenide in wireless communication devices.

Germanium is still used in diodes to transform alternating current into direct current and in transistors, although increasingly it is being replaced by silicon.

Other Fields

Germanium oxide is used in catalysts for polymerization in the production of polyethylene terephthalate (PET).

Pure germanium crystals are used in the detection of gamma rays.

22.4 Recycling

Approximately 30% of the germanium consumed worldwide is from recycling the equipment that contains it. In the manufacture of optical devices, more than 60% of germanium is reused.

Further Readings

Critical Mineral Resources of the United States (2017) Economic and environmental geology and prospects for future supply. URL: https://pubs.usgs.gov/pp/1802/i/pp1802i.pdf

Gray T, Mann N (2009) The elements. Black Dog & Leventhal Publishers Inc., New York

Mata JM, Sanz J (2007) Guia d'identificació de minerals. Manresa, Catalonia. Edicions UPC/Parcir (Catalan 2nd paper edition). 262 p. ISBN: 9788483019023. http://hdl.handle.net/2117/90445

Quadbeck-Seeger H-J (2007) World of the elements: elements of the world. Wiley-VCH Verlag GmbH & Co, Germany

Sanz J, Tomasa O (2017) Elements i Recursos minerals: aplicacions i reciclatge. Manresa, Catalonia. Zenobita Edicions/Iniciativa Digital Politècnica (Catalan 3rd digital edition). http://hdl.handle.net/2117/105113

Sanz J, Tomasa O (2018) Elementos y Recursos minerales: aplicaciones y reciclaje. Manresa, Catalonia. Zenobita Edicions/Iniciativa Digital Politècnica (Spanish 1st digital edition). http://hdl.handle.net/2117/123674

Stwertka A (2018) A guide to the elements, 4th edn. Oxford University Press, England

Umicore (2021) Germanium. URL: https://www.umicore.com/en/about/our-metals/germanium/. Last accessed May 2021

USGS (2021) Commodity statistics and information. Germanium. URL: https://www.usgs.gov/centers/nmic/ germanium-statistics-and-information. Last accessed May 2021

Gold (Au) [Z = 79]

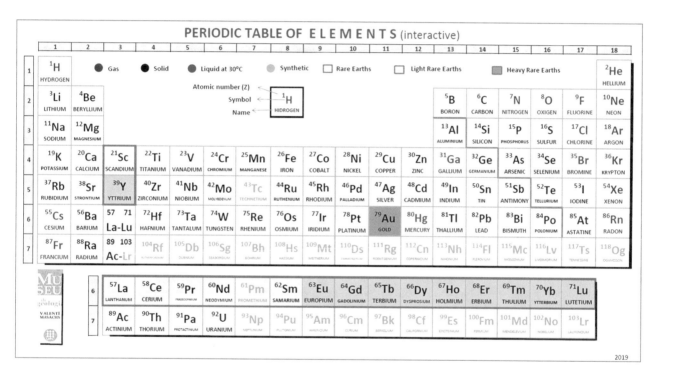

- A noble metal
- The most ductile and malleable metal known
- Very good heat and light reflector
- Excellent electrical conductor
- Does not rust
- One of the most stable metals
- Associated with other minerals in primary and alluvial deposits (Fig. 23.1). Also obtained as a by-product of copper metallurgy
- Found both pure and naturally alloyed with silver as 'electrum'.

23.1 Geology

Gold was mainly concentrated in the Earth's solid and compacted Fe–Ni core during the planet's accretionary stage, along with other highly siderophile elements. During partial melting of the mantle, gold is derived from the magmatic fluids that circulated to the surface in various tectonic contexts (cratons, ocean basins, divergent margins, convergent basins, and transform boundaries). Tectonic models and related geological processes control the genesis of ores not only of gold but of other metal deposits in Earth's geology and history, including the opening and closing of

J. Sanz et al., *Elements and Mineral Resources*, Springer Textbooks in Earth Sciences,
Geography and Environment, https://doi.org/10.1007/978-3-030-85889-6_23

Fig. 23.1 Native gold. *Segre River (Catalonia)* (*Photo* Joaquim Sanz. MGVM)

ocean basins, orogenesis of mountain ranges and geological structures, the distribution of mineral resources, and paleoclimates. Deposit types are grouped by geological setting, host rock, associated minerals, and depth of emplacement: volcanic-hosted massive sulfides (VMS); mesothermal orebodies; intrusion-related porphyry and non-porphyry deposits; iron oxide–copper–gold (IOCG); Carlin-type deposits; and epithermal deposits. Residual and detrital deposits are developed wherever the unroofing of a sufficiently large primary gold ore body has contributed gold to the regolith under stable conditions of weathering erosion and deposition.

23.2 Producing Countries

The country with the largest current gold reserves is Australia (10,000 t), followed by Russia (5300 t), South Africa (3200 t), the United States (3000 t), Indonesia and Brazil (2600 t), and Canada and China (2000 t), followed by other countries with smaller amounts (Fig. 23.2).

The gold reserves contained in porphyry copper deposits, as a by-product, are not quantified.

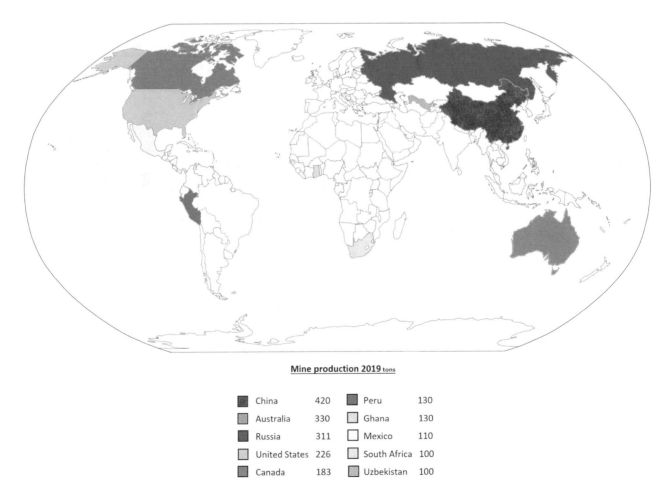

Mine production 2019 tons

China	420	Peru	130
Australia	330	Ghana	130
Russia	311	Mexico	110
United States	226	South Africa	100
Canada	183	Uzbekistan	100

Fig. 23.2 List of producing countries based on the US Geological Survey, Mineral Commodity Summaries

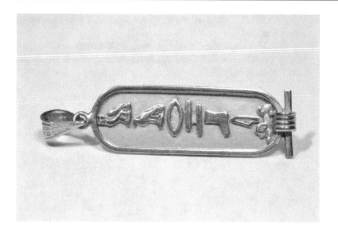

Fig. 23.3 Gold pendant. (*Photo* Joaquim Sanz. MGVM)

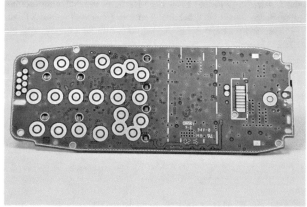

Fig. 23.4 Printed circuit board with gold-plated contacts. (*Photo* Joaquim Sanz. MGVM)

23.3 Applications

Jewelry

The main applications of gold are in jewelry (Fig. 23.3), investments, central bank reserves, and minting commemorative coins. Pure gold (24 ct) is soft and is used only in the form of ingots for investment and state funds. In jewelry gold is used in the form of alloys with other metals, such as 18 ct gold, an alloy of gold (75%), silver (12.5%), and copper (12.5%). There is also 18 ct white gold, an alloy of gold (75%), palladium (15%), and silver (10%).

Investments

At times of stock market instability due to anomalous social and economic situations, such as the COVID-19 pandemic, gold is a metal that for many investors represents sure value, inviting its purchase.

Electric and Electronic Industry

Because it is an excellent conductor of electricity and does not oxidize, gold is used in the manufacture of electrical and electronic connectors, gold-plated printed circuits, relays with gold-plated contacts, and microchips, as it guarantees good electrical contact every time (Fig. 23.4).

Other Fields

Recently, as it is not considered a toxic metal, thin gold leaf has become a decorative ingredient in select foods and beverages. It is also used in cosmetics.

Gold leaf is a very thin sheet of beaten gold (0.006 mm) long used to gild various art objects.

23.4 Recycling

Gold is 100% reusable, therefore all old jewelry and dental prostheses, offcuts, and new jewelry-making powder are recycled.

There are substantial amounts of gold in scrap metal from old electrical and electronic circuits.

According to Gold Info, 1 ton of rock in the mine can produce between 3 and 5 g of gold; 1 ton of electronic scrap can produce about 200 g of gold and involves more than significant savings in terms of mining resources, energy, and pollution.

Extracting metals from electronic scrap is known as 'urban mining', yielding highly valued metals such as palladium, silver, tantalum, gallium, copper, and tin, besides gold. Currently, several university and company laboratories are undertaking chemical tests to further this type of metal recovery (Biometallum project—Universitat Politècnica de Catalunya, Manresa).

The world's level of gold recycling is over 87%.

Further Readings

Gray T, Mann N (2009) The elements. Black Dog & Leventhal Publishers Inc., New York

Macdonald EH (ed) (2007) Geology of gold ore deposits. In: Handbook of gold exploration and evaluation. Woodhead Publishing, pp 62–133

Mata JM, Sanz J (2007) Guia d'identificació de minerals. Manresa, Catalonia. Edicions UPC/Parcir (Catalan 2nd paper edition). 262 p. ISBN: 9788483019023. http://hdl.handle.net/2117/90445

Quadbeck-Seeger H-J (2007) World of the elements: elements of the world. Wiley-VCH Verlag GmbH & Co, Germany

Sanz J, Tomasa O (2017) Elements i Recursos minerals: aplicacions i reciclatge. Manresa, Catalonia. Zenobita Edicions/Iniciativa Digital Politècnica (Catalan 3rd digital edition). http://hdl.handle.net/2117/105113

Sanz J, Tomasa O (2018) Elementos y Recursos minerales: aplicaciones y reciclaje. Manresa, Catalonia. Zenobita Edicions/Iniciativa Digital Politècnica (Spanish 1st digital edition). http://hdl.handle.net/2117/123674

Stwertka A (2018) A guide to the elements, 4th edn. Oxford University Press, England

USGS (2021) Commodity statistics and information. Gold. URL: https://www.usgs.gov/centers/nmic/gold-statistics-and-information. Last accessed May 2021

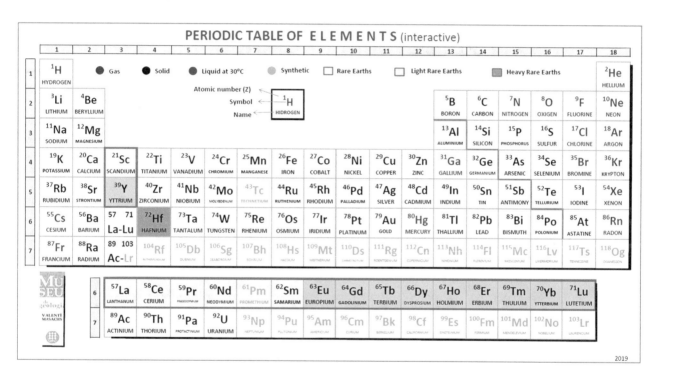

- A rare metal
- Ductile and a shiny silver color
- High melting point (2233 °C)
- Highly resistant to corrosion
- Always attached to zirconium (Fig. 24.1), from which it is difficult to separate
- Several researchers, including Mendeleyev, predicted its existence, but its similarity to zirconium interfered with its identification. George Charles de Hevesy (a Hungarian chemist) and Dirk Coster (a Dutch physicist) discovered it in the zircon structure using x-ray diffractometry
- Reacts with oxygen to form a protective layer

- Assigned the status of a strategic metal by the EU in 2017
- Obtained from zircon.

24.1 Geology

Hafnium is always found associated with zirconium, because of their similar geochemical behavior. Zircon ($ZrSiO_4$) is the most common naturally occurring zirconium- and hafnium-bearing mineral. Most zircon forms as a product of primary crystallization in igneous rocks.

Fig. 24.1 Zircon (zirconium silicate with hafnium). *Madagascar* (*Photo* Joaquim Sanz. MGVM)

There are two main types of deposits associated with zirconium and hafnium. The main ore deposits worldwide that are economically viable are: heavy-mineral sands produced by the weathering and erosion of pre-existing rocks; and the concentration of zircon and other economically important heavy minerals, such as ilmenite and rutile (for titanium), chromite (for chromium), and monazite (for rare earth elements), in sedimentary systems, particularly in coastal environments. In coastal deposits, heavy-mineral enrichment occurs where sediment is repeatedly reworked by wind, waves, currents, and tides. The resulting heavy-mineral sand deposits, called placers or paleoplacers, form preferentially at relatively low latitudes on passive continental margins and supply the entirety of the world's zircon. Zircon makes up a relatively small percentage of the economic heavy minerals in most deposits and is primarily a by-product of mining heavy-mineral sands for titanium minerals.

In addition, there are alkaline igneous rock deposits on the Kola Peninsula of Murmanskaya Oblast, Russia, where baddeleyite (zirconium oxide) is recovered as a by-product of apatite and magnetite mining. However, at present there are few primary igneous deposits of zirconium- and hafnium-bearing minerals that are of economic value.

24.2   Producing Countries

There is no information on hafnium reserves; however, as it is associated with zirconium, the world reserves of that metal give an indication (see: zirconium) (Fig. 24.2).

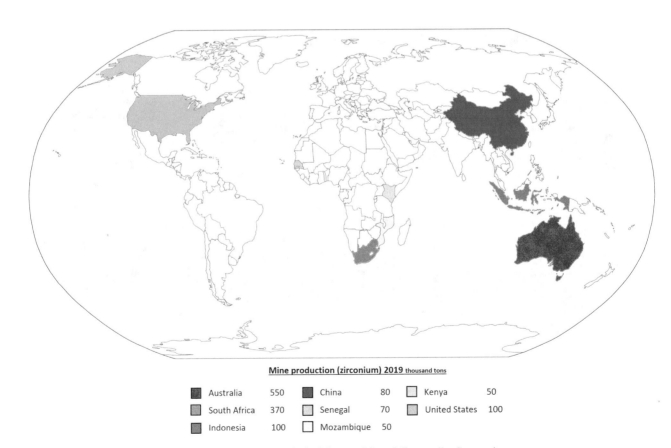

Mine production (zirconium) 2019 thousand tons

■ Australia	550	■ China	80	▫ Kenya	50
■ South Africa	370	▫ Senegal	70	▫ United States	100
■ Indonesia	100	□ Mozambique	50		

Fig. 24.2 List of producing countries based on the US Geological Survey, Mineral Commodity Summaries

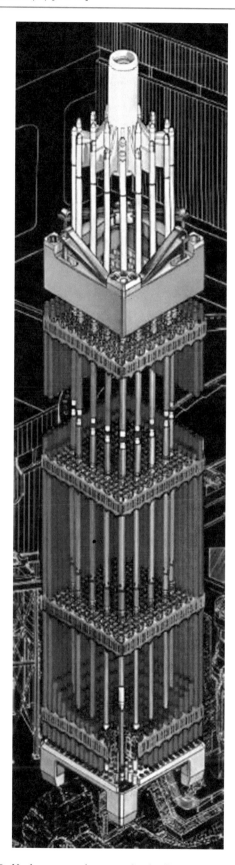

Fig. 24.3 Nuclear power plant control rods. (*Image courtesy of* Javier Castelo)

Fig. 24.4 Plasma cutting electrode. (*Photo* Joaquim Sanz. MGVM)

24.3 Applications

Nuclear Energy

Hafnium is used in the production of control rods for nuclear power plants because of its high capacity to absorb thermal neutrons. It is used in the reactors of nuclear submarines. Due to its mechanical properties and resistance to corrosion, it is used in pressurized water reactors (PWR), the system used in one of the reactors at the Ascó nuclear power plant (Catalonia), where water is used as a coolant (Fig. 24.3).

Metallurgical Industry

Hafnium is added to alloys of iron, titanium, niobium, and tantalum because it enhances their mechanical properties and corrosion resistance. These alloys are used in the space industry for manufacturing rockets and satellites.

60% of world demand is focused on nickel-hafnium alloys, which are used in the manufacture of gas turbines for electricity generators and jet engines due to their high resistance and great stability when operating at high temperatures.

Hafnium, due to its high melting point and resistance to degradation in oxygen-rich environments and very high temperatures, is ideal for use in high-power plasma cutting electrodes, which provide a higher speed of cutting of steels, greater precision, and better finish than that of oxygen cutting electrodes (oxyfuel) (Fig. 24.4).

Other Fields

It is used in the manufacture of projection incandescent lamps as an absorber of gases such as hydrogen and nitrogen, because of its affinity with these gases.

Hafnium carbide (HfC) has the highest melting point of all substances (3890 °C) and is used in the manufacture of the ceramics for thermal shielding of space rocket nozzles.

24.4   Recycling

The recycling of this element is insignificant.

Further Readings

Alkane.com.au. (2021) Hafnium. Available at: https://asm-au.com/products/hafnium/. Last accessed May 2021

Avalon Advanced Materials (2021) Hafnium. Available at: http://avalonadvancedmaterials.com/rare_metals/hafnium/. Last accessed May 2021

Critical Mineral Resources of the United States—Economic and Environmental Geology and Prospects for Future Supply. Available at: https://pubs.usgs.gov/pp/1802/v/pp1802v.pdf

Gray T, Mann N (2009) The elements. Black Dog & Leventhal Publishers Inc., New York

Mata JM, Sanz J (2007) Guia d'identificació de minerals. Manresa, Catalonia. Edicions UPC/Parcir (Catalan 2nd paper edition). 262 p. ISBN: 9788483019023. http://hdl.handle.net/2117/90445

Quadbeck-Seeger H-J (2007) World of the elements: elements of the world. Wiley-VCH Verlag GmbH & Co, Germany

Sanz J, Tomasa O (2017) Elements i Recursos minerals: Aplicacions i reciclatge. Manresa, Catalonia. Zenobita Edicions /Iniciativa Digital Politècnica (Catalan 3rd digital edition). http://hdl.handle.net/2117/105113

Sanz J, Tomasa O (2018) Elementos y Recursos minerales: Aplicaciones y reciclaje. Manresa, Catalonia. Zenobita Edicions/ Iniciativa Digital Politècnica (Spanish 1st digital edition). http://hdl.handle.net/2117/123674

Stwertka A (2018) A guide to the elements, 4th edn. Oxford University Press, England

USGS (2021) Commodity Statistics and Information. Zirconium. Available at: https://www.usgs.gov/centers/nmic/zirconium-and-hafnium-statistics-and-information. Last accessed May 2021

25

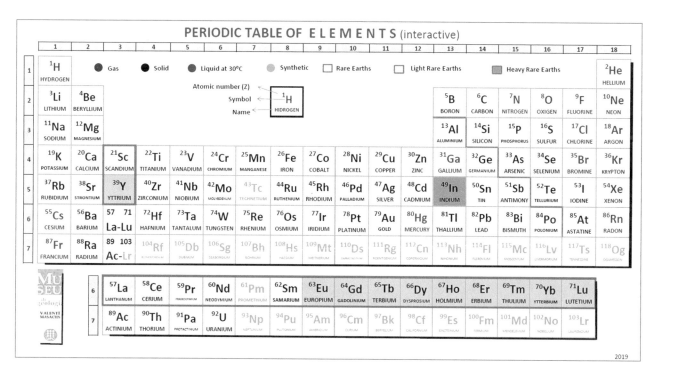

- Highly resistant to corrosion
- Light, soft, ductile, and malleable metal
- Good electrical conductor
- Melts at low temperature (157 °C)
- Stays malleable and ductile below −150 °C
- Named after its characteristic spectral line—the 'indi' is from the color indigo
- Emits a characteristic crunching sound when folded
- Has played a key role in technological advances since Dr. William S. Murray, Utica (New York), who in 1934 founded the Indium Corporation, starting investigations into its multiple applications ongoing to the present

- Assigned the status of a strategic metal by the EU in 2017
- Main ore is sphalerite (Fig. 25.1), as a by-product of zinc refinery.

25.1 Geology

Indium tends to occur in nature with base metals such as copper, silver, zinc, cadmium, tin, lead, and bismuth. As a mineral, indium can be found in sphalerite (ZnS), galena (PbS), cassiterite (SnO_2), the stannite group, and roquesite ($CuInS_2$).

Fig. 25.1 Sphalerite (zinc sulfide with indium). *Santander (Spain)* (*Photo* Joaquim Sanz. MGVM)

According to USGS (2021), indium is recovered as a by-product of refinery of other base-metal ores and concentrates, most commonly the zinc ore mineral, sphalerite. Several complex and proprietary methods have been developed to extract indium from various source materials. Waste products generated during zinc refinery, such as dusts, fumes, residues, and slag, are collected and treated to recover the indium. The materials are first leached with hydrochloric or sulfuric acid to dissolve the indium into an aqueous solution. The solution then undergoes a solvent extraction process to increase its concentration. Next, the indium is removed from solution by means of cementation, and the resulting sponge is cast into anodes for electrolytic refining to produce a metal of standard-grade purity (99.97 or 99.99%).

Approximately half of the world's by-product indium is produced at smelters in southern China, but the Republic of Korea, Japan, Canada, Belgium, and Peru also produce significant amounts. More than half of the material for smelters in China is from VMS (volcanogenic massive sulfides) and SEDEX (sedimentary exhalative) deposits, while much of the remaining production is from MVT (Mississippi Valley-type) deposits. It is difficult to decipher which of the world's VMS, SEDEX, and MVT deposits provide the zinc feedstock processed at Chinese smelters.

25.2 Producing Countries

The world's reserves of indium are related to its zinc sulfide, copper, and tin deposits (Fig. 25.2).

The percentage of indium content in zinc sulfides ranges from 1 to 100 parts per million (see: zinc).

25.3 Applications

Electronics Industry

Indium-tin oxide (ITO) is used in electrical circuits that are to be invisible in liquid crystal displays (LCDs), plasma displays, and touch screens of smartphones or tablets, since it is an optically transparent electrical conductor that provides an electrical supply to the pixels (Fig. 25.3).

Indium-doped germanium is used to manufacture integrated circuits and electrical components such as rectifiers and photoconductors.

The indium compounds of InAs, InGaAs, and InGaN are used in the manufacture of LEDs, laser diodes (LD), and integrated circuits.

Power Generation

Indium is used in the control rods of nuclear power plants to absorb neutrons.

Copper indium gallium selenide (CIGS) is a semiconductor used in the manufacture of the conductive thin layer of thin and flexible solar cells (Fig. 25.4).

Glass and Ceramic Industry

Indium can be used in the manufacture of mirrors of the same quality as those of silver, and with a greater resistance to corrosion.

Medicine

The alloy of gallium, indium, and tin (galinstan) is used in the new clinical thermometers.

The gamma rays emitted by indium-111 are employed in the detection and evaluation of neuroendocrine tumors in combination with PET, with fluorine-18 (see: fluorine).

Other Applications

Indium metal is used to prepare welding alloys that melt at temperatures of between 6.5 and 310 °C. These have less tendency to crack and better resistance to thermal fatigue than Pb–Sn solder.

Another property of indium is that it is the ideal adhesive to join quartz, glass, and glazed ceramics.

25.4 Recycling

The recovery of indium is from the ITO in electronic equipment such as LCD screens and touch screens.

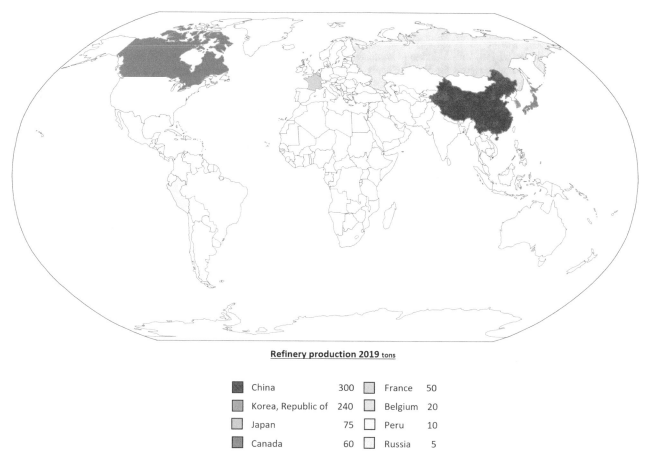

Refinery production 2019 tons

■	China	300	■ France	50
■	Korea, Republic of	240	■ Belgium	20
■	Japan	75	□ Peru	10
■	Canada	60	□ Russia	5

Fig. 25.2 List of producing countries based on the US Geological Survey, Mineral Commodity Summaries

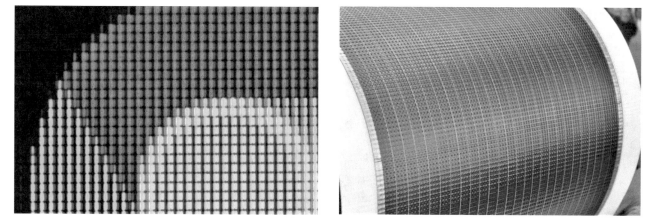

Fig. 25.3 Pixels of an LCD display (*Photo* Joaquim Sanz. MGVM)   **Fig. 25.4** Flexible solar CIGS (*Photo* Joaquim Sanz. MGVM)

Further Readings

Critical Mineral Resources of the United States—Economic and Environmental Geology and Prospects for Future Supply (2017) URL: https://pubs.usgs.gov/pp/1802/i/pp1802i.pdf

Gray T, Mann N (2009) The elements. Black Dog & Leventhal Publishers Inc., New York

Indium Corporation (2021) Metals. Indium. Available at: https://www.indium.com/metals/indium/. Last accessed May 2021

Jiménez-Franco A (2017) Geochemical and metallogenic model of the Santa Fe Sn-Zn-Pb-Ag-(In) deposit in the Central Andean tin belt. Doctoral thesis. Universitat Politècnica de Catalunya, Spain

Mata JM, Sanz J (2007) Guia d'identificació de minerals. Manresa, Catalonia. Edicions UPC/Parcir (Catalan, 2nd paper edition). 262 p. ISBN: 9788483019023. http://hdl.handle.net/2117/90445

Quadbeck-Seeger H-J (2007) World of the elements: elements of the world. Wiley-VCH Verlag GmbH & Co, Germany

Sanz J, Tomasa O (2017) Elements i Recursos minerals: Aplicacions i reciclatge. Manresa, Catalonia. Zenobita Edicions/Iniciativa Digital Politècnica (Catalan, 3rd digital edition). http://hdl.handle.net/2117/105113

Sanz J, Tomasa O (2018) Elementos y Recursos minerales: Aplicaciones y reciclaje. Manresa, Catalonia. Zenobita Edicions/Iniciativa Digital Politècnica (Spanish, 1st digital edition). http://hdl.handle.net/2117/123674

Stwertka A (2018) A guide to the elements, 4th edn. Oxford University Press, England

USGS (2021) Commodity statistics and information. Indium. Available at: https://www.usgs.gov/centers/nmic/indium-statistics-and-information. Last accessed May 2021

Iodine (I) [Z = 53]

PERIODIC TABLE OF E L E M E N T S (interactive)

A solid element of a very dark violet color at room temperature, But upon applying slight heat it sublimes to a violet vapor with an irritating odor

An essential element for humans and animals

Uses are focused on the thyroid gland

Antiseptic properties

Mainly obtained from *caliche* (rock with saltpeter [sodium potassium nitrate and calcium iodate]) in northern Chile (Fig. 26.1), and from iodides in brine from the natural gas fields of Minami Kanto (Japan).

26.1 Geology

Iodine is rarely found independently in nature, and is often combined with other elements, basically forming inorganic salts. According to the World Iodine Association (WIA), seawater is the world's greatest iodine reserve, with around 34.5 million tons; however, no economic extraction is feasible directly because of the low concentrations of less than 0.05 ppm.

Iodine ore deposits are found only in certain places, such as the underground water from drilling certain deep oil wells in Japan, mostly located in five zones: Chiba, Niigata, Sadowara, Okinawa, and Oshamambe. Of these, only the

Fig. 26.1 *Caliche*—rock with saltpeter and iodates. *Pedro de Valdivia Plant* (*Chile*) (*Photo* Joaquim Sanz. MGVM)

first three zones are producing, and Chiba alone is responsible for 80% of Japan's total production.

In the Atacama Desert of northern Chile and west of the Andes, *caliche* occupies an area averaging 700 km north–south by 30 km east–west. It is a whitish-gray, calcareous soil mixed with gypsum, sodium chloride, and other salts and sand, characteristic of arid and sem-iarid zones. However, the Chilean *caliche* is associated with salitre, a composite of sodium nitrate ($NaNO_3$) and potassium nitrate (KNO_3) enriched in iodine. These deposits are the largest known natural source of nitrates worldwide, containing up to 25% sodium nitrate and 3% potassium nitrate, as well as iodate minerals, sodium chloride, sodium sulfate, and sodium borate (borax). The *caliche* beds are from 0.2 to 5.0 m thick. They are mined and refined to produce a variety of products, including sodium nitrate (for agriculture or industry uses), potassium nitrate, sodium sulfate, iodine, and iodine derivatives.

According to Pueyo et al. (1998), the brines that reach the ore deposits originate in one of two ways: leaching of altered rocks; or pre-existing saline material. The main part of the saline deposits in northern Chile's arid zone is of volcanic origin, aside from a few cases of marine evaporitic formations. The leaching of volcanic materials would have been promoted by thermal processes related to the volcanic arc of the Middle Tertiary (Chong 1991). However, the evolution of brines interceded in the evaporation mechanisms, since they are remarkably effective in precipitating minerals with a high percentage of solubility. On the other hand, the brine fraction has been affected by repeated precipitation–evaporation during the long hydrologic circulation from the Andean range to the western coastal area. Moreover, according to other studies (e.g. Álvarez 2016), there is iodine enrichment due to the interaction both between underground waters and the enriched sedimentary biologic rock and with leached iodine.

In addition, iodine is obtained as a by-product of processing of sodium alginate, mainly in China. No more than 2% of total iodine consumption is derived from this source.

26.2   Producing Countries

The world's most notable reserves are in Minami Kanto (Japan), at 5 Mt (natural gas deposits), followed by Chile with 610,000 t, the United States with 250,000 t, Azerbaijan with 170,000 t, Russia with 120,000 t, Indonesia with 100,000 t, and Turkmenistan with 70,000 t (Fig. 26.2).

26.3   Applications

Chemical Industry

Iodine is used as a catalyst in the production of acetic acid and as a disinfectant in water treatments. It is also used as a biocide in paint, adhesives, and wood treatments.

Medicine

Povidone-iodine (polymer combined with iodine) is the basic component of many antiseptics and disinfectants for treating minor cuts and skin wounds.

Iodine is used as a contrast agent in x-rays of certain parts of the body, since it is an element that is opaque to x-rays (Fig. 26.3).

In cardiac catheterization, the presence of iodine in the circulatory system allows the detection of any narrowing of arteries and/or veins so that, if necessary, stents can be fitted to facilitate the passage of blood.

Iodine-131 (a half-life of 8 days) is used in therapy to treat hyperthyroidism. In 'seed' form (see iridium), iodine-125 (a half-life of 59 days) is used in brachytherapy to treat prostate tumors, as well as to help surgeons to locate small non-palpable tumors in the breast, thyroid, and lungs. Iodine-123 (a half-life of 13 h) is used in the diagnosis of Parkinson's disease.

According to Ajay-SQM, potassium iodide can mitigate the effects of radioactive iodine on the human body after exposure to high levels of radioactivity as a result of accident or nuclear conflict. However, this compound does not offer full protection to radioactive iodine: its extent is a function of the level of radiation and the time that has elapsed before treatment. All in all, although iodine can protect the thyroid from damage associated with radiation, it cannot protect other organs. In any case, this chemical must be taken in accordance with medical protocols established by health authorities.

Electrical Industry

Iodine is part of some types of halogen lamps.

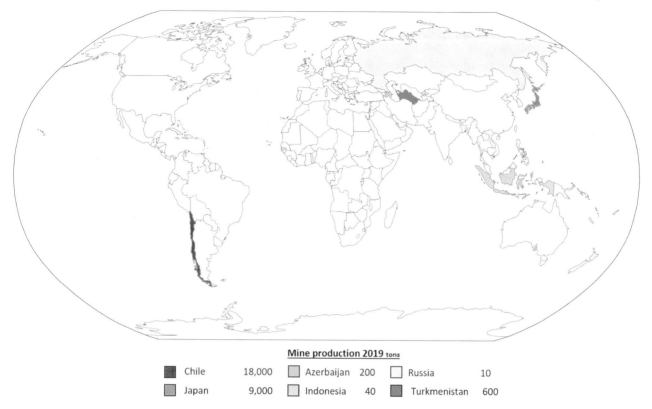

Mine production 2019 tons					
Chile	18,000	Azerbaijan	200	Russia	10
Japan	9,000	Indonesia	40	Turkmenistan	600

Fig. 26.2 List of producing countries based on the US Geological Survey, Mineral Commodity Summaries

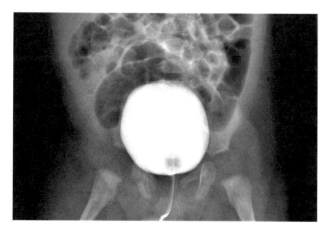

Fig. 26.3 X-ray of bladder, with iodine contrast. (*Image courtesy of* Hospital Sant Pau, Barcelona)

Other Fields

Iodized salt is used in many countries to correct dietary deficiencies of this element and to prevent possible diseases of the human body, such as goiter (Fig. 26.4).

Fig. 26.4 Iodized kitchen salt (*Photo* Joaquim Sanz. MGVM)

Silver iodide is combined with silver bromide as an active base for the photographic film still used by some practitioners, even in this digital age.

Iodine is a key component of polarizing screen films (LCD/LED) in smartphones, tablets, and televisions.

26.4   Recycling

Small amounts of iodine are recycled.

References

Álvarez Amado F (2015) Fuentes y reservorios de yodo natural y antropogénico en el Desierto de Atacama, norte de Chile. PhD thesis. Universidad de Chile, Santiago de Chile

Chong G (1991) Geología de los yacimientos de nitratos de Chile. Antecedentes para establecer una teoría sobre su génesis. In: Alonso RN et al (eds) Genesis de formaciones evaporiticas. Modelos andinos e ibericos. University of Barcelona, pp 377–417

Pueyo JJ, Chong G, Vega M (2010) Mineralogy and parental brine evolution in the Pedro de Valdivia nitrate deposit, Antofagasta Chile. Andean Geol 25(1):3–15. https://doi.org/10.5027/andgeoV25n1-a01

Further Reading

Ajay-SQM (2018) What role does iodine play in radiation? Available at: https://www.ajay-sqm.com/what-role-does-iodine-play-in-radiation/. Last accessed May 2021

Gray T, Mann N (2009) The elements. Black Dog & Leventhal Publishers Inc., New York

Mata JM, Sanz J (2007) Guia d'identificació de minerals. Manresa, Catalonia. Edicions UPC/Parcir (Catalan, 2nd paper edition). 262 p. ISBN: 9788483019023. URL: http://hdl.handle.net/2117/90445

Quadbeck-Seeger H-J (2007) World of the elements: elements of the world. Wiley-VCH Verlag GmbH & Co Germany

Roskill (2013). Market reports. Iodine. Available at: https://roskill.com/market-report/iodine/. Last accessed May 2021

SQM (2021) Solutions for human progress. Iodine. Available at: https://www.sqm.com/en/productos/yodo-y-derivados/. Last accessed May 2021

Sanz J, Tomasa O (2017) Elements i Recursos minerals: Aplicacions i reciclatge. Manresa, Catalonia. Zenobita Edicions/Iniciativa Digital Politècnica (Catalan, 3rd digital edition). http://hdl.handle.net/2117/105113

Sanz J, Tomasa O (2018) Elementos y Recursos minerales: Aplicaciones y reciclaje. Manresa, Catalonia. Zenobita Edicions/Iniciativa Digital Politècnica (Spanish 1st digital edition). http://hdl.handle.net/2117/123674

Stwertka A (2018) A guide to the elements, 4th edn. Oxford University Press, England

USGS (2021) Commodity Statistics and Information. Iodine. Available at: https://www.usgs.gov/centers/nmic/iodine-statistics-and-information. Last accessed May 2021

WIA (2021) World Iodine Association. https://www.worldiodineassociation.com/. Last accessed May 2021

Iridium (Ir) [Z = 77]

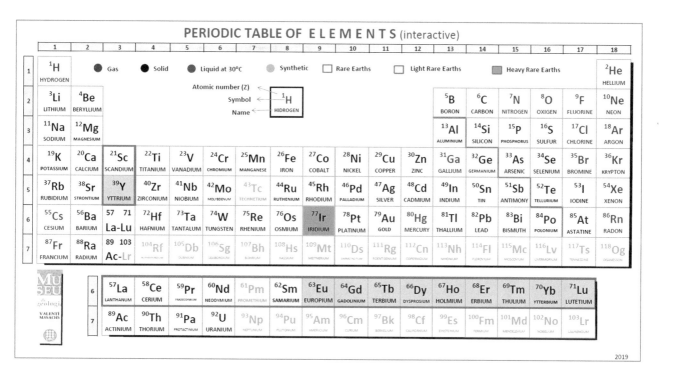

- A very hard, dense, and fragile noble metal
- The densest element known: 22.59 g/cm³
- Discovered in 1803 by English chemist Smithson Tennant, with same procedure as he used to discover osmium
- Has a high gloss and is very expensive
- The most resistant of all metals to corrosion
- Resistant to high temperatures (2443 °C)
- A platinum-group metal (PGM) ore
- Assigned the status of a strategic metal by the EU in 2017
- Found in PGM deposits—separation is difficult and laborious, performed only for expensive noble metals

- Also obtained from osmiridium and sperrylite, and as a by-product of electrorefinery of nickel and copper (Fig. 27.1)

27.1 Geology

The platinum-group metals (PGMs) are found almost exclusively in ores associated with mafic and ultramafic rocks at very low concentrations. They can be subdivided into two main groups: the Ir-subgroup (IPGE: Os, Ir, and Ru); and the Pt-subgroup (PPGE: Pt, Pd, and Rh). The most

Fig. 27.1 Nickel and copper sulfides with PGMs. *Norilsk (Russia)* (*Photo* Joaquim Sanz. MGVM)

common iridium mineral is irarsite, (Ir,Ru,Rh,Pt)AsS. This is commonly hosted in chromitites, forming layered intrusions as in the Bushveld Complex), or in ophiolitic podiform chromitites as in the Al'Ays ophiolite complex.

The laurite–erlichmanite series represents 75% of the total PGM present in this type of ore and occurs as mineral inclusions in chromites commonly associated with other base-metal sulfides, forming anhedral to euhedral crystals and in isolated or composite grains.

27.2 Producing Countries

The world's most important iridium reserves are in: South Africa (Bushveld) for platinum, rhodium, and palladium deposits; Norilsk (Russia) for copper and nickel; and Sudbury basin (Canada) for copper, nickel, and palladium deposits (Fig. 27.2).

27.3 Applications

Metallurgical Industry

Iridium's properties, such as its resistance to corrosion, high temperature, and hardness, make the manufacture of alloys

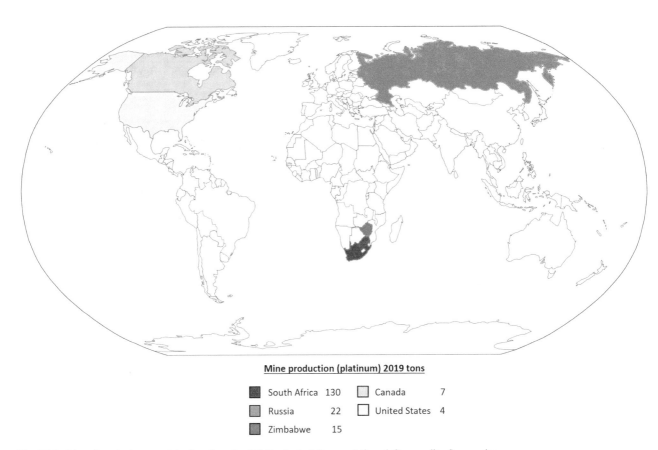

Mine production (platinum) 2019 tons

■	South Africa	130	☐ Canada	7
■	Russia	22	☐ United States	4
■	Zimbabwe	15		

Fig. 27.2 List of producing countries based on the US Geological Survey, Mineral Commodity Summaries

the main application of this metal. By adding iridium to platinum, the resulting alloy is harder than platinum alone.

Medicine

The radioisotope iridium-192, deposited inside small capsules known as 'seeds', is the source of gamma radiation in the treatment of cancer by brachytherapy (Fig. 27.3).

The tips of the electrodes in a Pacemaker device contain iridium and titanium.

Chemical Industry

In electrochemistry, iridium is used together with ruthenium to coat the electrodes in the production of chlorine and caustic soda by electrolysis.

Iridium can be used as a catalyst in electrolysis to split water into oxygen and hydrogen, increasingly in demand as a source of power cleaner than fossil fuels.

Compounds of iridium are used as catalysts in the production of acetic acid.

Electronic Industry

Since the beginning of 2021, iridium prices been increasing. This is due to growing demand for crucibles for forming the lithium tantalate crystals necessary in 5G technology. Crucibles of iridium are also used to grow corundum crystals (sapphire) for making electroluminescent diodes (LEDs).

Automotive Industry

Iridium is used in the manufacture of long-life spark plug electrodes, which last in the order of 120,000 km of driving (Fig. 27.4).

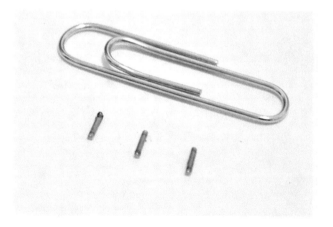

Fig. 27.3 Ir-192 'seeds' for brachytherapy (5 mm)—paper clip shown for scale. (*Photo* Joaquim Sanz.MGVM)

Fig. 27.4 Iridium spark plug electrode. (*Photo* Joaquim Sanz. MGVM)

Iridium will play an important role in the so-called 'hydrogen economy', obtaining that gas for the engines of many vehicles in a sustainable way through hydrolysis of water.

Other Fields

Because it is so hard, some fountain pen nibs are coated in iridium to make them resistant to wear.

The measurement standard of the international standard meter, in use from 1889 to 1960 and now kept in Paris, is a rod of platinum (90%) and iridium (10%).

27.4 Recycling

All metal scraps and remnants of objects containing iridium are recycled.

According to S&P Global Platts (March 2021), the amount of iridium that is recycled annually is between 7 and 8 thousand tons.

Further Reading

Gray T, Mann N (2009) The elements. Black Dog & Leventhal Publishers Inc., New York

International Platinum Group Metals Association (2021) Iridium. Available at: https://ipa-news.com/index/platinum-group-metals/the-six-metals/iridium.html. Last accessed May 2021

JM (2020) Platinum group metals. Market report. Available at: http://www.platinum.matthey.com/services/market-research/may-2020-pgm-market-report. Last accessed May 2021

Mata JM, Sanz J (2007) Guia d'identificació de minerals. Manresa, Catalonia. Edicions UPC/Parcir (Catalan 2nd paper edition). 262 p. ISBN: 9788483019023. URL: http://hdl.handle.net/2117/90445

NGK (2021) Iridium. Available at: https://ngksparkplugs.com/en/products/ignition-parts/spark-plugs/iridium-spark-plugs. Last accessed May 2021

Nornickel (2021) Iridum. Available at: https://www.nornickel.com/business/products/iridium. Last accessed May 2021

O'Driscoll B, González-Jiménez, JM (2016) Petrogenesis of the platinum-group minerals. Rev Miner Geochem 81(1):489–578. URL: https://doi.org/10.2138/rmg.2016.81.09

Pirajno F (2009) Hydrothermal processes and mineral systems. Springer, Berlin

Quadbeck-Seeger H-J (2007) World of the elements: elements of the world. Wiley-VCH Verlag GmbH & Co, Germany

Sanz J, Tomasa O (2017) Elements i Recursos minerals: Aplicacions i reciclatge. Manresa, Catalonia. Zenobita Edicions/Iniciativa Digital Politècnica (Catalan 3rd digital edition). URL: http://hdl.handle.net/2117/105113

Sanz J, Tomasa O (2018) Elementos y Recursos minerales: Aplicaciones y reciclaje. Manresa, Catalonia. Zenobita Edicions/Iniciativa Digital Politècnica (Spanish 1st digital edition). URL: http://hdl.handle.net/2117/123674

Stwertka A (2018) A guide to the elements, 4th edn. Oxford University Press, England

USGS (2021) Commodity statistics and information. Platinum group metals. Available at: https://www.usgs.gov/centers/nmic/platinum-group-metals-statistics-and-information. Last accessed May 2021

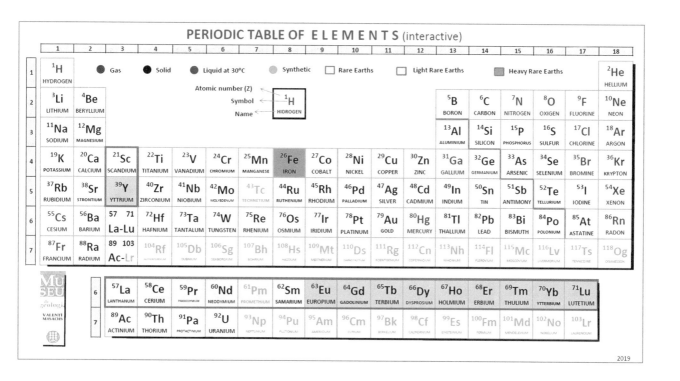

- Fourth most common element in the Earth's crust
- Essential for all living organisms, as it is part of the hemoglobin in blood, responsible for the transport of O_2
- Soft and fragile in its pure state
- When carbon is added, it becomes harder and more resistant
- During the Iron Age it was the metal most used to make weapons and tools
- Oxidizes in air to forms iron oxides
- The world's most widely used metal
- Alloyed with manganese, nickel, tungsten, and chrome, among others, makes steels of extreme hardness, great tensile strength, the capacity to compensate for vibrations, etc.
- Obtained from hematite (Fig. 28.1) and magnetite.

28.1 Geology

The primary mineral sources of iron (Fe) are magnetite (Fe_3O_4), hematite (Fe_2O_3), goethite (FeO(OH)), limonite (FeO(OH)·n(H_2O)), and siderite ($FeCO_3$). Iron mineralization is associated with several hydrothermal systems, such as banded-iron formations (BIF). BIF are iron deposits limited

J. Sanz et al., *Elements and Mineral Resources*, Springer Textbooks in Earth Sciences, Geography and Environment, https://doi.org/10.1007/978-3-030-85889-6_28

Fig. 28.1 Hematite (iron oxide). *Llucena (Spain)* (*Photo* Joaquim Sanz.MGVM)

to the Paleoproterozoic, in an enrichment associated with an evolving oxygenated atmosphere and the delivery of large quantities of ferrous iron from seafloor hydrothermal venting, possibly related to mantle superplume events. Therefore, the precipitation of Proterozoic iron formations reflects

a change from a predominantly anoxic and low pH acidic ocean to an oxic and alkaline ocean at about 2.5 Ga. Occasionally, granite and ultrapotassic igneous rocks segregate magnetite crystals to form masses of magnetite suitable for economic concentration. A few iron-ore deposits were formed from volcanic flows with magnetite. In environments like the Atacama Desert, alluvial accumulations of magnetite are common in the streams leading from these volcanic formations. Some magnetite skarn and hydrothermal deposits have been exploited. Other magnetite iron-ore sources include metamorphic accumulations of massive magnetite ore related to ophiolite ultramafic sequences.

28.2 Producing Countries

The most important iron-ore reserves in the world are in Australia (48,000 Mt), followed by Brazil (29,000 Mt), Russia (25,000 Mt), China (20,000 Mt), Canada (6000 Mt), the Ukraine (6500 Mt), and India (5500 Mt), followed at a distance by countries such as South Africa, the United States, Iran, and Kazakhstan (Fig. 28.2).

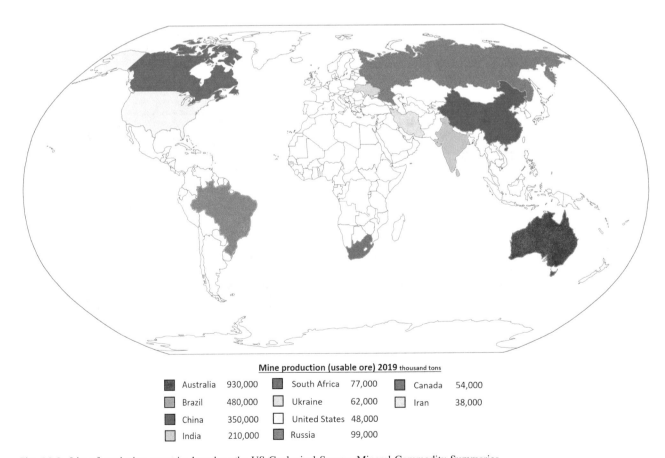

Mine production (usable ore) 2019 thousand tons

Australia	930,000	South Africa	77,000	Canada	54,000
Brazil	480,000	Ukraine	62,000	Iran	38,000
China	350,000	United States	48,000		
India	210,000	Russia	99,000		

Fig. 28.2 List of producing countries based on the US Geological Survey, Mineral Commodity Summaries

28.3 Applications

Metallurgical Industry

Iron is the most widely used metal in the manufacture of all types of steel (combined with other metals and carbon) to build machinery, railways, cars, structures, tools, kitchen utensils, beverage cans, etc. (Figs. 28.3 and 28.4).

Iron is vital in the manufacture of magnets and electro-magnets, due to its magnetic properties.

Terphenol-D is an alloy of iron, terbium, and highly magnetostrictive dysprosium, used in the construction of naval sonar systems and magnetomechanical sensors.

Chemical Industry

Iron catalysts are used in the production of ammonia, the raw material for fertilizers that supply one-third of the world's population.

Iron (III) chloride is used as a water purifier and in the treatment of wastewater. It is also used as a red pigment in paint and fabrics.

Iron (III) chloride is a reducing agent in the reduction of chromates in cement.

Ferrous (II) sulfate is used to treat iron deficiency (anemia), a disease that arises when there is insufficient iron in the body.

Other Applications

Due to their color stability, low cost and non-toxicity, iron oxides are widely used as natural coloring pigments in building materials, the chemical industry, plastics, and cosmetics.

28.4 Recycling

The main source of recycled iron is scrap and steels, mainly from scrapped cars, forming the raw material to produce new steel and gray cast-iron products (cast iron).

Recycling steel saves 62% of energy compared to producing it from iron ore, as well as saving 85% of the water and 95% of the coal that would otherwise be used; moreover, it reduces emissions from CO_2 by 80%.

The consumption of scrap for obtaining new steel is increasing, to the detriment of iron-ore extraction, due to the enormous amount of steel waste on the market and the energy savings that it represents.

Fig. 28.3 Excavator shovel (*Photo* Joaquim Sanz. MGVM)

Fig. 28.4 Iron railway bridge (*Photo* Joaquim Sanz. MGVM)

Further Reading

Gray T, Mann N (2009) The elements. Black Dog & Leventhal Publishers Inc., New York

Mata JM, Sanz J (2007) Guia d'identificació de minerals. Manresa, Catalonia. Edicions UPC/Parcir (Catalan 2nd paper edition). 262 p. ISBN: 9788483019023. URL: http://hdl.handle.net/2117/90445

Pirajno F (2009) Hydrothermal processes and mineral systems. Springer

Quadbeck-Seeger H-J (2007) World of the elements: elements of the world. Wiley-VCH Verlag GmbH & Co, Germany

Sanz J, Tomasa O (2017) Elements i Recursos minerals: Aplicacions i reciclatge. Manresa, Catalonia. Zenobita Edicions/Iniciativa Digital Politècnica (Catalan 3rd digital edition). URL: http://hdl.handle.net/2117/105113

Sanz J, Tomasa O (2018) Elementos y Recursos minerales: Aplicaciones y reciclaje. Manresa, Catalonia. Zenobita Edicions/Iniciativa Digital Politècnica (Spanish 1st digital edition). URL: http://hdl.handle.net/2117/123674

Stwertka A (2018) A guide to the elements, 4th edn. Oxford University Press, England

USGS (2021) Commodity Statistics and Information. Iron Ore. Available at: https://www.usgs.gov/centers/nmic/iron-ore-statistics-and-information. Last accessed May 2021

PERIODIC TABLE OF E L E M E N T S (interactive)

- A soft, ductile, malleable, and very heavy metal
- Highly resistant to corrosion
- Used for sewage pipes by the Romans; some in Rome are in perfect working order after 2000 years
- A good absorber of sound and radioactivity
- Due to its toxicity, its applications are reduced
- Obtained from galena (Fig. 29.1) and the treatment of other copper, zinc, and nickel sulfides.

29.1 Geology

Lead is most commonly found in hydrothermal veins. Lead and zinc are often found together in several kinds of ore deposits, and less so copper and iron. Metals precipitate from ore fluids by various processes, depending on the specific local conditions. The most common processes are cooling, mixing with other fluids, and changes in pH. Lead is most commonly obtained from galena (PbS) in assemblages with other sulfides such as sphalerite (ZnS), chalcopyrite

Fig. 29.1 Galena (lead sulfide). *El Molar (Catalonia)* (*Photo* Joaquim Sanz. MGVM)

(CuFeS$_2$), pyrite (FeS$_2$), and pyrrhotite (FeS). Basinal hydrothermal systems include Mississippi Valley-type (MVT) and sedimentary exhalative (SEDEX) deposits. The MVT deposits are so named because they were first found in mid-continent United States in the valley of the Mississippi River. These deposits are tectonically related to enriched high-salinity metal fluids from sedimentary basins going into carbonate platforms. Sedimentary exhalative (SEDEX) deposits consist of layers of lead–zinc-iron sulfides, a product of sedimentary processes, and are found within large ancient sedimentary basins. Famous SEDEX deposits include Broken Hill, Mount Isa, and McArthur River in Australia, and Sullivan in British Columbia. Lead is also found in volcanogenic massive sulfide (VMS) deposits and skarn deposits. Typical lead–zinc VMS and skarn deposits are in Mexico, Honduras, and Peru.

29.2 Producing Countries

The world reserves of galena are in Australia (36,000 t), China (18,000 t), Russia and Peru (6400 t), Mexico (5600 t), and the United States (5000 t), among others. However, abundant lead reserves are found in zinc, silver, and copper deposits in several countries around the world, including Australia, China, Ireland, Mexico, Peru, Portugal, Russia, and the United States (Fig. 29.2).

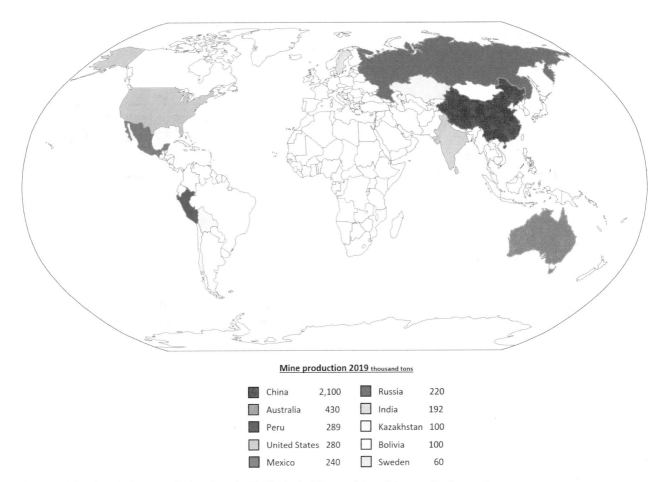

Mine production 2019 thousand tons

China	2,100	Russia	220
Australia	430	India	192
Peru	289	Kazakhstan	100
United States	280	Bolivia	100
Mexico	240	Sweden	60

Fig. 29.2 List of producing countries based on the US Geological Survey, Mineral Commodity Summaries

29.3 Applications

Battery Industry

The main application of lead is in lead-acid batteries, widely used in most vehicles, in powering electronic equipment, and as an emergency power source in companies, communications, and calculation centers (Fig. 29.3).

Construction

Lead is used in the manufacture of plates and bricks for shielding against ionizing radiation in hospital wards and centers where x-rays, CT scans, and nuclear medicine are undertaken, and in companies or scientific institutions with equipment that generates ionizing radiation.

Glass and Ceramics Industry

Lead oxide, added to glass for jewelry and for making 'crystal' glass, gives it a higher refractive index, more brilliance, and a pleasant sound when struck gently. It is also used in the manufacture of lead glass, which protects against ionizing radiation.

Lead is used to join the pieces of glass in the windows of many churches (leaded glass and stained glass).

Metallurgical Industry

Alloyed with tin and silver, lead is used for solders that need good mechanical resistance; however, due to its toxicity, it is being replaced by other metals.

Other Fields

Lead is used as a protective cover for underwater power cables and in the manufacture of bullets, pellets, and counterweights.

Lead containers are used to transport radioactive isotopes for nuclear medicine.

29.4 Recycling

In the European Union and the United States, the lead recycling rate is above 70 to 90%, mainly sourced from lead-acid batteries, scrap, and offcuts.

In the rest of the world, the average recovery rate for lead is above 80% from recycling vehicle batteries and battery equipment from emergency power supply facilities in industry and telecommunication and computing centers.

Further Reading

Eurometaux (2021) https://eurometaux.eu/about-our-industry/introducing-metals/. Last accessed May 2021

Gray T, Mann N (2009) The elements. Black Dog & Leventhal Publishers Inc., New York

International Lead Association (ILA) (2021) Lead. Available at: http://www.ila-lead.org. Last accessed May 2021

Mata JM, Sanz J (2007) Guia d'identificació de minerals. Manresa, Catalonia. Edicions UPC/Parcir (Catalan 2nd paper edition). 262 p. ISBN: 9788483019023. URL: http://hdl.handle.net/2117/90445

Pirajno F (2009) Hydrothermal processes and mineral systems. Springer

Quadbeck-Seeger H-J (2007) World of the elements: elements of the world. Wiley-VCH Verlag GmbH & Co, Germany

Sanz J, Tomasa O (2017) Elements i Recursos minerals: Aplicacions i reciclatge. Manresa, Catalonia. Zenobita Edicions/Iniciativa Digital Politècnica (Catalan 3rd digital edition). URL: http://hdl.handle.net/2117/105113

Sanz J, Tomasa O (2018) Elementos y Recursos minerales: Aplicaciones y reciclaje. Manresa, Catalonia. Zenobita Edicions/Iniciativa Digital Politècnica (Spanish 1st digital edition). URL: http://hdl.handle.net/2117/123674

Stwertka A (2018) A guide to the elements, 4th edn. Oxford University Press, England

USGS (2021) Commodity statistics and information. Lead. Available at: https://www.usgs.gov/centers/nmic/lead-statistics-and-information. Last accessed May 2021

Fig. 29.3 Lead-acid car battery (*Photo* Joaquim Sanz. MGVM)

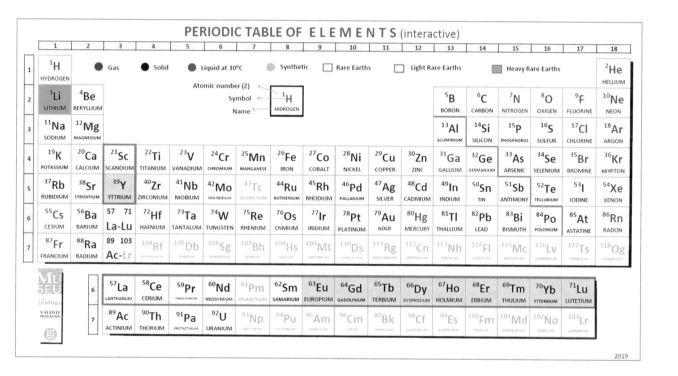

- The lightest metal known
- Extremely reactive with air and water
- Discovered in 1817 by Swedish chemist Johan August Arfwedson during routine chemical investigations of Swedish minerals
- Named after the Greek *lithos* (stone)
- Very high electrochemical potential
- Its presence in the human body is central to psychiatric health
- Obtained from spodumene (Fig. 30.1), amblygonite, lepidolite (in pegmatites), and brines that are lithium-rich but low in magnesium
- Assigned to the Critical Raw Materials list by the EU in September 2020.

30.1 Geology

Lithium is mostly found in granites and pegmatite rocks, brines (salt solution), seawater, and clays. It typically occurs as a minor component in minerals. There are more than 100 minerals known potentially to contain lithium, but only a few are currently economic to exploit. The most common lithium-bearing minerals are

Fig. 30.1 Spodumene (aluminum and lithium silicate). *Namibe (Moçâmedes) (Angola)* (*Photo* Joaquim Sanz. MGVM)

lepidolite (K(Li,Al)$_3$(Si,Al)$_4$O$_{10}$(F,OH)$_2$), petalite (LiAl-Si$_4$O$_{10}$), amblygonite (Li,Na)Al(PO$_4$)(F,OH), and spodumene (LiAlSi $_2$O$_6$), and the most abundant is spodumene. However, subsurface brines are a common source of lithium since the extraction process is easier than from igneous rocks.

Adam Webb (S&P Global, 2019) states: 'The cost of producing concentrates at hard-rock (pegmatites) mines is generally lower than that of producing lithium chemical products from brines operations. For brine assets, the biggest cost component is reagents (sodium carbonate and lime) for downstream processing, and royalty costs are also notably high at brine operations.... Although hard-rock (pegmatites) producers have lower cost, the price they receive for their end product, usually spodumene concentrate, is significantly lower than received for lithium carbonate, chloride and hydroxide, which are produced at brine.'

According to Dominic Wells (Industry News, 2020, October): 'In the transition towards a more sustainable lower-carbon future, reducing the CO$_2$ intensity of the lithium supply chain is crucial, with demand set to soar over the next decade. In recent years, the lithium sector has been dogged by allegations of excessive water use, chemical contamination incidents and poor relationships with local indigenous groups. However, as an industry, perhaps one of the most concerning aspects is the CO$_2$-intensive nature of its supply chain, particularly from hard rock sources.'

Also according to Dominic Wells (Industry News, 2020, November): 'Brine producers have an average carbon footprint of just 2.8t of CO$_2$ per tonne of lithium carbonate equivalent (LCE), compared to 9.6t of CO$_2$ of mineral producers.' Roskill has calculated that CO$_2$ emissions from lithium production are set to triple by 2025 and grow to six times current levels by 2030 (October 2020).

30.2 Producing Countries

The world's lithium reserves are distributed as follows:

Pegmatites (rock): The chief reserve is Australia (2.8 Mt), followed by China (1 Mt), Canada (0.3 Mt), Namibia, Zimbabwe, Brazil, and the United States (Fig. 30.2).

Brines: The main resources are in Bolivia (21 Mt) (according to SRK), followed by Argentina (17 Mt), Chile (9 Mt), China, and the United States (Fig. 30.2).

30.3 Applications

Battery Industry

Li-ion batteries are the main consumers of this element.

In 2019, Roskill estimated that the Li-ion rechargeable battery industry consumed 54% of all lithium mined in the world.

Li-ion batteries are more efficient and lighter than traditional lead-acid batteries. They have a higher energy density, longer life, lower memory effect, and lower charge loss when not being used, so they are used in electric vehicles, drills, portable tools, etc. (Fig. 30.3).

They are also manufactured in small sizes: button cells, digital watches, Pacemakers, and lithium-polymer batteries for smartphones, small motors, laptops, etc. (Fig. 30.4) These batteries must not be electrically overcharged as they can be ignited by an increase in temperature.

In the portable electronic devices industry, the most widely used cathode is the LCO (lithium cobalt oxide), while in the automotive industry it is the NMC (nickel, manganese, cobalt).

According to Roskill, the current trend (2020) is towards cathodes with a high nickel level (NMC 6: 2: 2 and NMC 8: 1: 1) in order to increase the energy density of the battery, even at the cost to its life and more expensive battery management systems. On the other hand, anode materials are evolving to maximize battery performance and battery longevity, with increasing amounts of silicon doping.

The lithium iron phosphate (LIP) cathode for Li-ion batteries is being introduced, especially in China. Despite its lower energy density than nickel manganese cobalt (NMC) batteries, it is usually more durable and cheaper because it is cobalt-free. While cobalt is necessary in vehicles with a range of more than 200 km and for portable electronics, the LFP cathode can be used for vehicles with a lesser range (vans and city buses) (Matthew Bohlson, Investor Intel, 2019, April).

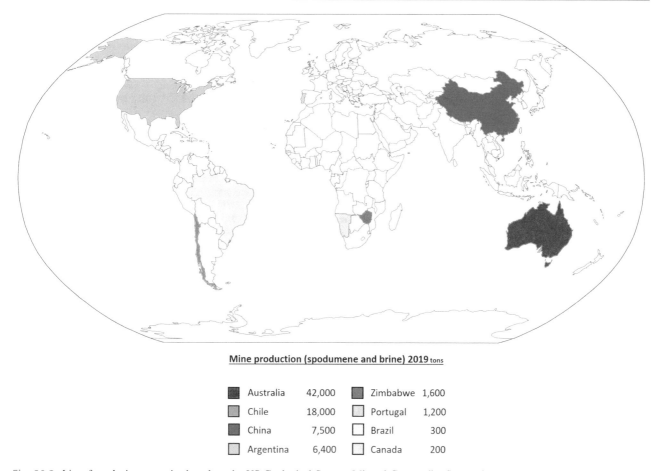

Mine production (spodumene and brine) 2019 tons

■	Australia	42,000	■	Zimbabwe	1,600
■	Chile	18,000	□	Portugal	1,200
■	China	7,500	□	Brazil	300
□	Argentina	6,400	□	Canada	200

Fig. 30.2 List of producing countries based on the US Geological Survey, Mineral Commodity Summaries

Fig. 30.3 Electric car (*Photo* Joaquim Sanz. MGVM)

Fig. 30.4 Li-ion battery (*Photo* Joaquim Sanz. MGVM)

Glass and Ceramics Industry

The consumption of lithium minerals in the glass, ceramic, and glass–ceramic industries is notable. Lithium added to glass and ceramics provides greater resistance to high temperatures, greater hardness, better resistance to impact and changes in temperature, and enhanced color and luster.

Medicine

Lithium carbonate is used as a drug in psychiatry to regulate the presence of lithium in the bodies of bipolar people. It is also used in the preparation of dental prostheses based on lithium disilicate.

Chemical Industry

Lithium stearate is used in the manufacture of synthetic rubber, greases, and other lubricants.

Metallurgical Industry

Lithium is used in alloys with aluminum, copper, or manganese in the aeronautical industry.

30.4 Recycling

There is already sufficient technology to recover lithium from used batteries (end of life—EOL), and there are some companies that recycle them. However, the current price of lithium makes this economically unfeasible, and as cobalt is a scarcer element and more expensive it is this element in used batteries, as well as nickel, that is most recycled.

Tesla, Volkswagen, BMW-Umicore-Northvolt, and Fujian Evergreen have started recycling used lithium batteries (EOL) from electric vehicles (EV) and thus recover, in addition to lithium, metals such as cobalt, nickel, aluminum, copper, and steel. However, according to Roskill, until there are no further used electric vehicle (EV) batteries on the market to purchase, giving a second life to a secondhand battery is more cost-effective than recycling the lithium that it contains.

Further Reading

Gray T, Mann N (2009) The elements. New York. Black Dog & Leventhal Publishers, Inc.

Intel Investor. Intel Technology Metals (2019) Lithium. Available at: https://investorintel.com/sectors/technology/technology-intel/nano-one-is-excited-by-their-lfp-battery-opportunity-with-pulead-technology/. Last accessed May 2021

Mata JM, Sanz J (2007) Guia d'identificació de minerals. Manresa, Catalonia. Edicions UPC/Parcir (Catalan 2nd paper edition). 262 p. ISBN: 9,788,483,019,023. URL: http://hdl.handle.net/2117/90445

Quadbeck-Seeger H-J (2007) World of the elements: elements of the world. Germany. Wiley–VCH Verlag GmbH & Co.

Roskill (2020) Market reports. Lithium. Available at: https://roskill.com/market-report/lithium/. Last accessed May 2021

Roskill (2020) Market reports. Lithium-ion Batteries. Available at: https://roskill.com/market-report/lithium-ion-batteries/. Last accessed May 2021

Roskill (2020) Market reports. Lithium sustainability. Available at: https://roskill.com/news/lithium-sustainability-limited-improvement-expected-from-shipping-emissions-over-coming-decade/. Last accessed May 2021

Sanz J, Tomasa O (2017) Elements i Recursos minerals: Aplicacions i reciclatge. Manresa, Catalonia. Zenobita Edicions/Iniciativa Digital Politècnica (Catalan 3rd digital edition). URL: http://hdl.handle.net/2117/105113

Sanz J, Tomasa O (2018) Elementos y Recursos minerales: Aplicaciones y reciclaje. Manresa, Catalonia. Zenobita Edicions/Iniciativa Digital Politècnica (Spanish 1st digital edition). URL: http://hdl.handle.net/2117/123674

Stwertka A (2018) A guide to the elements, 4th edn. Oxford University Press, England

USGS (2021) Commodity STATISTICS AND INFORMAtion. Lithium. Available at: https://www.usgs.gov/centers/nmic/lithium-statistics-and-information. Last accessed May 2021

Magnesium (Mg) [Z = 12]

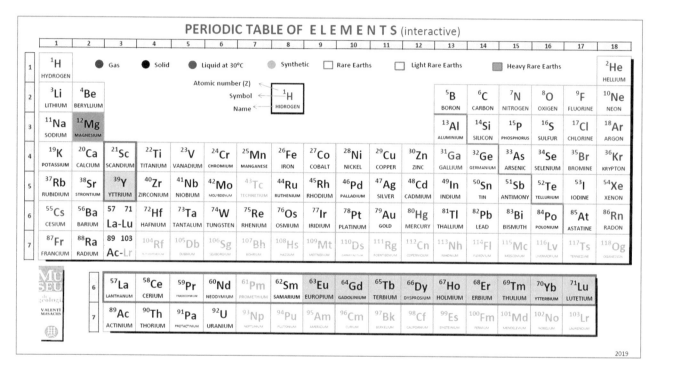

- An alkaline earth metal
- Very light (1.74 gr/cm^3), resistant, and easy to machine
- Lighter than aluminum and highly resistant to corrosion
- Highly flammable (in tape form, it can be lit with a match)
- In powder form it is highly explosive (early flash photography sprayed magnesium powder into a candle flame)

- In large pieces, it is difficult to ignite: the mass of the metal dissipates the heat fast enough to prevent it from burning.
- Seventh most abundant element in the Earth's crust
- Essential element for the human organism and for chlorophyll in plants to photosynthesize
- Assigned the status of a strategic metal by the EU in 2017
- Obtained from magnesite (Fig. 31.1) and magnesium chloride from brine, seawater, and salt lakes that contain it.

Fig. 31.1 Magnesite (magnesium carbonate). *Eugui (Spain)* (*Photo* Joaquim Sanz. MGVM)

31.1 Geology

Magnesium is the eighth-most abundant element in the Earth's crust, constituting about 2%, and in seawater it is the third most abundant element, excluding hydrogen and oxygen, after sodium salts.

Basically it is found forming carbonates, silicates, chlorides, and sulfates. The main source of magnesium is magnesite ($MgCO_3$), but minerals such as dolomite CaMg $(CO_3)_2$, carnallite ($KMgCl_3.6H_2O$), brucite ($Mg(OH)_2$), and olivine (($Mg,Fe)_2(SiO_4)$) are also of commercial importance.

Magnesite usually forms during alteration of magnesium-rich or carbonate rocks by metamorphism or chemical weathering. Magnesites can be divided into three categories on the basis of their crystal characteristics and metallogenic environment (Zheng et al. 2015): sparry magnesite; aphanitic magnesite deposits; and sedimentary metamorphic-hydrothermal metasomatic magnesite.

Sparry magnesite deposits, formed in sedimentary or metamorphic magnesium carbonate rocks, are mainly layered or lenticular bodies in Precambrian dolomite marble formations. They mostly formed in continental platform regions, such as the Haicheng-Ashiqiao magnesite deposit of the China and North Korea paleo-continent (Hurai et al. 2011).

Aphanitic magnesite deposits are found commonly related to ultramafic rocks, mainly in serpentinized rocks such as the Serbian deposits in Sumadija district. They mostly formed in the shallow lacustrine sediments of the Tertiary period and are usually large in size, occurring in secondary sediments in superficial strata, such as those in Australia (Zheng et al. 2015).

Sedimentary metamorphic-hydrothermal metasomatic magnesite deposits are generally layered or lenticular orebodies mostly in rocks formed during the Pre-Ediacaran and Ediacaran periods, such as dolomite or marble. These represent major types of large and supersized deposits. Typical examples are those in the Liaodong region and the Tianshan Mountain area of Xinjiang (Dong et al. 2014).

China has dominated the world's magnesium supply for decades. Over the years, it has discovered new sources across the country, such as Haicheng in Liaoning province, Dahe in Hebei province, and Basha in Tibet (Zheng et al. 2015). Nowadays, the main magnesium-producing area is Liaoning province, focused on two areas: Dashiqiao in Yingkou; and Haicheng in Anshan (Roskill 2021). Moreover, China has further magnesium-enriched mineral sources, such as dolomite resources, with total proven reserves of more than 3 billion tonnes, in addition to Qinghai Salt Lake that contains 3.2 billion tonnes of magnesium chloride and 1.6 billion tonnes of magnesium sulfate (Zheng et al. 2015).

Australia and Canada are developing two new processes in order to use minerals that have already been extracted, such as by-products or tailings. The Australian process is a hydrothermal and metallurgic process based on producing magnesium from by-products of fly ash. The Canadian process is based on reusing tailings from abandoned asbestos (calcium magnesium silicates) mining operations.

31.2 Producing Countries

The world's reserves of magnesium are secure due to its extensive sources: brines, seawater, dolostone, magnesite, and other minerals of evaporite origin (Fig. 31.2).

31.3 Applications

Metallurgical Industry
The main application of magnesium is in the manufacture of alloys with aluminum. For example, alloy car wheels are made of magnesium or a combination of magnesium and aluminum. Also, magnesium provides strength and helps to dissipate the heat produced by vehicles' braking system, as well as giving greater resistance to corrosion and achieving a decrease in weight (Fig. 31.3).

Beverage cans are generally made of an aluminum-magnesium alloy.

Magnesium is one of the alloys used in the manufacture of very lightweight bicycles.

The elektron ZRE1 alloy combines magnesium with zirconium, zinc, and rare earths, and the result is a very strong product for Eurocopter-type helicopter parts and the aerospace industry.

Magnesium is used as a reducing agent in the production of titanium and other metals.

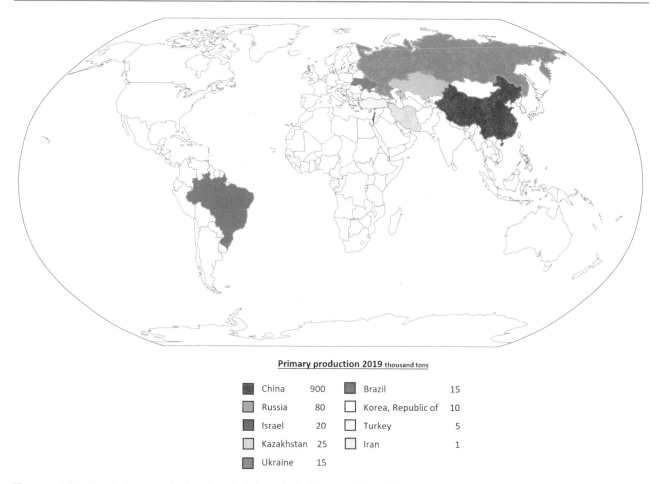

Primary production 2019 thousand tons

■	China	900	■	Brazil	15
■	Russia	80	☐	Korea, Republic of	10
■	Israel	20	☐	Turkey	5
☐	Kazakhstan	25	☐	Iran	1
■	Ukraine	15			

Fig. 31.2 List of producing countries based on the US Geological Survey, Mineral Commodity Summaries

Fig. 31.3 Magnesium alloy car wheel (*Photo* Joaquim Sanz. MGVM)

Electronics Industry

Magnesium is used in the manufacture of chassis for mobile phones, computers, cameras, etc., due to its light weight and good mechanical properties.

Steel Industry

Magnesium oxide is used as a refractory material in steel furnaces in the production of steels, non-ferrous metals, glass, and cement.

Other Fields

Magnesium is essential to plants (it is the main element in chlorophyll), as a supplement to livestock feed (Fig. 31.4), and for the recovery of contaminated soil.

Magnesium carbonate is a natural antacid that for decades has contributed to neutralizing excess acid in the stomach, constituting a healthy supplement. It is also used as a desiccant for athletes' and gymnasts' hands to improve their grip on equipment, and for rock-climbers.

Ferrocerium is a material containing iron, magnesium, cerium, lanthanum, neodymium, and praseodymium, and it is used in the manufacture of sparking lighter 'flints'.

In chip and powder form, magnesium is highly flammable and explosive. It produces an intense white light and a great deal of noise, so it is much used in fireworks.

Fig. 31.4 Livestock feed (*Photo* Joaquim Sanz. MGVM)

The main body of many metal pencil sharpeners is of magnesium.

31.4 Recycling

Recycling of magnesium metal is directly linked to the aluminum and steel industry. All magnesium metal offcuts from the various manufacturing processes are recovered, as well as from cast alloys and disused objects containing it.

The process of obtaining magnesium by recycling saves 50% of the energy consumed by obtaining it from ore.

References

Dong A, Zhu X, Li S, Wang Y, Gao Z (2014) Genesis of Precambrian Strata-bound Magnesite Deposit in NE China. In: Acta Geol Sin—English Edn 88(s2): 1559–1560). URL: https://doi.org/10.1111/1755-6724.12384_4

Hurai V, Huraiová M, Koděra P, Prochaska W, Vozárová A, Dianiška I (2011) Fluid inclusion and stable CO isotope constraints on the origin of metasomatic magnesite deposits of the Western Carpathians, Slovakia. Russ Geol Geophys 52(11):1474–1490. URL: https://doi.org/10.1016/j.rgg.2011.10.015

Zheng Z, Cui X, Denghong W; Yuchuan C, Ge B, Jiankang L, Xinxing L (2015) Review of the metallogenic regularity of magnesite deposits in China. Acta Geol Sin-English Edn 89(5):1747–1761. URL: https://doi.org/10.1111/1755-6724.12576

Further Reading

Gray T, Mann N (2009) The elements. Black Dog & Leventhal Publishers Inc., New York

Luxfer Mel Technologies (2020) Extruded Magnesium. https://www.luxfermeltechnologies.com/. Last accessed May 2021

Mata JM, Sanz J (2007) Guia d'identificació de minerals. Manresa, Catalonia. Edicions UPC/Parcir (Catalan 2nd paper edn). 262 p. ISBN: 9788483019023. URL: http://hdl.handle.net/2117/90445

Quadbeck-Seeger H-J (2007) World of elements: elements of the world. Wiley-VCH Verlag GmbH & Co, Germany

Roskill (2020) Market reports. Magnesium Metal. https://roskill.com/market-report/magnesium-metal/. Last accessed May 2021

Sanz J, Tomasa O (2017) Elements i Recursos minerals: Aplicacions i reciclatge. Manresa, Catalonia. Zenobita Edicions/Iniciativa Digital Politècnica (Catalan 3rd digital edn). URL: http://hdl.handle.net/2117/105113

Sanz J, Tomasa O (2018) Elementos y Recursos minerales: Aplicaciones y reciclaje. Manresa, Catalonia. Zenobita Edicions/Iniciativa Digital Politècnica (Spanish 1st digital edition). URL: http://hdl.handle.net/2117/123674

Stwertka A (2018) A guide to the elements, 4th edn. Oxford University Press, England

USGS (2021) Commodity statistics and information. Magnesium metal. https://www.usgs.gov/centers/nmic/magnesium-statistics-and-information. Last accessed May 2021

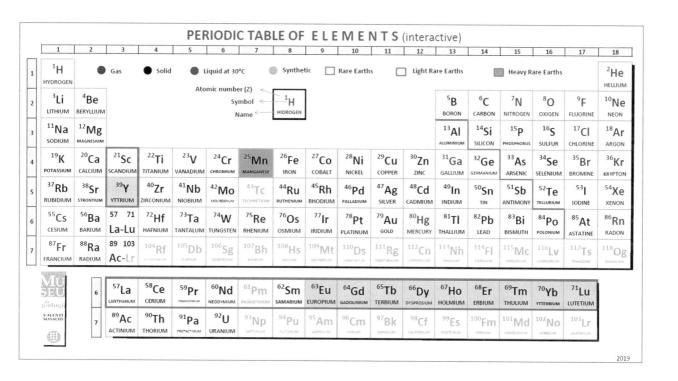

PERIODIC TABLE OF ELEMENTS (interactive)

- Very hard and fragile metal
- Refractory and easily oxidized
- The black of manganese oxide was being used to paint images in caves 17,000 years ago
- Obtained from pyrolusite (Fig. 32.1) and other manganese oxides such as manganite
- Manganese nodules with other elements (nickel, copper, iron, and silicon) are found on many seabeds at varying depths, from the shallow Baltic Sea to 4000 and 6000 m in the Pacific Ocean, and they may represent a future source of manganese.

32.1 Geology

According to USGS (2021), the main minerals from which to obtain manganese are pyrolusite (MnO_2) and manganite (MnO(OH). Manganite is a manganese hydroxide that forms commonly by oxidation at or near Earth's surface as a result of circulating groundwater. It is a significant mineral in major supergene deposits, such as those in Gabon and Ghana, as well as in parts of the Kalahari district in South Africa.

Pyrolusite is a compositionally simple manganese oxide. It forms in a variety of oxidizing conditions, especially in cases of higher acidity. It is a major ore mineral in many

Fig. 32.1 Pyrolusite (manganese oxide). *Tosa d'Alp (Catalonia)* (*Photo* Joaquim Sanz. MGVM)

significant mining districts, including in Brazil, Gabon, Ghana, the Republic of Georgia, and South Africa.

The most important manganese ores are land-based deposits consisting mostly of ancient marine sediments and zones of secondary enrichment. There is a wide range of deposits of this type, as shown below:

- Manganese deposits in marine sedimentary rocks, such as the vast Kalahari deposits of South Africa that occur as interlayers in a banded-iron formation (BIF).
- Manganese deposits without iron enrichments. The largest of such deposits include Molango in Mexico, Groote Eylandt in Australia, the deposits of the Black Sea region, and many deposits in China.
- Iron-related manganiferous sedimentary deposits. Many occurrences of manganiferous sedimentary deposits are interlayered with iron-rich strata. These include the vast manganese deposits of the Kalahari district in South Africa and the Urucum mining district in Brazil, and many deposits in India.
- Secondary enrichment (supergene) deposits. These deposits form where chemical reactions takes place within tens of meters of the surface, redistributing manganese at a local scale and also leaching out non-manganese components, resulting in a residual enrichment. Significant deposits of these types are the Moanda deposit in Gabon, the Azul and the Serra do Navio deposits in Brazil, and numerous deposits in India.

32.2 Producing Countries

The most important reserves in the world are in South Africa (260 Mt), followed by the Ukraine and Brazil (140 Mt each), Australia (100 Mt), and then countries such as Gabon (61 Mt), China (54 Mt), and India (34 Mt) (Fig. 32.2).

32.3 Applications

Metallurgical Industry
Manganese is essential in the manufacture of steels because it improves their mechanical properties and gives them resistance to wear, characteristics that make them suitable for the manufacture of strongboxes, bearings, railway tracks, cutting tools, etc. (Fig. 32.3).

Manganese is used in aluminum alloys for soft drink cans, because it increases the resistance of the alloy against corrosion.

Chemical Industry
Manganese oxide is used as a reagent in the oxidation of benzyl alcohols. An organic derivative of manganese is used as an additive in unleaded gasoline to increase the octane.

Potassium permanganate is used in water purification by absorbing toxic gases from the water.

Battery Industry
Manganese dioxide reacts with zinc in a potassium hydroxide medium, and this reaction produces the electrical energy in alkaline batteries (Fig. 32.4). These batteries have a higher energy density and are more durable than the classic zinc and carbon batteries from Leclanché that are still manufactured, and they are very economical.

The combination of manganese dioxide and lithium, nickel, and cobalt (NCM) is already being used to produce batteries with more power, greater thermal stability, greater safety, and lower cost than Li-ion batteries.

Other Fields
Another application of manganese is as an agricultural fertilizer to increase crop yields by stimulating the process of photosynthesis.

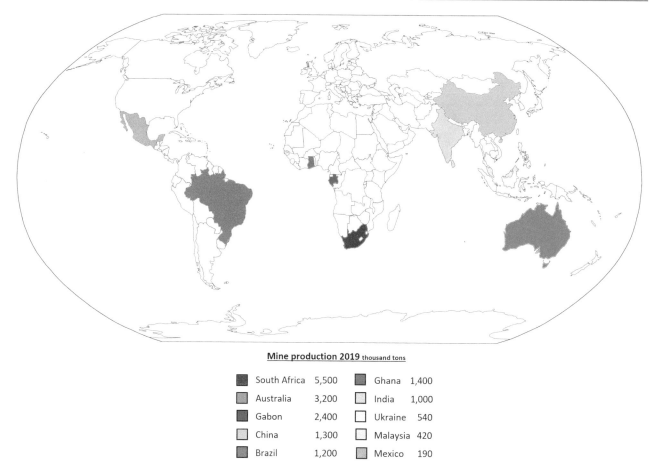

Mine production 2019 thousand tons

■	South Africa	5,500	Ghana	1,400
	Australia	3,200	India	1,000
	Gabon	2,400	Ukraine	540
	China	1,300	Malaysia	420
	Brazil	1,200	Mexico	190

Fig. 32.2 List of producing countries based on the US Geological Survey, Mineral Commodity Summaries

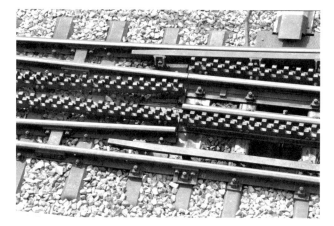

Fig. 32.3 Manganese steel railway tracks (*Photo* Joaquim Sanz. MGVM)

This metal is used as a brown pigment in the manufacture of paint, and the black powder of manganese oxides is a pigment for certain ceramics.

Fig. 32.4 Alkaline battery (*Photo* Joaquim Sanz. MGVM)

32.4 Recycling

The main source of manganese recovery is the offcuts obtained during the manufacture of steels.

Further Reading

Gray T, Mann N (2009) The elements. Black Dog & Leventhal Publishers Inc., New York

Mata JM, Sanz J (2007) Guia d'identificació de minerals. Manresa, Catalonia. Edicions UPC/Parcir (Catalan 2nd paper edn). 262 p. ISBN: 9788483019023. URL: http://hdl.handle.net/2117/90445

Pirajno F (2009) Hydrothermal processes and mineral systems. Springer

Quadbeck-Seeger H-J (2007) World of the elements: elements of the world. Wiley-VCH Verlag GmbH & Co, Germany

Roskill (2020) Market reports. Manganese. URL: https://roskill.com/market-report/manganese/. Last accessed May2021

Sanz J, Tomasa O (2017) Elements i Recursos minerals: Aplicacions i reciclatge. Manresa, Catalonia. Zenobita Edicions/Iniciativa Digital Politècnica (Catalan 3rd digital edn). URL: http://hdl.handle.net/2117/105113

Sanz J, Tomasa O (2018) Elementos y Recursos minerales: Aplicaciones y reciclaje. Manresa, Catalonia. Zenobita Edicions/Iniciativa Digital Politècnica (Spanish 1st digital edn). URL: http://hdl.handle.net/2117/123674

Stwertka A (2018) A guide to the elements, 4th edn. Oxford University Press, England

USGS (2017) Manganese. Chapter L of critical mineral resources of the United States. Economic and environmental geology and prospects for future supply. Available at: https://doi.org/10.3133/pp1802L. Last accessed May 2021

USGS (2021) Commodity statistics and information. Manganese. Available at: https://www.usgs.gov/centers/nmic/manganese-statistics-and-information. Last accessed May 2021

PERIODIC TABLE OF E L E M E N T S (interactive)

	1	2	3	4	5	6	7	8	9	10	11	12	13	14	15	16	17	18
1	^{1}H HYDROGEN	● Gas ● Solid ● Liquid at 30°C ○ Synthetic ☐ Rare Earths ☐ Light Rare Earths ☐ Heavy Rare Earths			Atomic number (Z) Symbol ← ^{1}H Name ← HIDROGEN													^{2}He HELLIUM
2	^{3}Li LITHIUM	^{4}Be BERYLLIUM											^{5}B BORON	^{6}C CARBON	^{7}N NITROGEN	^{8}O OXIGEN	^{9}F FLUORINE	^{10}Ne NEON
3	^{11}Na SODIUM	^{12}Mg MAGNESIUM											^{13}Al ALUMINIUM	^{14}Si SILICON	^{15}P PHOSPHORUS	^{16}S SULFUR	^{17}Cl CHLORINE	^{18}Ar ARGON
4	^{19}K POTASSIUM	^{20}Ca CALCIUM	^{21}Sc SCANDIUM	^{22}Ti TITANIUM	^{23}V VANADIUM	^{24}Cr CHROMIUM	^{25}Mn MANGANESE	^{26}Fe IRON	^{27}Co COBALT	^{28}Ni NICKEL	^{29}Cu COPPER	^{30}Zn ZINC	^{31}Ga GALLIUM	^{32}Ge GERMANIUM	^{33}As ARSENIC	^{34}Se SELENIUM	^{35}Br BROMINE	^{36}Kr KRYPTON
5	^{37}Rb RUBIDIUM	^{38}Sr STRONTIUM	^{39}Y YTTRIUM	^{40}Zr ZIRCONIUM	^{41}Nb NIOBIUM	^{42}Mo MOLYBDENUM	^{43}Tc TECHNETIUM	^{44}Ru RUTHENIUM	^{45}Rh RHODIUM	^{46}Pd PALLADIUM	^{47}Ag SILVER	^{48}Cd CADMIUM	^{49}In INDIUM	^{50}Sn TIN	^{51}Sb ANTIMONY	^{52}Te TELLURIUM	^{53}I IODINE	^{54}Xe XENON
6	^{55}Cs CESIUM	^{56}Ba BARIUM	57 71 La-Lu	^{72}Hf HAFNIUM	^{73}Ta TANTALUM	^{74}W TUNGSTEN	^{75}Re RHENIUM	^{76}Os OSMIUM	^{77}Ir IRIDIUM	^{78}Pt PLATINUM	^{79}Au GOLD	^{80}Hg MERCURY	^{81}Tl THALLIUM	^{82}Pb LEAD	^{83}Bi BISMUTH	^{84}Po POLONIUM	^{85}At ASTATINE	^{86}Rn RADON
7	^{87}Fr FRANCIUM	^{88}Ra RADIUM	89 103 Ac-Lr	^{104}Rf	^{105}Db	^{106}Sg	^{107}Bh	^{108}Hs	^{109}Mt	^{110}Ds	^{111}Rg	^{112}Cn	^{113}Nh	^{114}Fl	^{115}Mc	^{116}Lv	^{117}Ts	^{118}Og

6	^{57}La LANTHANUM	^{58}Ce CERIUM	^{59}Pr PRASEODYMIUM	^{60}Nd NEODYMIUM	^{61}Pm PROMETHIUM	^{62}Sm SAMARIUM	^{63}Eu EUROPIUM	^{64}Gd GADOLINIUM	^{65}Tb TERBIUM	^{66}Dy DYSPROSIUM	^{67}Ho HOLMIUM	^{68}Er ERBIUM	^{69}Tm THULIUM	^{70}Yb YTTERBIUM	^{71}Lu LUTETIUM
7	^{89}Ac ACTINIUM	^{90}Th THORIUM	^{91}Pa PROTACTINIUM	^{92}U URANIUM	^{93}Np NEPTUNIUM	^{94}Pu PLUTONIUM	^{95}Am AMERICIUM	^{96}Cm CURIUM	^{97}Bk BERKELIUM	^{98}Cf CALIFORNIUM	^{99}Es EINSTEINIUM	^{100}Fm FERMIUM	^{101}Md MENDELEVIUM	^{102}No NOBELIUM	^{103}Lr LAURENCIUM

MUSEU de geologia VALENTÍ MASACHS

2019

- The only metal to remain liquid at room temperature (20 °C)
- Very high density (13.6 gr/cm^3): lead (11.34 g/cm^3) floats on it
- Named after the planet, its chemical symbol (Hg) is from the Latin *hydrargyrum*
- Known in ancient China and found in Egyptian tombs dating to 1500 BC
- Forms amalgams with other metals, such as gold and silver

- Highly toxic: in January 2010 its use and trade were banned in the EU, as in other countries
- Not very good at conducting electricity
- Obtained from cinnabar (Fig. 33.1)
- One of world's most important deposits of cinnabar is in Spain, exploited for 2000 years until 2001/03, and large reserves remain (see link)
- Called *azogue* (see link) in the ancient Arabic-Hispanic language.

© The Author(s), under exclusive license to Springer Nature Switzerland AG 2022
J. Sanz et al., *Elements and Mineral Resources*, Springer Textbooks in Earth Sciences,
Geography and Environment, https://doi.org/10.1007/978-3-030-85889-6_33

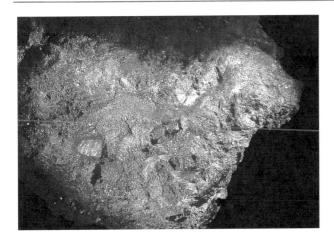

Fig. 33.1 Cinnabar (mercury sulfide). *Almadén (Spain)* *(Photo* Joaquim Sanz. MGVM)

33.1 Geology

Metallic mercury (Hg) is obtained mainly from cinnabar (HgS), which has a characteristic bright scarlet color. This generally occurs as a vein-filling related to magmatic (volcanic) activity. The Almadén district of Spain is a major source of mercury, with stratiform mineralization hosted by the Criadero quartzite. Almadén contains other structural types, some of economic importance. Las Cuevas is a fully discordant deposit stratigraphically located at the top of the Silurian sequence. The district lies in the Central-Iberian zone of the Iberian Massif. There are three cycles of deposition: Late Precambrian, Paleozoic, and Late Cenozoic. Mercury ore is found in Paleozoic rocks.

33.2 Producing Countries

The world's reserves of cinnabar are plentiful for industry's needs for the years to come. The values are unknown; however, it is estimated that China, Kyrgyzstan, and Peru have the most notable reserves (Fig. 33.2).

33.3 Applications

Medicine
In the EU, mercury is now scarcely used for dental fillings, the vast majority of which are of resin.

Mercury was used in thermometers and blood pressure monitors, but these are gradually being replaced by digital displays or in the case of thermometers, galinstan (see: gallium).

Electrical Industry
Mercury vapor is used in the manufacture of fluorescent and energy-saving lamps, where it produces ultraviolet light when excited by an electric discharge. Depending on the color of the glass, the fluorescent powder on the walls of the tube reacts with the vapor to produce white or colored light (Fig. 33.3).

Chemical Industry
Mercury is still used in some countries as a catalyst in the manufacture of polyvinyl chloride (PVC), chlorine, and caustic soda, particularly in China, although this country is in the process of switching to membrane cell systems.

Other Fields
Mercury is still used in the manufacture of some type of button batteries, although its use is decreasing. Unfortunately, in many countries in South America, Africa, Asia, and Australia it is still used to extract gold. When dissolved into an amalgam it leaves a yellowish-gray paste that, when heated, gives off mercury vapor yet leaves any gold. People who work with it inhale the vapor, causing disease; moreover, the process pollutes the surrounding environment.

NGOs such as Mineria en Acción (UPC, Polytechnic University of Catalonia) collaborate annually with mining cooperatives in South America to devise methods to extract gold without using mercury, to avoid polluting the environment.

Mercury fulminate is a highly dangerous and unstable explosive still used as a detonator in certain countries.

Cautions
Methyl mercury is a compound derived from this metal. When it enters the food chain (through environmental contamination) it accumulates and is concentrated in large marine animals such as tuna fish (see link: Society of Environmental Toxicology and Chemistry).

Mercury poisoning is caused by ingesting large amounts, damaging the central nervous system.

33.4 Recycling

The main sources of recycled mercury are the chemical industry and batteries, fluorescent lamps, medical instruments, electrical switches, and thermometers that contain the liquid metal.

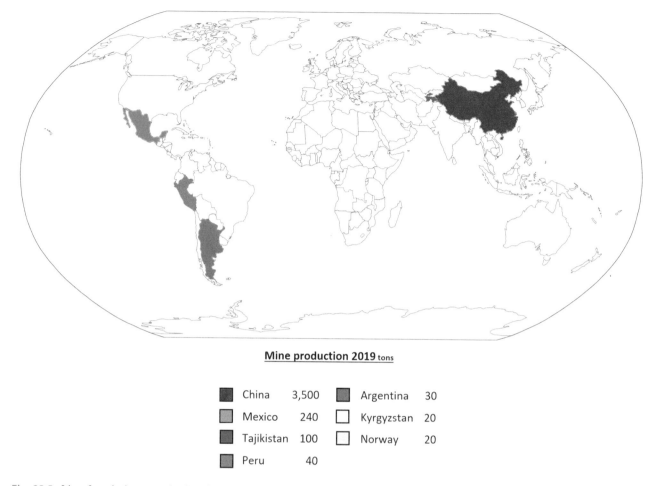

Mine production 2019 tons

■ China	3,500	■ Argentina	30
■ Mexico	240	□ Kyrgyzstan	20
■ Tajikistan	100	□ Norway	20
■ Peru	40		

Fig. 33.2 List of producing countries based on the US Geological Survey, Mineral Commodity Summaries

Fig. 33.3 Fluorescent lamp (*Photo* Joaquim Sanz. MGVM)

Further Reading

Almadén (2021) Mercury. Available at: https://www.parquemine rodealmaden.es/index.php?idioma=in. Last accessed May 2021

Almadén (2021) Mercury (Azogue). Available at: http://www. loscaminosdelazogue.org/. Last accessed May 2021

Gray T, Mann N (2009) The elements. Black Dog & Leventhal Publishers Inc., New York

Hernández A et al (1999) The Almadén mercury mining district. Spain. Miner Deposita 34(5–6):539–548. https://doi.org/10.1007/s001260050219

Mata JM, Sanz J (2007) Guia d'identificació de minerals. Manresa, Catalonia. Edicions UPC/Parcir (Catalan 2nd paper edn). 262 p. ISBN: 9788483019023. URL: http://hdl.handle.net/2117/90445

Quadbeck-Seeger H-J (2007) World of the elements: elements of the world. Wiley-VCH Verlag GmbH & Co, Germany

Sanz J, Tomasa O (2017) Elements i Recursos minerals: Aplicacions i reciclatge. Manresa, Catalonia. Zenobita Edicions/Iniciativa Digital Politècnica (Catalan 3rd digital edn). URL: http://hdl.handle.net/2117/105113

Sanz J, Tomasa O (2018) Elementos y Recursos minerales: Aplicaciones y reciclaje. Manresa, Catalonia. Zenobita Edicions/Iniciativa Digital Politècnica (Spanish 1st digital edn). URL: http://hdl.handle.net/2117/123674

Stwertka A (2018) A guide to the elements, 4th edn. Oxford University Press, England

Society of Environmental Toxicology and Chemistry (2019) Mercury toxicology. Available at: https://setac.onlinelibrary.wiley.com/doi/abs/https://doi.org/10.1002/etc.4513. Last accessed May 2021

USGS (2021) Commodity statistics and information. Mercury. Available at: https://www.usgs.gov/centers/nmic/mercury-statistics-and-information. Last accessed May 2021

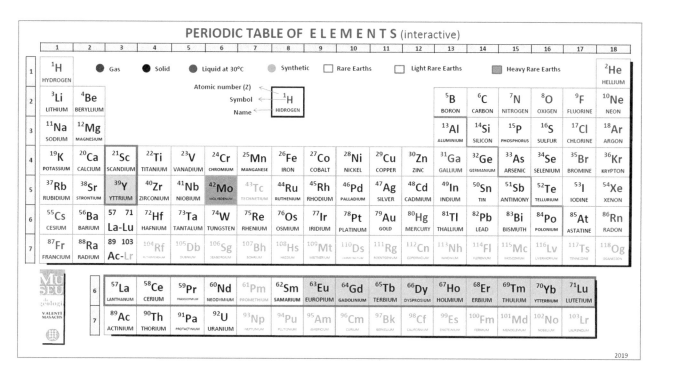

- A good thermal and electrical conductor
- Has a low coefficient of expansion
- A refractory metal (melts at 2625 °C)
- Swedish chemist Carl Wilhelm Scheele isolated and identified this element in molybdenite in 1778
- Obtained from molybdenite (Fig. 34.1) and as a by-product of copper mining, especially in South America.

34.1 Geology

The main molybdenum ores are molybdenite (MoS_2), wulfenite ($PbMoO_4$), and powellite ($CaMoO_4$). Porphyry systems are the main source of molybdenum minerals. Porphyry ore systems take their name from the porphyritic texture of mineralized intrusions that originate from high-temperature magmatic-hydro-thermal fluids. They are characterized by large tonnage yet low grade.

Fig. 34.1 Molybdenite (molybdenum sulfide). *Gualba (Catalonia)* (*Photo* Joaquim Sanz. MGVM)

Molybdenum's mineralization typically occurs as disseminations, stockworks, and veins of sulfides of Fe, Pb, Zn, W, Bi, and Sn, and native Au. Chile is one of the world's main producers of cobalt-molybdenum. In 2019, the

Chuquicamata and Radomiro Tomic mines produced nearly 12 thousand tonnes of molybdenum, while the El Teniente mine produced 7.5 thousand tonnes.

34.2   Producing Countries

The most notable reserves in the world are in China (8.3 Mt), followed by Peru (2.9 Mt), the United States (2.7 Mt), Chile (1.4 Mt), Russia (1 Mt, and Turkey (0.7 Mt), among others (Fig. 34.2).

34.3   Applications

Metallurgical Industry
Molybdenum is used in stainless steels, high-speed cutting steels, and super alloys for knives (Fig. 34.3), bearings, corrosion- and heat-resistant steels for the manufacture of aircraft parts (Fig. 34.4), electrical contacts, industrial engines, and oil and gas drilling equipment.

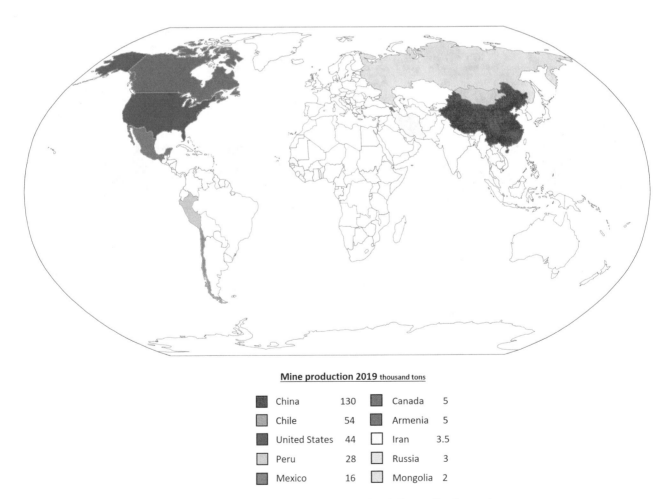

Mine production 2019 thousand tons

China	130	Canada	5
Chile	54	Armenia	5
United States	44	Iran	3.5
Peru	28	Russia	3
Mexico	16	Mongolia	2

Fig. 34.2 List of producing countries based on the US Geological Survey, Mineral Commodity Summaries

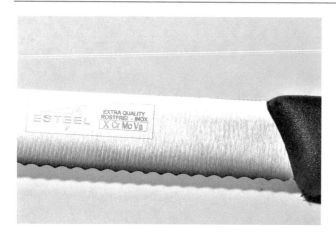

Fig. 34.3 Knife blade of alloy steel with molybdenum, chrome, and vanadium (*Photo* Joaquim Sanz. MGVM)

Fig. 34.4 Helicopter exhaust gas nozzle made of molybdenum steel (*Photo* Joaquim Sanz. MGVM)

Ferro-molybdenum compounds impart multiple properties to steel, one of which is to improve the penetration of heat treatments and to facilitate hardening.

Oil Industry

Molybdenum is used as a catalyst in the oil industry to remove sulfur.

Medicine

Molybdenum, alloyed with cobalt and chromium, is currently used in medical stainless steels to make coronary stents to reduce the stenosis of coronary arteries (among others) that could lead to a heart attack.

Titanium-molybdenum alloy wire (TMA), nickel free, is widely used as an orthodontic archwire. It has good properties between stainless steel and Ni–Ti archwires.

The Mo-99 isotope is used in nuclear medicine as a generator of technetium-99, which is one of the most widely used radioisotopes in radiography (see: technetium).

Other Fields

Molybdenum disulfide is used as a natural lubricating grease, because it resists high mechanical working temperatures without losing its lubricating characteristics and consistency.

Sodium molybdate is used as a fertilizer for broccoli and cauliflower crops.

34.4 Recycling

Molybdenum is recycled from the various alloys, super alloys, and steel-cutting tools that contain it, as well as from catalysts used in the oil industry.

Further Reading

Gray T, Mann N (2009) The elements. Black Dog & Leventhal Publishers Inc., New York

International Molybdenum Association (2021) https://www.imoa.info/index.php. Last accessed May 2021

Mata JM, Sanz J (2007) Guia d'identificació de minerals. Manresa, Catalonia. Edicions UPC/Parcir (Catalan 2nd paper edn). 262 p. ISBN: 9788483019023. URL: http://hdl.handle.net/2117/90445

Pirajno F (2009) Hydrothermal processes and mineral systems. Springer

Quadbeck-Seeger H-J (2007) World of the elements: elements of the world. Wiley-VCH Verlag GmbH & Co, Germany

Roskill (2020) Market reports. Molybdenum (2021) https://roskill.com/market-report/molybdenum/. Last accessed May 2021)

Sanz J, Tomasa O (2017) Elements i Recursos minerals: Aplicacions i reciclatge. Manresa, Catalonia. Zenobita Edicions/Iniciativa Digital Politècnica (Catalan 3rd digital edn). URL: http://hdl.handle.net/2117/105113

Sanz J, Tomasa O (2018) Elementos y Recursos minerales: Aplicaciones y reciclaje. Manresa, Catalonia. Zenobita Edicions/Iniciativa Digital Politècnica (Spanish 1st digital edn). URL: http://hdl.handle.net/2117/123674

Stwertka A (2018) A guide to the elements, 4th edn. Oxford University Press, England

USGS (2021) Commodity statistics and information. Molybdenum. https://www.usgs.gov/centers/nmic/molybdenum-statistics-and-information. Last accessed May 2021

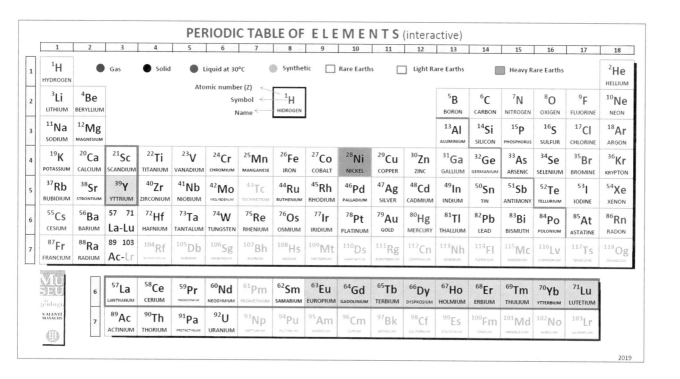

- A ductile and malleable metal
- Slightly ferromagnetic at room temperature
- Melts at 1455 °C
- Extremely resistant to corrosion: it does not rust
- Many scientists believe that Earth's core is chiefly iron and nickel. This could explain why nickel is present in many meteorites, both metallic and non-metallic
- Extracted from nickel-rich laterites (Fig. 35.1) and as a by-product of cobalt and copper ores.

35.1 Geology

The most common nickel ores are pentlandite [$(Ni,Fe)_9S_8$] and garnierite [$(Ni,Mg)_6[(OH)_8Si_4O_{10}]$]. Garnierite is the general name for the nickel-magnesium hydrosilicates that occur in many nickel–cobalt laterite deposits. Most economic nickel deposits occur in one of two types of geological environments: magmatic sulfide deposits or laterite deposits. Operational mines are divided equally between both; however, according to the USGS, laterites account for about 70% of known nickel–cobalt resources. They occur due to chemical weathering in wet tropical and subtropical conditions (warm temperature and high rainfall).

Fig. 35.1. Nepouite (nickel silicate). *Népoui (New Caledonia) (France)* (*Photo* Joaquim Sanz. MGVM)

Lateritic deposits of hydrosilicates are characterized by a serpentine-saprolite horizon normally covered by a layer of limonite (lateritic horizon, sensu stricto). In these deposits, nickel is mostly found forming veins in the serpentinite zone or at the base of the laterite profile (saprolite + garnierite). These deposits are formed under tectonic uplift conditions with a restricted water column.

35.2   Producing Countries

World reserves of nickel are large and are mostly concentrated in three countries: Indonesia (21 Mt), Australia (20 Mt), and Brazil (11 Mt). Next come Russia (6.9 Mt), Cuba (5.5 Mt), Philippines (4.8 Mt), China (2.8 Mt), and Canada (2.6 Mt), among others (Fig. 35.2).

35.3   Applications

Metallurgical Industry
The main use of this nickel is in the manufacture of austenitic stainless steels (with more than 7% nickel).

Classic 18/8 stainless-steel cutlery comprises—apart from iron—chromium (18%) and nickel (8%) (Fig. 35.3).

Nickel gives the copper-nickel alloy better resistance to marine corrosion, biological contamination, and improved ductility. It is widely used in both underwater and pipeline applications, in the manufacture of platform and hull linings, in cages for aquaculture, and also to make heat exchangers and condensers. This metal is used in antioxidant electrolytic coatings (nickel plating), including nickel plating to protect

Mine production 2019 thousand tons

Indonesia	800	Canada	180	Cuba	51
Philippines	420	China	110	Guatemala	49
Russia	270	Brazil	67		
New Caledonia	220	Australia	180		

Fig. 35.2 List of producing countries based on the US Geological Survey, Mineral Commodity Summaries

Fig. 35.3 Stainless-steel elevator (*Photo* Joaquim Sanz. MGVM)

Fig. 35.4 Coin with nickel, copper, and zinc (*Photo* Joaquim Sanz. MGVM)

neodymium magnets (see neodymium). It is also used in the manufacture of crucibles for chemical laboratories.

Nickel forms an alloy with titanium that has a 'memory effect' and is used in dental braces. In small quantities, it is in the composition of coronary stents.

Battery Industry

Nickel hydroxide was used in the manufacture of rechargeable nickel–cadmium (Ni–Cd) batteries, but they were heavy and could not store as much energy as Ni-MH (nickel-metal hydride) batteries, which have superseded them because they are cheaper and less contaminating (see: Rare Earths: lanthanum).

Nickel sulfate is fundamental as a cathode for lithium-ion batteries (NCM and NCA, and its consumption will increase as the electric vehicle market grows, also because it improves the electrical density of the batteries and increases vehicles' range.

Nickel is used in the manufacture of sodium and nickel batteries (zebra batteries), which work at high temperatures.

Manufacture of Magnets

Nickel is involved in the manufacture of Al–Ni–Co permanent magnets, which consist of sintered aluminum, nickel, and cobalt. They were the strongest magnets known until the introduction of neodymium and samarium-cobalt magnets (see: Rare Earths: neodymium and samarium). However, neodymium magnets are more expensive than Al–Ni–Co magnets.

Other Fields

Nickel is used in minting coins, such as €1 and €2 coins, as well as the US coin known as 'a nickel' (five cents), which contains both nickel and copper (Fig. 35.4).

The nickel–chromium-iron alloy constitutes nicrom, used as electrical resistance for toasters, ovens, industrial cutters, etc. It can withstand high temperatures (over 500 °C).

35.4 Recycling

Nickel is recovered, mainly from stainless-steel scrap and other steels containing nickel. The stainless-steel market has an average of 68% recycled nickel (Nickel Institute 2010).

Nickel is also recovered from old Ni-MH rechargeable batteries.

Further Reading

Gray T, Mann N (2009) The elements. Black Dog & Leventhal Publishers Inc., New York

Mata JM, Sanz J (2007) Guia d'identificació de minerals. Manresa, Catalonia. Edicions UPC/Parcir (Catalan 2nd paper edn). 262 p. ISBN: 9788483019023. URL: http://hdl.handle.net/2117/90445

Nickel Institute (2021) http://www.nickelinstitute.org. Last accessed May 2021

Nornickel (2021) Nickel. Available at: https://ar2019.nornickel.com/commodity-market-overview/nickel. Last accessed May 2021

Quadbeck-Seeger H-J (2007) World of the elements: elements of the world. Wiley-VCH Verlag GmbH & Co, Germany

Roskill (2020) Market reports. Nickel. Available at: https://roskill.com/market-report/nickel/. Last accessed May 2021

Roskill (2020) Market reports. Nickel Sulphate. Available at: https://roskill.com/market-report/nickel-sulphate/. Last accessed May 2021

Sanz J, Tomasa O (2017) Elements i Recursos minerals: Aplicacions i reciclatge. Manresa, Catalonia. Zenobita Edicions/Iniciativa Digital Politècnica (Catalan 3rd digital edition). URL: http://hdl.handle.net/2117/105113

Sanz J, Tomasa O (2018) Elementos y Recursos minerales: Aplicaciones y reciclaje. Manresa, Catalonia. Zenobita Edicions/Iniciativa Digital Politècnica (Spanish 1st digital edn). URL: http://hdl.handle.net/2117/123674

Stwertka A (2018) A guide to the elements, 4th edn. Oxford University Press, England

USGS (2021) Commodity statistics and information. Nickel. Available at: https://www.usgs.gov/centers/nmic/ nickel-statistics-and-information. Last accessed May 2021

Villanova-de-Benavent C et al (2014) Garnierites and garnierites: textures, mineralogy and geochemistry of garnierites in the Falcondo Ni-laterite deposit, Dominican Republic. Ore Geol Rev 58:91–109. https://doi.org/10.1016/j.oregeorev.2013.10.008

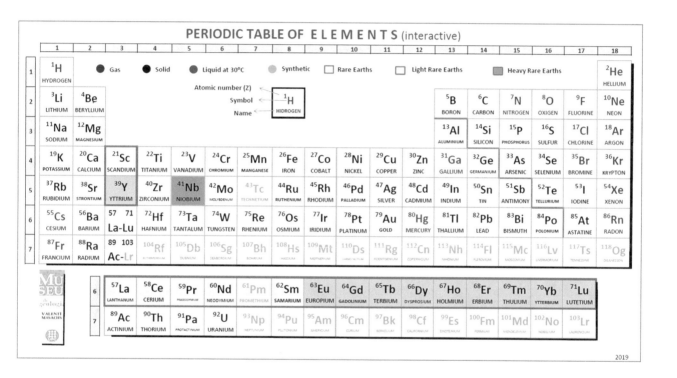

- Corrosion resistant
- Melting point of 2468 °C
- Great hardness
- Oxidizes in air to form a self-protecting layer
- Originally called 'columbium' (Charles Hatchett, 1801), but the German chemist Heinrich Rose called it niobium, from Niobe, daughter of Tantal (son of Zeus in mythology), and this was accepted in 1949
- A rare metal
- Assigned the status of a strategic metal by the EU in 2017
- Found in pyrochlore (Fig. 36.1), columbite-tantalite (coltan), and euxenite
- Very often appears next to tantalum
- Coltan and euxenite may contain small amounts of thorium and/or uranium and emit radioactivity.

36.1 Geology

The columbite group is the primary ore for niobium, referring back to its old name. Niobium and tantalum are usually related. The columbite group includes minerals of varying

Fig. 36.1 Pyrochlore (sodium and calcium niobate). *Oka (Canada)* (*Photo* Joaquim Sanz. MGVM)

1. Rare-metal pegmatites, in which the ores are usually represented by zonal vein bodies ranging in size from a few hundred meters to 1–2 km, with small amounts of spodumene or petalite. Rare-metal Ni-Ta-bearing granites (greisen) are represented by small pipes and granitic domes. A certain amount of Nb–Ta is also extracted from the Sn or W greisen deposits. Minas Gerais (Brazil) is one of the world's largest tantalum producers.
2. Placers related to pegmatitic sources, and which contain cassiterite and minerals of the columbite-tantalite group, and nepheline syenites.
3. Carbonatites, like the Aley deposit in northern British Columbia, Canada, hosted by metamorphosed calcite and dolomite. Primary niobium mineralization consists of pyrochlore ($A_2Nb_2(O,OH)_6Z$) and ferrocolumbite ($Fe^{2+}Nb_2O_6$).

proportions in their composition, with varying proportions of Nb or Ta in composition.

The main minerals from which Nb is extracted are columbite ($(Mn,Fe)(Ta,Nb)_2O_6$), hatchettolite ($(Ca,U,Ce)_2(Nb,Ti,Ta)_2O_6(OH,F)$, microlite $A_{2-m}Ta_2X_{6-w}Z_{-n}$, and ixiolite ($(Nb,Ta,Sn,W,Sc)_3O_6$).

The most common genetic types of Nb–Ta tantalum ore are:

36.2 Producing Countries

The world reserves of niobium are basically centered on two countries, Brazil (11.0 Mt); and Canada (1.6 Mt). The United States (0.2 Mt) follows a long way behind (Fig. 36.2).

They are sufficient reserves to cover the demands of industry for many years to come.

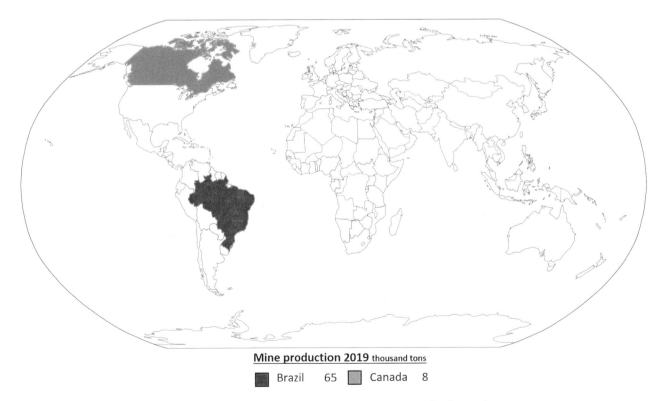

Mine production 2019 thousand tons

■ Brazil 65 ■ Canada 8

Fig. 36.2 List of producing countries based on the US Geological Survey, Mineral Commodity Summaries

36.3 Applications

Metallurgical Industry

Ferroniobium is an important iron-niobium alloy with a niobium content of the order of 60 to 70%. It is the main component of HSLA steels (high-strength, low-alloy).

Niobium is used in steel production because it makes the steel hard, improves weldability, and reduces its density. The suspension bridge at Millau (France), the highest in the world, is made of niobium steel and is 60% lighter as a result (Fig. 36.3).

Microalloyed stainless steels with niobium are used in refractory structural components of cars, aircraft, and rocket nozzles, in the manufacture of oil and gas pipelines in the Arctic, and in the manufacture of steels for turbines in hydroelectric and thermal power plants.

Electronic Industry

Niobium is used in the manufacture of capacitors because, thanks to its high dielectric constant, it can store an electrical charge. Lithium niobate is an electro-optical material used in the manufacture of optical communications equipment.

Medicine

Niobium is used to produce very strong magnetic fields, as at low temperatures it is a superconductor. The magnetic fields produced by the Nb-Ti alloy (niobium-titanium) make this material useful for magnetic resonance equipment (Fig. 36.4) and for the construction of particle accelerators.

Niobium is used in medical devices, such as Pacemakers, because it is a metal that is compatible with the human body.

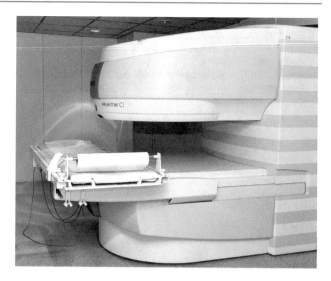

Fig. 36.4 Magnetic resonance equipment (*Photo* Joaquim Sanz. MGVM)

Glass and Ceramics Industry

Niobium is added to glass to achieve a higher refractive index and to allow corrective lenses in spectacles to be thinner and lighter.

36.4 Recycling

Niobium is recycled from steels and superalloys that contain it and from old superconducting magnets. Recycled niobium accounts for approximately 20% of apparent consumption.

Further Reading

Alkane Resources Ltd (2020) Niobium. Available at: http://www. alkane.com.au/products/niobium/. Last accessed May 2021

Chakhmouradian AR et al (2014) Carbonatite-hosted niobium deposit at Aley (British Columbia, Canada). Ore Geol Rev 64 (January):642–666. URL: https://doi.org/10.1016/j.oregeorev.2014. 04.020

Gray T, Mann N (2009) The elements. Black Dog & Leventhal Publishers Inc., New York

Mata JM, Sanz J (2007) Guia d'identificació de minerals. Manresa, Catalonia. Edicions UPC/Parcir (Catalan 2nd paper edn). 262 p. ISBN: 9788483019023. URL: http://hdl.handle.net/2117/ 90445

Quadbeck-Seeger H-J (2007) World of the elements: elements of the world. Wiley-VCH Verlag GmbH & Co Germany

Roskill (2020) Market reports. Niobium. Available at: https://roskill. com/market-report/niobium/. Last accessed May 2021

Sanz J, Tomasa O (2017) Elements i Recursos minerals: Aplicacions i reciclatge. Manresa, Catalonia. Zenobita Edicions/Iniciativa Digital Politècnica (Catalan 3rd digital edn). URL: http://hdl.handle.net/ 2117/105113

Fig. 36.3 Millau viaduct. (France) (*Image courtesy of* Xavier Lluis Tartera)

Sanz J, Tomasa O (2018) Elementos y Recursos minerales: Aplicaciones y reciclaje. Manresa, Catalonia. Zenobita Edicions/Iniciativa Digital Politècnica (Spanish 1st digital edition). URL: http://hdl.handle.net/2117/123674

Stwertka A (2018) A guide to the elements, 4th edn. Oxford University Press, England

Tantalum-Niobium International Study Center (2021) https://tanb.org/about-niobium. Last accessed May 2021

USGS (2021) Commodity statistics and information. Niobium. Available at: https://www.usgs.gov/centers/nmic/niobium-columbium-and-tantalum-statistics-and-information. Last accessed May 2021

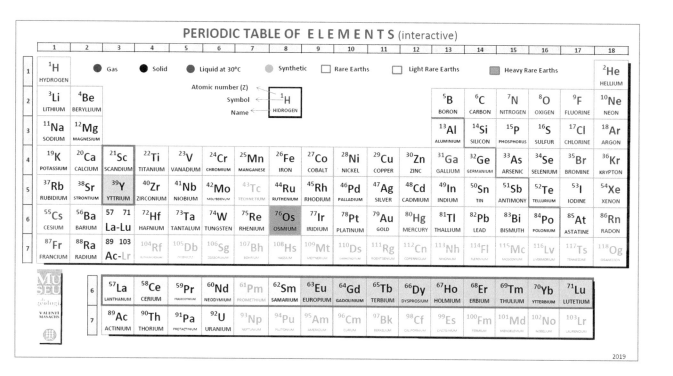

- Scarce noble metal, and extremely hard – more so than platinum
- The most corrosion-resistant element
- Melts at 3050 °C, the highest of the platinum group metals (PGMs)
- The densest PGM—twice as dense as lead
- Resistant to corrosion and high temperatures
- From 1906 it was used as filaments for incandescent bulbs, hence the name of the Osram brand

- Harmful to health, in must be handled in crystalline form: in combination with oxygen it forms osmium tetroxide, which at room temperature sublimates to a highly toxic vapor
- It appears associated with other PGMs, as a by-product of nickel and copper sulfides, and in minerals such as osmiridium (Fig. 37.1).
- Assigned the status of a strategic mineral by the EU in 2017.

© The Author(s), under exclusive license to Springer Nature Switzerland AG 2022
J. Sanz et al., *Elements and Mineral Resources*, Springer Textbooks in Earth Sciences,
Geography and Environment, https://doi.org/10.1007/978-3-030-85889-6_37

Fig. 37.1 Osmiridium (natural alloy of osmium, iridium, rhodium, and ruthenium). *Urals (Russia) (Photo Joaquim Sanz. MGVM)*

into the Ir-subgroup (IPGE: Os, Ir, and Ru) and the Pt-subgroup (PPGE: Pt, Pd, and Rh). The most abundant PGMs are in the laurite–erlichmanite (RuS_2–OsS_2) series. These are commonly hosted in chromitites forming layered intrusions as in the Bushveld Complex, or in ophiolitic podiform chromitites as in Al'Ays ophiolite complex.

The laurite–erlichmanite series represents 75% of the total PGM present in this type of ore and occurs as mineral inclusions in chromites commonly associated with other base-metal sulfides, forming anhedral to euhedral crystals and in isolate or composite grains.

37.2 Producing Countries

The world's largest reserves of osmium are in South Africa (Bushveld) for platinum, rhodium, and palladium deposits, Norilsk (Russia) for copper and nickel deposits, and Sudbury basin (Canada) for copper, nickel, and palladium deposits (Fig. 37.2).

37.3 Applications

General Industry

37.1 Geology

The platinum-group metals (PGMs) are found almost exclusively at very low concentrations in ores associated with mafic and ultramafic rocks. They can be subdivided

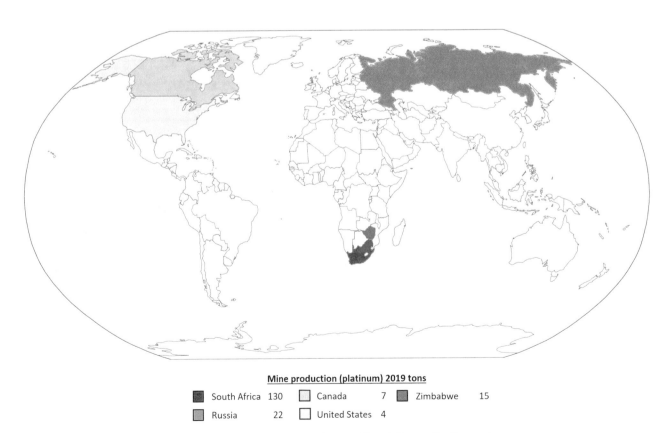

Mine production (platinum) 2019 tons

South Africa	130	Canada	7	Zimbabwe	15
Russia	22	United States	4	

Fig. 37.2 List of producing countries based on the US Geological Survey, Mineral Commodity Summaries

Fig. 37.3 Fountain pen nib. *(Photo Joaquim Sanz. MGVM)*

37.4 Recycling

All possible osmium, from offcuts and scrap alloys containing it, is recycled.

Osmium alloys with platinum and iridium are used for products where high resistance to friction is required, for certain electrical contacts, instrument pins (such as compasses), and the nibs of fountain pens (Fig. 37.3).

Electronics Industry

Osmium's high conductivity makes it an effective and durable alternative to gold or platinum in electronic products.

Chemical Industry

In the chemical industry it is used as an oxidation catalyst, as it is extremely efficient.

Other Fields

It is used as a stain for slides to be viewed by transmission electron microscopy and in certain chemical reactions.

Further Reading

Gray T, Mann N (2009) The elements. Black Dog & Leventhal Publishers Inc., New York

International Platinum Group Metals Association (2021) The six metals. Osmium. Available at http://ipa-news.com/index//platinum-group-metals/the-six-metals. Last accessed May 2021

Mata JM, Sanz J (2007) Guia d'identificació de minerals. Manresa, Catalonia. Edicions UPC/Parcir (Catalan 2nd paper edition). 262 p. ISBN: 9788483019023. http://hdl.handle.net/2117/90445

O'Driscoll B, González-Jiménez J-M (2016) Petrogenesis of the platinum-group minerals. Rev Mineral Geochem 81(1):489–578. https://doi.org/10.2138/rmg.2016.81.09

Pirajno F (2009) Hydrothermal processes and mineral systems. Springer, Berlin

Quadbeck-Seeger H-J (2007) World of the elements: elements of the world. Wiley-VCH Verlag GmbH & Co., Germany

Sanz J, Tomasa O (2017) Elements i recursos minerals: aplicacions i reciclatge. Manresa, Catalonia. Zenobita Edicions/Iniciativa Digital Politècnica (Catalan 3rd digital edition). http://hdl.handle.net/2117/105113

Sanz J, Tomasa O (2018) Elementos y recursos minerales: aplicaciones y reciclaje. Manresa, Catalonia. Zenobita Edicions/Iniciativa Digital Politècnica (Spanish 1st digital edition). http://hdl.handle.net/2117/123674

Stwertka A (2018) A guide to the elements, 4th edn. Oxford University Press, England

USGS (2021) Commodity statistics and information. Platinum-Group Metals. Available at https://www.usgs.gov/centers/nmic/platinum-group-metals-statistics-and-information. Last accessed May 2021

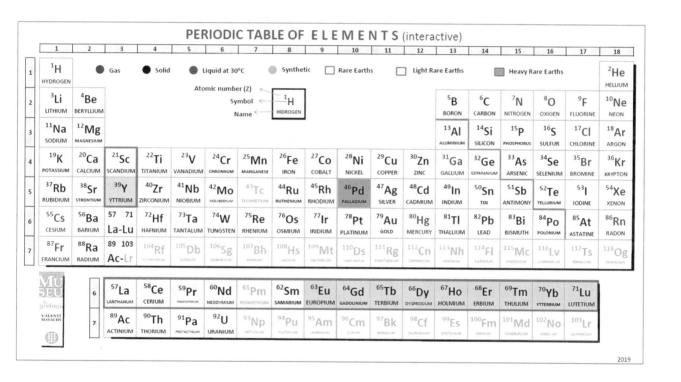

PERIODIC TABLE OF E L E M E N T S (interactive)

38.1 Geology

- Noble, light, ductile, and malleable metal
- Like gold, can be hammered into super-thin sheets of the order of microns
- Has the lowest melting point (1554 °C) of the platinum-group metals (PGMs)
- Good oxidation catalyst
- Capable of absorbing 900 times its own volume of hydrogen at room temperature
- Assigned the status of a strategic metal by the EU in 2017
- It is obtained as a by-product, mixed with other PGMs, in alluvial deposits and from mining nickel and copper sulfides (Fig. 38.1).

The platinum-group metals (PGMs) are almost exclusively found at very low concentrations in ores associated with mafic and ultramafic rocks. They can be divided into the Ir-subgroup (IPGE: Os, Ir, and Ru) and the Pt-subgroup (PPGE: Pt, Pd, and Rh). PGMs are a group of six minerals in which at least one of the six is essential to the composition of each mineral. A single major magmatic PGM ore deposit currently provides $\sim$ 80% of worldwide demand for PGM: the Bushveld Complex in South Africa, which hosts about 80% of the world's platinum resources. Several palladium minerals are known, and all are very rare. An example is braggite (Pt, Pd, Ni)S.

Fig. 38.1 Nickel and copper sulfides with PGMs. *Norilsk (Russia) (Photo Joaquim Sanz. MGVM)*

Most PGM deposits are related to high-temperature magmatic processes and are found associated with mafic or ultramafic igneous rocks. They result from various processes, including chromite deposits hosted in ophiolitic chromitites.

38.2   Producing Countries

The world's largest palladium reserves are in Norilsk (Russia) for copper and nickel deposits and PGM group metals, South Africa (Bushveld) for platinum, rhodium, and palladium deposits, and Sudbury basin (Canada) for copper, nickel, and palladium deposits (Fig. 38.2).

38.3   Applications

Automotive and Chemical Industry

The 2010 Nobel Prize in Chemistry was jointly awarded to Richard F. Heck, Ei-ichi Negishi, and i Akira Suzuki for palladium-catalyzed cross-couplings in organic synthesis.

The main application of palladium (65%), due to its being a good oxidation catalyst, is the manufacture of catalytic converters used in oil cracking and as a catalyst for gasoline vehicles, where pollutants (CO, NO_x, and C_xH_x) are converted into the less-harmful compounds of CO_2, N_2 (these are pollutants) and H_2O. This is in view of the increasingly stringent global legislation on permitted emissions from vehicles (for example, the EURO 6d-TEMP for Europe) (Fig. 38.3).

Palladium is a reagent in the membrane reactors used in the production of high-purity hydrogen. It is an important catalyst in the removal of toxic substances from aquifers.

Electronics Industry

The chemical stability and electrical conductivity of palladium make it an effective alternative to gold in the production of electronic components, such as in the multilayer

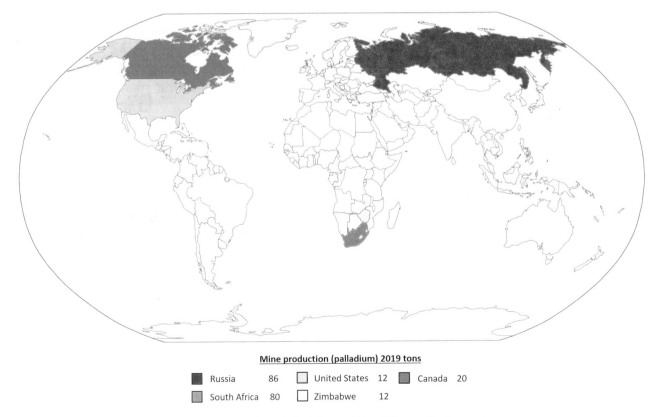

Mine production (palladium) 2019 tons

■ Russia	86	☐ United States	12	■ Canada	20
■ South Africa	80	☐ Zimbabwe	12		

Fig. 38.2 List of producing countries based on the US Geological Survey, Mineral Commodity Summaries

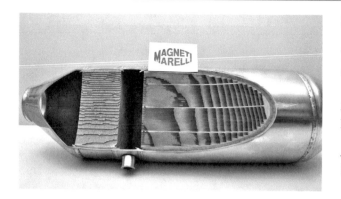

Fig. 38.3 Automotive catalytic converter. *(Photo Joaquim Sanz. MGVM)*

ceramic capacitors (MLCC) used in media players, smartphones, digital tablets, palladium and silver conductive tracks, hybrid-integrated circuits (HICs), and connectors.

Medicine

Palladium is used in odontology to make dental bridges and crowns, as it provides strength, rigidity, and durability and, when alloyed with silver, it becomes more malleable.

The radioisotope palladium-103 can be used in the treatment of prostate cancer (brachytherapy).

Jewelry

Palladium is used alone or combined with gold and silver to prepare the alloy known as white gold to make bracelets, necklaces, and rings (see: gold) (Fig. 38.4).

Given the steadily rising value of palladium in recent years, it is made into ingots as an investment.

Fig. 38.4 White gold ring. *(Photo Joaquim Sanz. MGVM)*

38.4 Recycling

The main sources of recycled palladium are scrap catalytic converters from cars, as well as jewelry and electronic scrap.

In 2019, recycled output grew by 12 t to 109 t as demand grew for catalytic converter scrap on the back of increased prices for palladium and steel scrap (Nornickel 2021).

Further Reading

Gray T, Mann N (2009) The elements. Black Dog & Leventhal Publishers Inc., New York

International Platinum Group Metals Association (2021) The six metals. Palladium. Available at https://ipa-news.de/index/platinum-group-metals/the-six-metals/palladium.html. Last accessed May 2021

JM (2020) Platinum group metals. Market report. Available at http://www.platinum.matthey.com/services/market-research/may-2020-pgm-market-report. Last accessed May 2021

Mata JM, Sanz J (2007) Guia d'identificació de minerals. Manresa, Catalonia. Edicions UPC/Parcir (Catalan 2nd paper edition). 262 p. ISBN 9788483019023. http://hdl.handle.net/2117/90445

Nornickel (2021) Palladium. Available at https://ar2019.nornickel.com/commodity-market-overview/palladium. Last accessed May 2021

O'Driscoll B, González-Jiménez J-M (2016) Petrogenesis of the platinum-group minerals. Rev Mineral Geochem 81(1):489–578. https://doi.org/10.2138/rmg.2016.81.09

Pirajno F (2009) Hydrothermal processes and mineral systems. Springer, Berlin

Quadbeck-Seeger H-J (2007) World of the elements: elements of the world. Wiley-VCH Verlag GmbH & Co., Germany

Sanz J, Tomasa O (2017) Elements i recursos minerals: aplicacions i reciclatge. Manresa, Catalonia. Zenobita Edicions/Iniciativa Digital Politècnica (Catalan 3rd digital edition). http://hdl.handle.net/2117/105113

Sanz J, Tomasa O (2018) Elementos y recursos minerales: aplicaciones y reciclaje. Manresa, Catalonia. Zenobita Edicions/Iniciativa Digital Politècnica (Spanish 1st digital edition).http://hdl.handle.net/2117/123674

Stwertka A (2018) A guide to the elements, 4th edn. Oxford University Press, England

USGS (2021) Commodity statistics and information. Platinum Group Metals. Available at https://www.usgs.gov/centers/nmic/platinum-group-metals-statistics-and-information. Last accessed May 2021

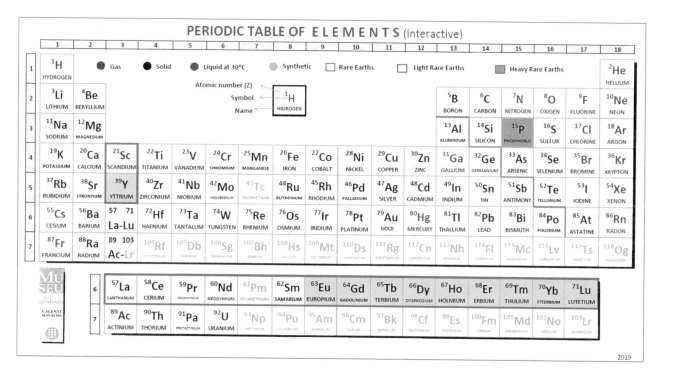

PERIODIC TABLE OF E L E M E N T S (interactive)

2019

- A highly reactive non-metal; oxidizes spontaneously on contact with oxygen and emits light (phosphorescence)
- Insoluble in water
- An essential nutrient for plants
- Cells have an energy storehouse of adenosine triphosphate (ATP)
- Elemental phosphorus can be white or red
- Obtained from phosphorite (cryptocrystalline variety of apatite) (Fig. 39.1) and from igneous deposits with apatite and other phosphates
- Assigned the status of a strategic mineral by the EU in 2017.

39.1 Geology

Phosphate deposits are widespread throughout the world, occurring on all continents and dating from the Precambrian to the Holocene. In terms of origin there are two main sources of phosphate deposit: sedimentary or igneous.

Phosphate rock resources are mainly from sedimentary marine phosphorites. Modern phosphorites are characterized by either grains of cryptocrystalline or amorphous carbonate (CO_3)-fluorapatite $(Ca_5(PO_4)_3F)$, variously referred to as collophane or francolite, occurring as beds varying in thickness from a few centimeters to tens of meters.

Phosphorus is derived from organic material, such as fecal matter, bone material, and decaying marine organisms,

Fig. 39.1 Phosphorite. *Logrosán (Spain)*. (*Photo* Joaquim Sanz. MGVM)

which are either accumulated in situ or carried into shallow coastal regions by upwellings. These nutrients enhance a diverse biota to flourish and finally produce organic-rich sediments. A changeable depositional environment with alternating reduced deposition and rebuilding sediments in shallow seas helps phosphogenesis. Nevertheless, settings and methods of deposition vary widely and are subject to debate (Dar et al. 2017).

Older phosphorites may have undergone diagenesis, deformation, and metamorphism. During the initial diagenesis stages, collophane precipitation occurs within the topmost layers of these sediments from pore waters rich in phosphorus, leached from organic remains. Precipitation improves where phosphatic nuclei are already present (Dar et al. 2017).

In the case of igneous ore deposits, the main ore is apatite $Ca_5(PO_4)_3(F, Cl, OH)$, and it can be found in a range of igneous ore deposits such as pegmatites, carbonatites, apatite-nepheline rock, and hydrothermal veins. Typically, apatite crystals are found within a hard-crystalline matrix of other igneous minerals. However, its matrix can be destroyed by intense weathering, producing a residual deposit. Therefore, its texture resembles an unconsolidated sediment more than a hard-rock igneous deposit. Furthermore, weathering of igneous deposits may be at an intermediate stage between an unaltered hard rock and the soft, unconsolidated material.

The largest sedimentary deposits are found in northern Africa and China. One of the world's largest phosphate deposits is in Morocco, where Late Cretaceous marine sediments occur on the flatlands in front of the Atlas Mountains. There are further notable economic occurrences of igneous origin.

39.2 Producing Countries

The world's most significant phosphorite reserves are in Morocco (50,000 Mt), China (3200 Mt), Algeria (2200 Mt), Syria (1800 Mt), Saudi Arabia (1400 Mt), Egypt (1300 Mt), Australia (1200 Mt), and the United States (1000 Mt), and other countries (Fig. 39.2).

Regarding phosphate reserves in igneous rocks, the chief countries are Brazil (1700 Mt), South Africa (1400 Mt), Finland (1000 Mt), and Russia (600 Mt), and other countries (Fig. 39.2).

39.3 Applications

Chemical Industry
The main use of phosphorus is in the manufacture of concentrated phosphoric and superphosphoric acid for the preparation of agricultural fertilizers and animal supplements (Fig. 39.3).

Phosphorus is basically obtained from phosphorite rock (see: phosphorite). In the manufacture of phosphoric acid, fluorosilicic acid (FSA) is obtained as a by-product, and is used in the extraction of aluminum from bauxite (see: fluorine).

Another utility of phosphorus was previously the manufacture of phosphates for detergents to decrease the hardness of the water and achieve more efficient cleaning. A European regulation banned it in detergents for washing machines and dishwashers from January 2017, due to the contamination of river waters, since treatment plants cannot eliminate them effectively.

Electronic Industry and Lamps
Phosphorus was used in the manufacture of cathode ray tubes, where it was excited by an electron beam to produce fluorescence.

Phosphorus, activated with rare earth metals, is a phosphorescent material that is used in the manufacture of plasma displays and LCDs (see: Rare Earths). It is also used in the manufacture of LEDs, x-ray equipment, ionizing radiation detectors, lasers, and infrared rays.

Phosphorus generates most of the light intensity in fluorescent lamps (Fig. 39.4).

Medicine
The potassium titanyl phosphate (KTP) crystal is used to treat benign prostate enlarged (BPE). The crystal is engaged by a beam generated by a neodymium: yttrium aluminum garnet (Nd: YAG) laser to produce a beam in the green

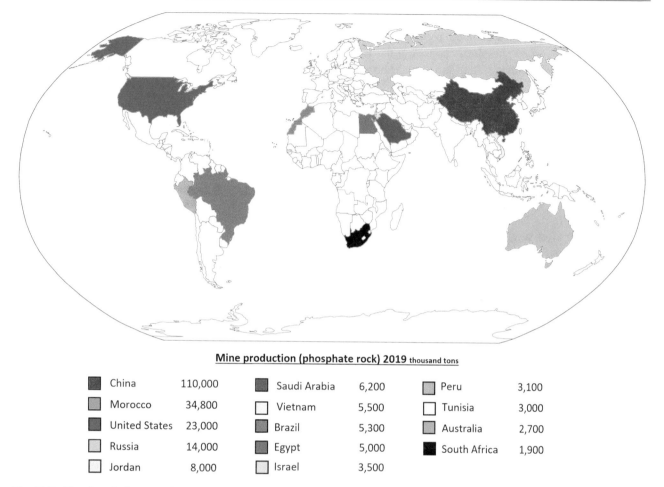

Mine production (phosphate rock) 2019 thousand tons

■	China	110,000	■	Saudi Arabia	6,200	■	Peru	3,100
■	Morocco	34,800	□	Vietnam	5,500	□	Tunisia	3,000
■	United States	23,000	■	Brazil	5,300	■	Australia	2,700
■	Russia	14,000	■	Egypt	5,000	■	South Africa	1,900
□	Jordan	8,000	■	Israel	3,500			

Fig. 39.2 List of producing countries based on the US Geological Survey, Mineral Commodity Summaries

Fig. 39.3 Wheat field fertilized with phosphates. (*Photo* Joaquim Sanz. MGVM)

Fig. 39.4 Fluorescent tube. (*Photo* Joaquim Sanz. MGVM)

visible spectrum with a wavelength of 532 nm. The KTP-green laser, which vaporizes tissue, has very good absorption by hemoglobin without bleeding or complications, allowing the patient to return home after a stay of just a few hours in hospital. However, it is not useful for breaking kidney stones or urinary bladder stones, because these have no hemoglobin (see: holmium and thulium). The process is also used in dermatology.

Other Fields

Phosphorus compounds are used in the preparation of paint and fluorescent inks. Red phosphorus is used in the manufacture of the red strip on the side of safety matchboxes.

39.4 Recycling

Recycling of phosphorus is unknown.

Reference

Dar SA, Khan KF, Birch WD (2017) Sedimentary rocks. Phosphates. Ref Module Earth Syst Environ Sci. https://doi.org/10.1016/B978-0-12-409548-9.10509-3

Further Reading

Birch WD (2005) Sedimentary rocks. Elsevier, Phosphates. https://doi.org/10.1016/B0-12-369396-9/00282-3

Gray T, Mann N (2009) The elements. Black Dog & Leventhal Publishers Inc., New York

Mata JM, Sanz J (2007) Guia d'identificació de minerals. Manresa, Catalonia. Edicions UPC/Parcir (Catalan 2nd paper edition). 262 p. ISBN 9788483019023. http://hdl.handle.net/2117/90445

Phosphor Technology (2021) http://www.phosphor-technology.com/. Last accessed May 2021

Quadbeck-Seeger H-J (2007) World of the elements: elements of the world. Wiley-VCH Verlag GmbH & Co., Germany

Sanz J, Tomasa O (2017) Elements i recursos minerals: aplicacions i reciclatge. Manresa, Catalonia. Zenobita Edicions/Iniciativa Digital Politècnica (Catalan 3rd digital edition).http://hdl.handle.net/2117/105113

Sanz J, Tomasa O (2018) Elementos y recursos minerales: aplicaciones y reciclaje. Manresa, Catalonia. Zenobita Edicions/Iniciativa Digital Politècnica (Spanish 1st digital edition).http://hdl.handle.net/2117/123674

Stwertka A (2018) A guide to the elements, 4th edn. Oxford University Press, England

USGS (2021) Commodity statistics and information. Phosphate Rock. Available at https://www.usgs.gov/centers/nmic/phosphate-rock-statistics-and-information. Last accessed May 2021

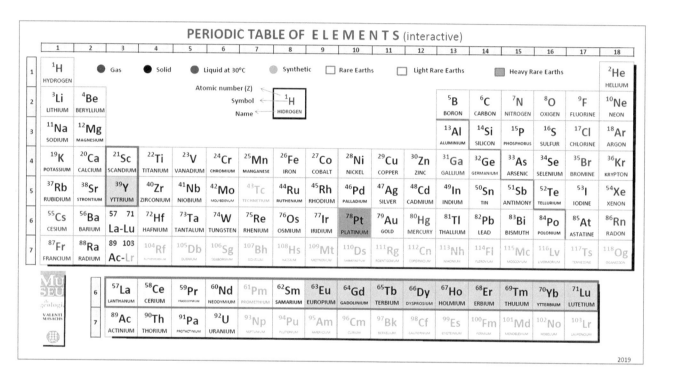

PERIODIC TABLE OF E L E M E N T S (interactive)

- Noble, dense, malleable, and ductile metal
- Does not rust
- The international standard kilogram is a cylinder of platinum with 10% iridium, manufactured in 1879 and kept in the Paris Museum of Measurements
- A good oxidation catalyst
- Melting point of 1772 °C
- Resistant to corrosion and high temperatures
- Assigned the status of a strategic metal by the EU in 2017
- Obtained as a by-product, mixed with other platinum-group metals (PGMs) (Fig. 40.1), in the mining of nickel and copper sulfides, and native in alluvial deposits
- In 1924, South Africa became the world's leading source when the German geologist Hans Merensky found the largest known deposits.

40.1 Geology

The platinum-group metals (PGMs) are found almost exclusively at very low concentrations in ores associated with mafic and ultramafic rocks. They can be divided into

Fig. 40.1 Rock with sulfide and PGMs. *(Merensky Reef) South Africa.* (*Photo* Joaquim Sanz. MGVM)

two main groups: the Ir-subgroup (IPGE: Os, Ir, and Ru) and the Pt-subgroup (PPGE: Pt, Pd, and Rh). The PGMs are a group for which at least one of the six minerals is essential to the composition. One major magmatic PGM ore deposit currently provides ~ 80% of worldwide demand for PGMs: the Bushveld Complex in South Africa (hosting about 80% of platinum resources). Sperrylite (PtAs$_2$) and Pt–Pd-Rh ± (base metal) alloys are the main platinum ores.

Most PGM deposits are related to high-temperature magmatic processes and are found associated with mafic or ultramafic igneous rocks. PGM deposit types originate from various processes, including chromite deposits hosted in ophiolitic chromitites.

40.2 Producing Countries

The world's largest platinum reserves are: South Africa (Bushveld) for platinum, rhodium, and palladium deposits; and Norilsk (Russia) and Sudbury basin (Canada) for copper, nickel, and PGM (Fig. 40.2).

40.3 Applications

Automotive Industry

The main application of platinum is the manufacture of catalytic convertors for vehicles, mainly with diesel, where the polluting compounds (CO, NO$_x$ and C$_x$H$_x$) are changed to less-harmful compounds: CO$_2$, and N$_2$ (which are pollutants), and H$_2$O, in view of the increasingly stringent global legislation on permitted emissions from

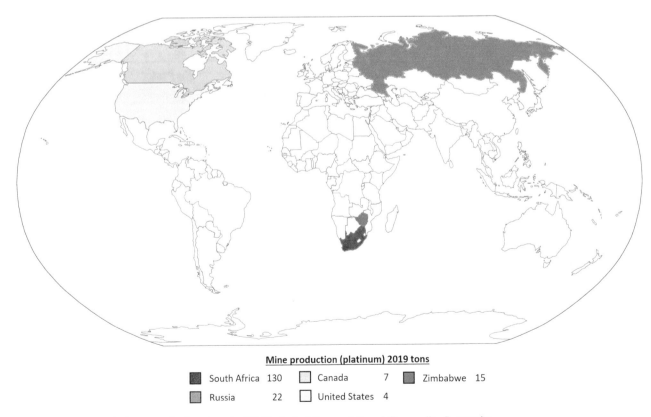

Mine production (platinum) 2019 tons
South Africa 130   Canada 7   Zimbabwe 15
Russia 22   United States 4

Fig. 40.2 List of producing countries based on the US Geological Survey, Mineral Commodity Summaries

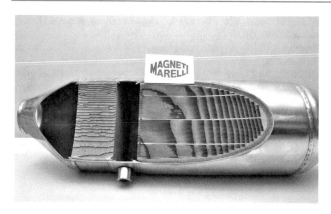

Fig. 40.3 Automotive catalytic converter. (*Photo* Joaquim Sanz. MGVM)

vehicles (for example the EURO 6d-TEMP for Europe) (Fig. 40.3).

Platinum is used in the manufacture of catalysts for fuel cells capable of generating energy in vehicles that use hydrogen as fuel. It is also used in the manufacture of long-life spark plugs.

Petrochemical Industry

Platinum catalysts are widely used in obtaining gasoline from petroleum, in the manufacture of high-octane fuels, and in obtaining nitric acid.

Fig. 40.4 Carboplatin (chemotherapy drug). (*Photo* Joaquim Sanz. MGVM)

Electronics Industry

Cobalt-alloyed platinum is used in the production of computer hard disks, as it improves their magnetic properties and allows a greater data storage capacity.

Medicine

Platinum compounds such as cisplatin and carboplatin can inhibit the growth of cancer cells and are therefore used in the treatment of certain tumors (chemotherapy) (Fig. 40.4).

Other Fields

Platinum is highly valued in jewelry making for its characteristic color, and it is also formed into ingots as an investment. Platinum is used in chemistry laboratories in the form of wire, crucibles, filters, and electrodes.

Platinum is also used for making silicone molds that need to go into the oven (up to 500 °C) and in the manufacture of nitric acid, as a catalyst.

40.4 Recycling

The main sources of recycled platinum are used exhaust gas catalytic converters and scrap jewelry. Recycled output grew by 6 t to 71 t in 2019. However, the growth of recycling is hampered by difficulties in using new-type silicon carbide-based diesel catalysts. Being a refractory material, this can damage furnaces that are unsuited to handling it. This involves employing processors to sort through catalytic converters and separate the material with a high silicon content, involving extra time and resources (Nornickel).

Further Reading

Gray T, Mann N (2009) The elements. Black Dog & Leventhal Publishers Inc., New York

International Platinum Group Metals Association (2021) The six metals. Platinum. Available at https://ipa-news.de/index/platinum-group-metals/the-six-metals/platinum.htm. Last accessed May 2021

JM (2020) Platinum Group Metals. Market report. Available at http://www.platinum.matthey.com/services/market-research/may-2020-pgm-market-report. Last accessed May 2021

Mata JM, Sanz J (2007) Guia d'identificació de minerals. Manresa, Catalonia. Edicions UPC/Parcir (Catalan 2nd paper edition). 262 p. ISBN 9788483019023. http://hdl.handle.net/2117/90445

NGK Spark Plugs (2021) NGK laser series. Available at https://ngksparkplugs.com/en/products/ignition-parts/spark-plugs/platinum-spark-plugs. Last accessed May 2021

Nornickel (2021) Platinum. Available at https://ar2019.nornickel.com/commodity-market-overview/platinum. Last accessed May 2021

O'Driscoll B, González-Jiménez JH (2016) Petrogenesis of the platinum-group minerals. Rev Mineral Geochem 81(1):489–578. https://doi.org/10.2138/rmg.2016.81.09

Pirajno F (2009) Hydrothermal processes and mineral systems. Springer, Berlin

Quadbeck-Seeger H-J (2007) World of the elements: elements of the world. Wiley-VCH Verlag GmbH & Co., Germany

Sanz J, Tomasa O (2017) Elements i Recursos minerals: aplicacions i reciclatge. Manresa, Catalonia. Zenobita Edicions/Iniciativa Digital Politècnica (Catalan 3rd digital edition). http://hdl.handle.net/2117/105113

Sanz J, Tomasa O (2018) Elementos y Recursos minerales: aplicaciones y reciclaje. Manresa, Catalonia. Zenobita Edicions/Iniciativa Digital Politècnica (Spanish 1st digital edition). http://hdl.handle.net/2117/123674

Stwertka A (2018) A guide to the elements, 4th edn. Oxford University Press, England

USGS (2021) Commodity statistics and information. Platinum group metals. Available at https://www.usgs.gov/centers/nmic/platinum-group-metals-statistics-and-information. Last accessed May 2021

PERIODIC TABLE OF E L E M E N T S (interactive)

- An alkaline metal
- Reacts with water to give hydrogen
- Oxidizes rapidly in air and vigorously in water, forming flames
- An essential element for human, animal, and plant life. Essential for cell transmission in the nervous system and the regulation of neuromuscular activity
- Obtained from sylvite (Fig. 41.1), sylvinite and carnallite, also from the water of certain seas (Dead Sea).

41.1 Geology

The characteristic mineral source of potassium is sylvite (KCl). Sylvite is frequently observed associated with halite (NaCl), forming the mix known as sylvinite. In most cases, relatively pure sylvinite exists with essentially no soluble sulfate or other salts, yet it can occasionally be associated with carnallite ($KMgCl_3 \cdot 6H_2O$), with a similar crystalline structure and nearly free from other soluble salts.

Potassium compounds, especially potassium carbonate, are usually called potash. The main potash deposits originate from evaporation of epicontinental sea basins lacking significant drainage, either incoming or outgoing. Optimal

Fig. 41.1 Sylvite (potassium chloride). *Sallent (Catalonia).* (*Photo* Joaquim Sanz)

evaporation occurs in arid conditions. A frequent mineral sequence found within depositional basins found as the evaporation of seawater increases is calcite (or dolomite), gypsum (or anhydrite), halite, and potash and, in some locations, soluble magnesium and sulfate salts.

41.2 Producing Countries

The world reserves of K_2O equivalent are in Canada (1000 Mt), Belarus (750 Mt), Russia (600 Mt), China (350 Mt), and the United States (220 Mt), while Israel and Jordan recover potash from the Dead Sea, which contains nearly 2 billion tons of KCl. Other countries with lesser reserves are Chile, Spain, Germany, and Brazil (Fig. 41.2).

41.3 Applications

Chemical Industry

Potassium is essential to the growth of plants and trees, therefore, potassium compounds such as potassium chloride (KCl) (popularly known as potash), potassium sulfate (SOP), potassium-magnesium sulfate (SOPM) and potassium muriate (MOP), with 95% KCl and NaCl, are used as fertilizers in agriculture (see: sylvite/sylvinite) (Fig. 41.3).

Potassium nitrate is used in the manufacture of gunpowder and matches.

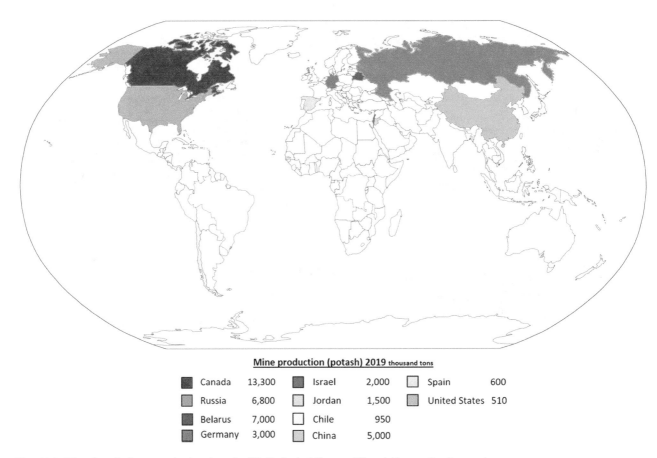

Mine production (potash) 2019 thousand tons					
Canada	13,300	Israel	2,000	Spain	600
Russia	6,800	Jordan	1,500	United States	510
Belarus	7,000	Chile	950		
Germany	3,000	China	5,000		

Fig. 41.2 List of producing countries based on the US Geological Survey, Mineral Commodity Summaries

Fig. 41.3 Cherries fertilized with potassium. (*Photo* Joaquim Sanz. MGVM)

The sodium and potassium (NaK) mixture is used as a heat-transfer medium in nuclear reactors.

Potassium hydroxide is used in the manufacture of soft soaps because of its mildness and because it is more soluble than sodium hydroxide. It is also used as an electrolyte in alkaline batteries.

Potassium chromate is used in the manufacture of explosives and as a bright red colorant in fireworks and matches.

Food Industry

Potassium chloride is used as a salt substitute to decrease sodium intake in the control of hypertension. Potassium bisulfite is used as a food preservative in alcoholic beverages such as wine and beer.

Medicine

To treat benign prostate enlargement (BPE), the Nd: YAG laser is used with a crystal of titanyl and potassium phosphate (KTP – green laser). This vaporizes tissue and has very good absorption by hemoglobin, with no bleeding or complications, so patients can return home after a stay of just a few hours in hospital. Potassium chloride is used in heart surgery, to stop the heart temporarily.

Other Fields

Potassium bisulfite is used as a fabric bleach and in leather treatments.

Potassium K-40 is a radioactive isotope and is present in rocks and in soils where vegetables and fruit are grown, constituting an important source of what is called natural background radiation. Therefore, many fruit (such as bananas and peaches) and vegetables (such as potatoes and carrots) are sources of healthy potassium and necessary to the proper functioning of the human body.

Potassium K-40 has a half-life of 1.25 billion years and, when it decays, it is transformed into argon-40 gas. For this reason, by looking at the amount of argon-40 that is still present, we can establish the age of the rocks and minerals that form them from the K-40/Ar-40 ratio.

41.4 Recycling

Potassium recycling is unknown.

Further Reading

Garrett DE (2012) Potash: deposits, processing, properties and uses. Springer Science & Business Media. https://doi.org/10.1007/978-94-009-1545-9

Gray T, Mann N (2009) The elements. Black Dog & Leventhal Publishers Inc., New York

ICL. Iberia (2021) Potassium. Available at https://www.icl-ip.com/specialty-minerals/sm-product/potassium/. Last accessed May 2021

Mata JM, Sanz J (2007) Guia d'identificació de minerals. Manresa, Catalonia. Edicions UPC/Parcir (Catalan 2nd paper edition). 262 p. 9788483019023. http://hdl.handle.net/2117/90445

Quadbeck-Seeger H-J (2007) World of the elements: elements of the world. Wiley-VCH Verlag GmbH & Co., Germany

Sanz J, Tomasa O (2017) Elements i recursos minerals: aplicacions i reciclatge. Manresa, Catalonia. Zenobita Edicions/Iniciativa Digital Politècnica (Catalan 3rd digital edition). http://hdl.handle.net/2117/105113

Sanz J, Tomasa O (2018) Elementos y recursos minerales: aplicaciones y reciclaje. Manresa, Catalonia. Zenobita Edicions/Iniciativa Digital Politècnica (Spanish 1st digital edition). http://hdl.handle.net/2117/123674

Stwertka A (2018) A guide to the elements, 4th edn. Oxford University Press, England

USGS (2021) Commodity statistics and information. Potash. Available at https://www.usgs.gov/centers/nmic/potash-statistics-and-information. Last accessed May 2021

Warren JK (2016) Potash resources: occurrences and controls. In Evaporites, pp 1081–1185. Cham. Springer. https://doi.org/10.1007/978-3-319-13512-0_11

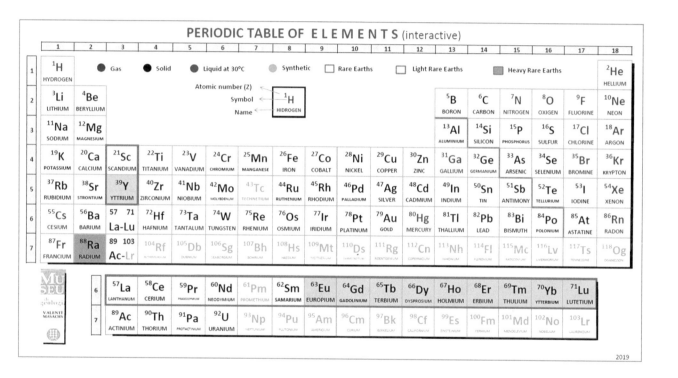

- The heaviest metal of the alkaline earths
- Highly radioactive, much more so than uranium
- Discovered in December 1898 by Marie Sklodowska Curie and husband Pierre, from the radioactive properties of pitchblende (uraninite-black color)
- Earlier, in July, they had identified polonium (named in honor of Marie's native Poland)
- In 1911 Marie Curie received the Nobel Prize in Chemistry for isolating pure radium
- Disintegrates to form radon gas

- Found in small amounts in uranium ores. According to the Musée Curie (Paris), Marie Curie extracted 1–2 mg of radium chloride from a ton of pitchblende (uraninite) (Fig. 42.1).
- Radium salts are isolated by processing the residues of uranium production from uraninite (UO_2) or carnotite ($K_2(UO_2)_2V_2O_8 \cdot 3H_2O$), according to Eckert and Ziegler. The pure element can be obtained by electrolysis of the $RaCl_2$ (see: uranium).

J. Sanz et al., *Elements and Mineral Resources*, Springer Textbooks in Earth Sciences,
Geography and Environment, https://doi.org/10.1007/978-3-030-85889-6_42

Fig. 42.1 Uraninite (uranium oxide). *Don Benito (Badajoz) (Spain).* (*Photo* Joaquim Sanz. MGVM)

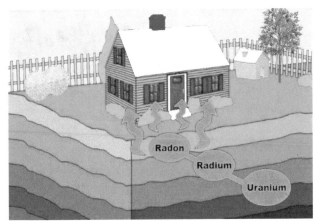

Fig. 42.2 Natural radon gas generation. (*Image courtesy of indoor Breathing*)

42.1 Applications

Medicine

The radium-223 isotope is made artificially from a generator containing actinium-227, similar to preparing technetium-99 m from molybdenum-99 (see: technetium).

Radium is used in the treatment of prostate tumors when there have been metastases to the bone. However, he European Medicines Agency (July 2018) recommends restricting the use of radium-223 dichloride to patients who have had two previous treatments for metastatic prostate cancer (it has spread to the bone) or who can have no other treatment.

Radon Generation

Radium-226 has very limited use. The main application is the preparation of radon gas in research laboratories.

Radon is a colorless, odorless, tasteless, radioactive, gaseous chemical element that comes from the natural decay of radium (uranium-radium-radon). It weighs more than air and, when it seeps out from rock and soil, tends to accumulate in high concentrations in the basements of houses and be harmful to health. Its concentration decreases to very low levels when such spaces are ventilated, thus avoiding harm (Fig. 42.2).

Its presence is directly linked to the type of terrain: granite generates more than clay. Radon gas is used to study the transport of air masses and physical phenomena of dispersion and to validate atmospheric transport models.

42.2 Recycling

Recycling of radium is unknown.

Further Reading

European Radon Association (2021) http://radoneurope.org/. Last accessed May 2021

Gray T, Mann N (2009) The elements. Black Dog & Leventhal Publishers Inc., New York

Mata JM, Sanz J (2007) Guia d'identificació de minerals. Manresa, Catalonia. Edicions UPC/Parcir (Catalan 2nd paper edition). 262 p. ISBN 9788483019023. http://hdl.handle.net/2117/90445

Musée Curie (2021) Pitchblende-radium. Available at https://musee.curie.fr/decouvrir/documentation/histoire-de-la-radioactivite. Last accessed May 2021

Quadbeck-Seeger H-J (2007) World of the elements: elements of the world. Wiley-VCH Verlag GmbH & Co., Germany

Sanz J, Tomasa O (2017) Elements i recursos minerals: aplicacions i reciclatge. Manresa, Catalonia. Zenobita Edicions/Iniciativa Digital Politècnica (Catalan 3rd digital edition). http://hdl.handle.net/2117/105113

Sanz J, Tomasa O (2018) Elementos y recursos minerales: aplicaciones y reciclaje. Manresa, Catalonia. Zenobita Edicions/Iniciativa Digital Politècnica (Spanish 1st digital edition). http://hdl.handle.net/2117/123674

Stwertka A (2018) A guide to the elements, 4th edn. Oxford University Press, England

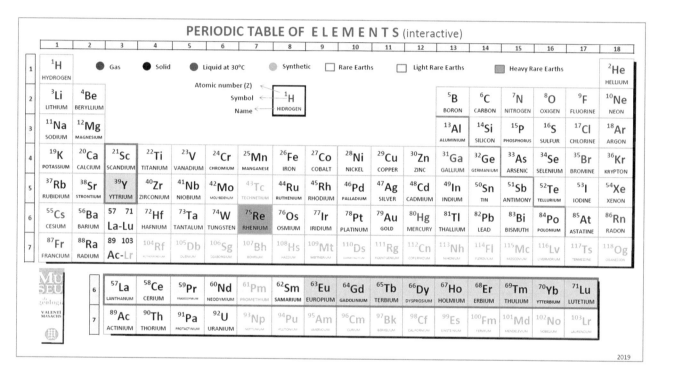

43.1 Geology

- Extremely hard and very dense
- Melting point of 3186 °C
- Corrosion- and wear-resistant
- Good conductor of electricity and heat
- Superconductor at low temperatures
- A very scarce and expensive metal
- Extracted from molybdenite (Fig. 43.1) and obtained as a by-product of copper and molybdenum sulfides.

Chuquicamata, in northern Chile, is the world's largest porphyry copper ore body with molybdenite. Roasting the molybdenite emits gases, and the renium can then be extracted.

The Merlin Project is an important rhenium-molybdenite deposit in northwestern Queensland, Australia. It is hosted in the Precambrian Mount Isa Inlier and is grouped within the Eastern Fold Belt. The host rocks are metasedimentary rocks of the Proterozoic Kuridala and Staveley Formations, adjacent to the Mount Dore Granite. The Merlin zone represents

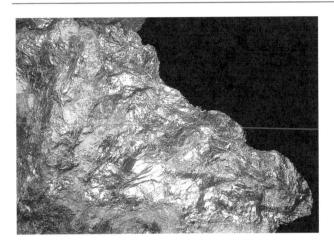

Fig. 43.1 Molybdenite (molybdenum sulfide). *Gualba (Catalonia).* (*Photo* Joaquim Sanz. MGVM)

part of the Mount Dore polymetallic deposit, containing copper, zinc, silver, gold, lead, and cobalt. The host rocks are of three main types: black shale; phyllite; and altered banded calc-silicate. The deposit was discovered in 2008 by routine infill reverse circulation sampling conducted for the Mount Dore copper deposit: one hole penetrated below the

main target copper horizon to intersect with visible molybdenite. During 2009 and 2010, diamond drilling delineated a total mineral resource of 6.7 Mt (1.3% Mo, 23 g/t Re, and 0.3% Cu).

43.2   Producing Countries

The world's rhenium reserves are associated with molybdenite deposits in porphyry copper deposits. The first country is Chile (1300 t), followed by the United States (400 t), Russia (310 t), Kazakhstan (190 t), Armenia (95 t), Peru (45 t), and Canada (32 t), and others (Fig. 43.2).

43.3   Applications

Metallurgical Industry

The main application of rhenium (70%) is in the manufacture of superalloys with iron, nickel, and cobalt, very resistant to both wear and the high working temperatures of turbines/reactors for aviation (Fig. 43.3) and industrial gas turbines (IGT).

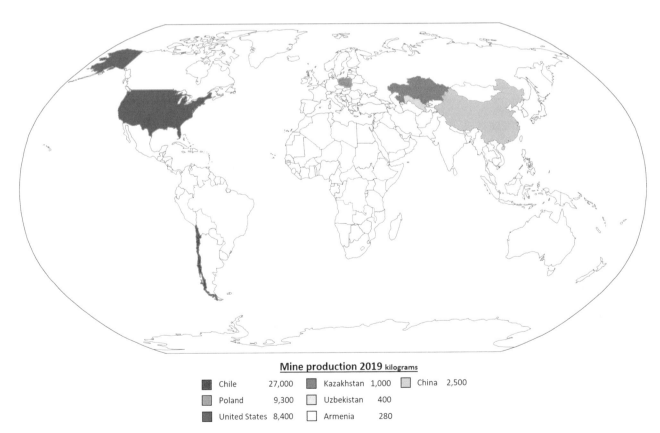

Mine production 2019 kilograms					
Chile	27,000	Kazakhstan	1,000	China	2,500
Poland	9,300	Uzbekistan	400		
United States	8,400	Armenia	280		

Fig. 43.2 List of producing countries based on the US Geological Survey, Mineral Commodity Summaries

Fig. 43.3 Aircraft engine blades. (*Photo* Carlos Domínguez Morano)

Unless the turbine blades of these engines contain rhenium in their composition, they cannot exceed temperatures of 1500 °C. With rhenium they can reach 1610 °C, improving fuel economy by 40–60%, and doubling the power and thrust, reducing emissions from CO_2 to 64%.

Petrochemical Industry
The rhenium-platinum combination is used as a catalyst in the process of distilling crude oil to produce high-octane and unleaded gasoline.

Other Fields
Rhenium is involved in the manufacture of crucibles and electrical contacts that must withstand very high temperatures.

In the manufacture of thermocouples for measuring temperatures up to 2760 °C, rhenium is used alongside tungsten. Rhenium is used in the manufacture of electrical contacts that have to withstand electric arcs, such as electrodes and filaments.

43.4 Recycling

As it is one of the most expensive metals, the rhenium present in old turbine blades of jet engines and manufacturing scrap is recycled.

Further Reading

Babo J, Spandler C, Oliver NH, Brown M, Rubenach MJ, Creaser RA (2017) The high-grade Mo-Re Merlin deposit, Cloncurry district, Australia: paragenesis and geochronology of hydrothermal alteration and ore formation. Econ Geol 112(2):397–422

Gray T, Mann N (2009) The elements. Black Dog & Leventhal Publishers Inc., New York

Mata JM, Sanz J (2007) Guia d'identificació de minerals. Manresa, Catalonia. Edicions UPC/Parcir (Catalan 2nd paper edition). 262 p. ISBN 9788483019023. http://hdl.handle.net/2117/90445

Molymet (2021) Metallic/super alloys industries. Rhenium. Available at https://molymet.com/mo-re/. Last accessed May 2021

Quadbeck-Seeger H-J (2007) World of the elements: elements of the world. Wiley-VCH Verlag GmbH & Co., Germany

Roskill (2020) Market reports. Rhenium. Available at https://roskill.com/market-report/rhenium/. Last accessed May 2021

Sanz J, Tomasa O (2017) Elements i recursos minerals: aplicacions i reciclatge. Manresa, Catalonia. Zenobita Edicions/Iniciativa Digital Politècnica (Catalan 3rd digital edition). http://hdl.handle.net/2117/105113

Sanz J, Tomasa, O (2018) Elementos y recursos minerales: aplicaciones y reciclaje. Manresa, Catalonia. Zenobita Edicions/Iniciativa Digital Politècnica (Spanish 1st digital edition). http://hdl.handle.net/2117/123674

Stwertka A (2018) A guide to the elements, 4th edn. Oxford University Press, England

USGS (2021) Commodity Statistics and Information. Rhenium. Available at: https://www.usgs.gov/centers/nmic/rhenium-statistics-and-information. Last accessed May 2021

PERIODIC TABLE OF E L E M E N T S (interactive)

44.1 Geology

- Noble metal, very scarce and expensive
- Dense, malleable, ductile, and very hard
- Resistant to corrosion and high temperatures (1960 °C)
- Assigned the status of a strategic metal by the EU in 2017
- The most expensive of the precious metals
- Appears associated with other platinum group metals (PGMs) in nickel and copper sulfides (Fig. 44.1).

The platinum-group metals (PGMs) are almost exclusively found at very low concentrations in ores associated with mafic and ultramafic rocks. They can be subdivided into two main groups: the Ir-subgroup (IPGE: Os, Ir, and Ru); and the Pt-subgroup (PPGE: Pt, Pd, and Rh). PGM minerals are a group in which at least one of the six PGM is essential to the mineral composition. Pt–Pd–Rh ± (base metals) alloys are the main platinum ore. Rh minerals are very rare, and zaccarinite (RhNiAs) is an example.

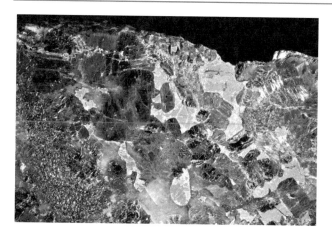

Fig. 44.1 Rock with nickel and copper sulfides and PGMs. *Merensky Reef (Bushveld) South Africa. (Photo* Joaquim Sanz. MGVM)

Most PGM deposits are related to high-temperature magmatic processes and are found associated with mafic or ultramafic igneous rocks. PGM deposit types result from various processes, including chromite deposits hosted in ophiolitic chromitites.

44.2 Producing Countries

The main reserves are in South Africa, followed some way behind by Russia, Zimbabwe, North America, and other countries (Fig. 44.2).

44.3 Applications

Automotive Industry

Rhodium is used in the manufacture of catalysts for vehicles, mainly diesel-powered, where the polluting compounds (CO, NO_x, and C_xH_x) are converted into less-harmful compounds: CO_2, N_2 (pollutants), and H_2O, in view of increasingly stringent global legislation regarding permitted emissions from vehicles (for example the EURO 6d-TEMP for Europe) (Fig. 44.3).

Jewelry

In silver and gray gold jewelry, a thin layer of rhodium (rhodium plating) is applied electrolytically, greatly improving the appearance, preventing oxidation of the silver

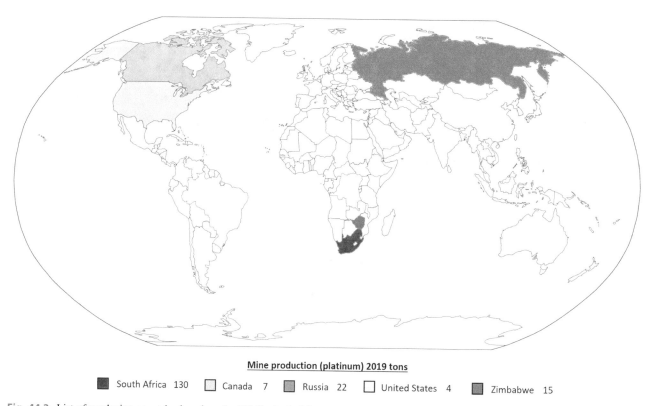

Mine production (platinum) 2019 tons

South Africa 130   Canada 7   Russia 22   United States 4   Zimbabwe 15

Fig. 44.2 List of producing countries based on the US Geological Survey, Mineral Commodity Summaries

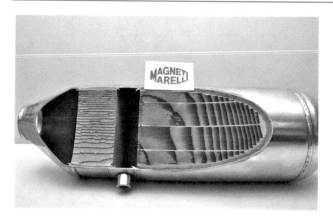

Fig. 44.3 Automotive catalytic converter. (*Photo* Joaquim Sanz. MGVM)

Fig. 44.4 Rhodium-plated silver pendant. (*Photo* Joaquim Sanz. MGVM)

and enhancing its wear resistance (Fig. 44.4). Alloyed with platinum, it hardens it.

Chemical Industry

As a catalyst, it is very useful in the process of distillation of hydrocarbons and in the manufacture of nitric acid.

Electrical Industry

Rhodium is used in the manufacture of electrical contacts that have to work very frequently and in the preparation of rhodium-platinum thermocouples.

Glass Industry

Rhodium, alloyed with platinum, is used in furnaces that make glass and fiberglass, due to its resistance to high temperatures.

44.4 Recycling

Most of the rhodium that is recycled is from car catalyst treatments.

According to S&P Global Platts (March 2021), the annual recycling of rhodium is 30 thousand tons.

Further Reading

Gray T, Mann N (2009) The elements. Black Dog & Leventhal Publishers Inc., New York

International Platinum Group Metals Association (2021) The six metals. Rhodium. https://ipa-news.de/index/platinum-group-metals/the-six-metals/rhodium.html. Last accessed May 2021

JM (2020) Platinum group metals. Market report. http://www.platinum.matthey.com/services/market-research/may-2020-pgm-market-report. Last accessed May 2021

Mata JM, Sanz J (2007) Guia d'identificació de minerals. Manresa, Catalonia. Edicions UPC/Parcir (Catalan 2nd paper edition). 262 p. ISBN 9788483019023. http://hdl.handle.net/2117/90445

Nornickel (2021) Rhodium. https://www.nornickel.com/business/products/rhodium. Last accessed May 2021

O'Driscoll B, González-Jiménez J-M (2016) Petrogenesis of the platinum-group minerals. Rev Mineral Geochem 81(1):489–578. https://doi.org/10.2138/rmg.2016.81.09

Pirajno F (2009) Hydrothermal processes and mineral systems. Springer, Berlin

Quadbeck-Seeger H-J (2007) World of elements: elements of the world. Wiley-VCH Verlag GmbH & Co., Germany

Sanz J, Tomasa O (2017) Elements i recursos minerals: aplicacions i reciclatge. Manresa, Catalonia. Zenobita Edicions/Iniciativa Digital Politècnica (Catalan 3rd digital edition). http://hdl.handle.net/2117/105113

Sanz J, Tomasa O (2018) Elementos y Recursos minerales: aplicaciones y reciclaje. Manresa, Catalonia. Zenobita Edicions/Iniciativa Digital Politècnica (Spanish 1st digital edition).http://hdl.handle.net/2117/123674

Stwertka A (2018) A guide to the elements, 4th edn. Oxford University Press, England

USGS (2021) Commodity statistics and information. Platinum group metals. https://www.usgs.gov/centers/nmic/platinum-group-metals-statistics-and-information. Last accessed May 2021

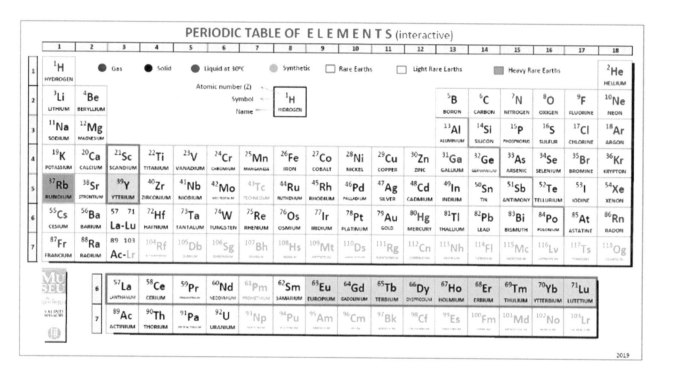

PERIODIC TABLE OF ELEMENTS (interactive)

2019

- Rare alkaline metal
- Associated with cesium and potassium
- One of the most reactive elements. In contact with water it reacts vigorously, giving hydrogen, and in air it ignites.
- Almost liquid at room temperature; at 39 °C it melts completely
- Discovered by the German chemists Bunsen and Kirchhoff in 1861 when checking the emission spectrum of lepidolite. Two dark red lines appeared, hence the name from the Latin *rubidus*, meaning dark red

- Found in lepidolite mica (Fig. 45.1), pollucite, and muscovite mica.

45.1 Geology

Rubidium replaces potassium or cesium in the composition of many minerals. Lepidolite $((KLi_2Al(Al,Si)_3O_{10}(F,OH)_2)$ is the most common source of rubidium. Rubidium also occurs in leucite $(KAlSi_2O_6)$, carnallite $(KCl.MgCl_2 \cdot 6(H_2O))$, zinnwaldite $(KLiFeAl(AlSi_3)O_{10}(OH,F)_2)$, and pollucite $(Cs(Si_2Al)O_6.nH_2O)$. It is considered to be an

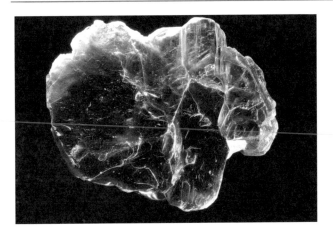

Fig. 45.1 Lepidolite (potassium-lithium aluminosilicate fluoride with rubidium) *Minas Geraes (Brasil). (Photo* Joaquim Sanz. MGVM)

incompatible element, meaning that during the magma crystallization process it is concentrated with other heavy elements like cesium in the liquid phase, crystallizing last. Therefore, the largest deposits of rubidium come from pegmatite orebodies. Two notable sources of rubidium are the Bernic Lake deposit in Manitoba (Canada) and the island of Elba (Italy), which is rich in rubicline ((Rb,K)AlSi$_3$O$_8$). The largest Canadian producers of cesium also produce rubidium as a by-product.

45.2   Producing Countries

No reliable data are available to determine reserves in specific countries; however, Australia, Namibia, Canada, Zimbabwe, and China are thought to have reserves totaling less than 200,000 tons (Fig. 45.2).

45.3   Applications

Electrical Industry
Rubidium, due to its high dielectric constant, is used in the manufacture of ceramic insulators for high voltage lines, as it greatly increases the insulation capacity and reduces voltage losses.

Electronics Industry
Like cesium, rubidium is used in solar cells because it converts light into a flow of electrons, i.e. electricity (Fig. 45.3).

Atomic clocks that use rubidium-87 to control frequency are cheaper yet less accurate and less compact than atomic clocks with cesium-133. They are used to control the frequency of television and telephone stations, and also in global positioning systems (GPS).

Mine production 2019 tons

(No official resources for production data)

■ Namibia   ▢ Zimbabwe   ▢ Canada

Fig. 45.2 List of producing countries based on the US Geological Survey, Mineral Commodity Summaries

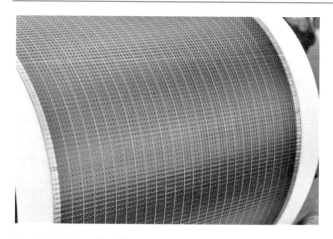

Fig. 45.3 Solar cells. (*Photo* Joaquim Sanz. MGVM)

Rubidium carbonate, as a reducer of electrical conductivity, is used to improve the stability and durability of the optical fibers in communication networks.

Rubidium is being investigated for its involvement in the future quantum computers, planned for 2025.

Medicine

The radioactive isotope rubidium-82 is rapidly absorbed by cells in the heart and can be used to identify regions of this muscle with a poor blood flow and therefore at risk of heart attack, by means of positron emission tomography (PET) (see: lutetium). Compared to technetium-99 m, rubidium-82 allows a more detailed diagnosis and a lower level of radiation exposure, but it is more expensive.

Rubidium drugs are used to treat epilepsy and thyroid abnormalities.

Other Fields

The radioisotope rubidium-87 is used in geochronological determinations of rock (its half-life is 50×10^9 years).

Rubidium is involved in the manufacture of special glasses for optoelectronic components, such as night-vision equipment.

This metal is constantly being researched for further applications.

45.4 Recycling

Recycling of this metal is unknown.

Further Reading

Gray T, Mann N (2009) The elements. Black Dog & Leventhal Publishers Inc., New York

Live Science. Facts about rubidium. Available at https://www.livescience.com/34519-rubidium.html. Last accessed May 2021

Mata JM., Sanz J (2007) Guia d'identificació de minerals. Manresa, Catalonia. Edicions UPC/Parcir (Catalan 2nd paper edition). 262 p. ISBN: 9788483019023. http://hdl.handle.net/2117/90445

Quadbeck-Seeger H-J (2007) World of the elements: elements of the world. Wiley-VCH Verlag GmbH & Co., Germany

Sanz J, Tomasa O (2017) Elements i recursos minerals: aplicacions i reciclatge. Manresa, Catalonia. Zenobita Edicions/Iniciativa Digital Politècnica (Catalan 3rd digital edition). http://hdl.handle.net/2117/105113

Sanz J, Tomasa O (2018) Elementos y recursos minerales: aplicaciones y reciclaje. Manresa, Catalonia. Zenobita Edicions/Iniciativa Digital Politècnica (Spanish 1st digital edition).http://hdl.handle.net/2117/123674

Stwertka A (2018) A guide to the elements, 4th edn. Oxford University Press, England

USGS. Commodity statistics and information. Rubidium. Available at https://www.usgs.gov/centers/nmic/cesium-and-rubidium-statistics-and-information. Last accessed May 2021

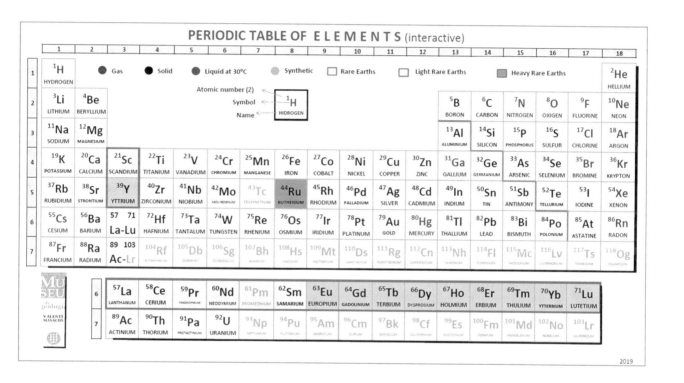

PERIODIC TABLE OF ELEMENTS (interactive)

- Noble, dense, malleable, and ductile metal
- Resistant to corrosion and high temperatures (2310 °C)
- Estonian chemist Karl Karlovitch Klaus discovered it, naming it after the Latin name *Ruthenia*, or Russia
- Presents great difficulty in being worked. Even at 1500 °C it remains tough and brittle
- Assigned the status of a strategic metal by the EU in 2017
- Appears associated with other platinum-group metals (PGMs) in nickel and copper sulfides (Fig. 46.1).

46.1 Geology

The platinum-group metals (PGMs) are found almost exclusively at very low concentrations in ores associated with mafic and ultramafic rocks. They can be subdivided into the Ir-subgroup (IPGE: Os, Ir, and Ru) and the Pt-subgroup (PPGE: Pt, Pd, and Rh). The most abundant PGM are the laurite–erlichmanite (RuS_2–OsS_2) series. These are commonly hosted in chromitites, forming layered intrusions as in the Bushveld Complex) or in ophiolitic podiform chromitites as in the Al'Ays ophiolite complex.

The laurite–erlichmanite series represents 75% of the total PGM present in this type of ore and occurs as mineral

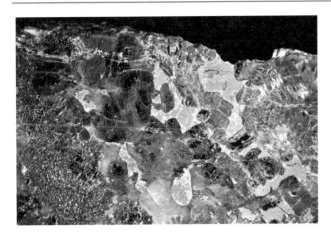

Fig. 46.1 Rock with nickel and copper sulfides and with PGMs. *Merensky Reef (Bushveld) South Africa. (Photo* Joaquim Sanz. MGVM)

inclusions in chromites commonly associated with other base-metal sulfides, forming anhedral to euhedral crystals and in isolated or composite grains.

46.2   Producing Countries

The most important world reserves are in South Africa, followed at some distance by Russia (Fig. 46.2).

46.3   Applications

Metallurgical and Electrical Industry

Ruthenium is used in platinum and palladium alloys to improve wear resistance; these alloys are used as coatings for electrical contacts.

The addition of 0.1% ruthenium to titanium improves corrosion resistance by a factor of 100.

Ruthenium, alloyed with molybdenum, becomes superconductive.

Automotive Industry

Ruthenium is used in the manufacture of long-life spark plug electrodes, which last in the order of 120,000 km of driving distance.

Ruthenium will play an important role in the so-called 'hydrogen economy', to obtain that gas in a sustainable way through the hydrolysis of water for the engines of many vehicles.

Chemical Industry

Ruthenium, associated with iridium, is used in coating electrodes used in the production of chlorine and caustic soda, replacing the old mercury electrodes. It is also a very good catalyst in many types of chemical and industrial processes.

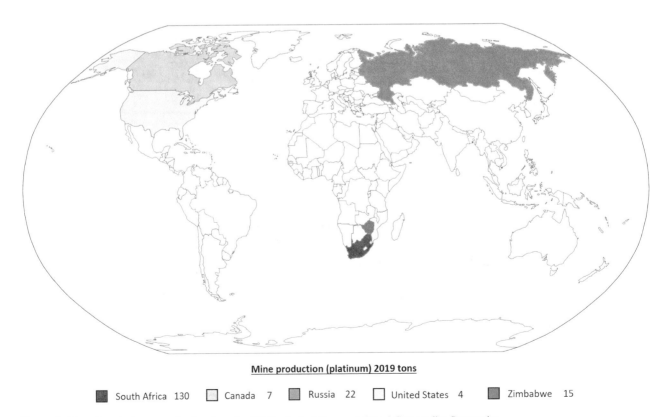

Mine production (platinum) 2019 tons

South Africa   130   Canada   7   Russia   22   United States   4   Zimbabwe   15

Fig. 46.2 List of producing countries based on the US Geological Survey, Mineral Commodity Summaries

Fig. 46.3 Jewelry made of palladium (95%) and ruthenium (5%). (*Photo* Joaquim Sanz. MGVM)

Jewelry

Small amounts of ruthenium are used to increase the stability of gold in jewelry. The alloy of palladium and ruthenium is used to make jewelry (Fig. 46.3).

Other Fields

Ruthenium is used in the center of electrodes to improve the ignition as well as the durability of vehicle spark plugs.

46.4 Recycling

All the ruthenium that can be recovered from the electrical and chemical industries using this metal is recycled.

According to S&P Global Platts (March 2021), annual recycling of ruthenium is between 35 and 40 Mt.

Further Reading

Gray T, Mann N (2009) The Elements. Black Dog & Leventhal Publishers Inc., New York

International Platinum Group Metals Association (2021) The six metals. Ruthenium. https://ipa-news.de/index/platinum-group-metals/the-six-metals/ruthenium.html. Last accessed May 2021

JM. Platinum Group Metals (2020) Market report. http://www.platinum.matthey.com/services/market-research/may-2020-pgm-market-report. Last accessed May 2021

JM. NGK. Spark plugs ruthenium. Available at https://ngksparkplugs.com/en/products/ignition-parts/spark-plugs/ruthenium-spark-plugs. Last accessed May 2021

Mata JM, Sanz J (2007) Guia d'identificació de minerals. Manresa, Catalonia. Edicions UPC/Parcir (Catalan 2nd paper edition). 262 p. ISBN 9788483019023. http://hdl.handle.net/2117/90445

Nornickel (2021) Ruthenium. https://www.nornickel.com/business/products/ruthenium. Last accessed May 2021

O'Driscoll B, González-Jiménez JM (2016) Petrogenesis of the platinum-group minerals. Rev Mineral Geochem 81(1):489–578. https://doi.org/10.2138/rmg.2016.81.09

Pirajno F (2009) Hydrothermal processes and mineral systems. Springer, Berlin

Quadbeck-Seeger, H.-J. (2007) World of the elements: elements of the world. iley-VCH Verlag GmbH & Co., Germany

Sanz J, Tomasa O (2017) Elements i recursos minerals: aplicacions i reciclatge. Manresa, Catalonia. Zenobita Edicions/Iniciativa Digital Politècnica (Catalan 3rd digital edition). http://hdl.handle.net/2117/105113

Sanz J, Tomasa O (2018) Elementos y recursos minerales: aplicaciones y reciclaje. Manresa, Catalonia. Zenobita Edicions/Iniciativa Digital Politècnica (Spanish 1st digital edition). http://hdl.handle.net/2117/123674

Stwertka A (2018) A guide to the elements, 4th edn. Oxford University Press, England

USGS (2021) Commodity statistics and information. Platinum group metals. https://www.usgs.gov/centers/nmic/platinum-group-metals-statistics-and-information. Last accessed May 2021

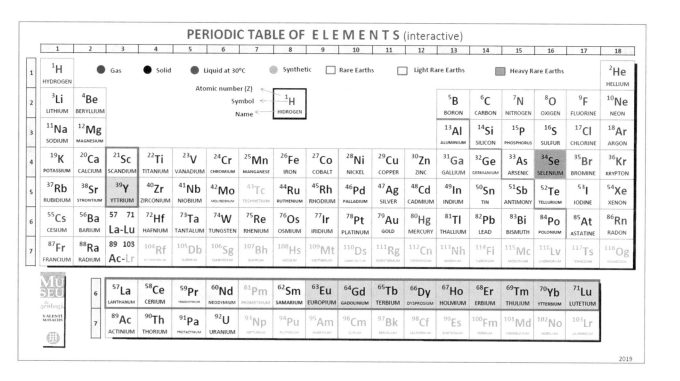

47.1 Geology

- Non-metal semiconductor
- Photoconductive (electrical conductivity increases when exposed to light)
- Discovered by J. Jacob Bercellus in 1817
- Name derived from *selene*, the Ancient Greek word for moon.
- Obtained mainly from electrolytic copper refinery of seleniferous copper sulfide ores, like chalcopyrite (Fig. 47.1).

Selenium is found in several deposit types, such as magmatic, volcanic, hydrothermal, and exogenic. It is found substituted into sulfide minerals such as chalcopyrite, cinnabar, cobaltite (CoAsS), galena, molybdenite, pentlandite, pyrite, pyrrhotite, sphalerite, and stibnite. Selenide minerals are less common than selenium-bearing sulfides, but phases such as clausthalite (PbSe), guanajuatite (Bi_2Se_3), naumannite (Ag_2Se), and penroseite (blockite) ($(Ni,Co,Cu)Se_2$) are found in the veins and stockworks of hydrothermal deposits.

Fig. 47.1 Chalcopyrite (copper and iron sulfide). *El Brull (Catalonia).* (*Photo* Joaquim Sanz. MGVM)

More than 80% of global production of selenium is obtained as a by-product of copper refinery; it is recovered from the anode slimes generated in the electrolytic production of copper. Blister copper (the partly refined copper formed during smelting) that serves as the anode in an electrolytic cell contains an average of 0.05% selenium (0.5 kg of selenium per metric tonne of copper), but

recovery ranges from 0.02% to 0.038% (0.2–0.38 kg of selenium per metric tonne of copper).

Because selenium production is mainly a by-product of copper refinery, countries with selenium resources and reserves are those with resources and reserves of copper, including Canada, Chile, China, the Congo (Kinshasa), Mexico, Peru, the Philippines, Russia, the United States, and Zambia. Minor amounts of selenium have also been recovered from lead, nickel, and zinc ores.

47.2   Producing Countries

The world's reserves are in China (26,000 t), Russia (20,000 t), Peru (13,000 t), Canada (6000 t), and Poland (3000 t), among other countries (Fig. 47.2).

47.3   Applications

Glass and Ceramics Industry

Selenium is used in glass production, where it bleaches the green tone caused by the iron impurities that it contains.

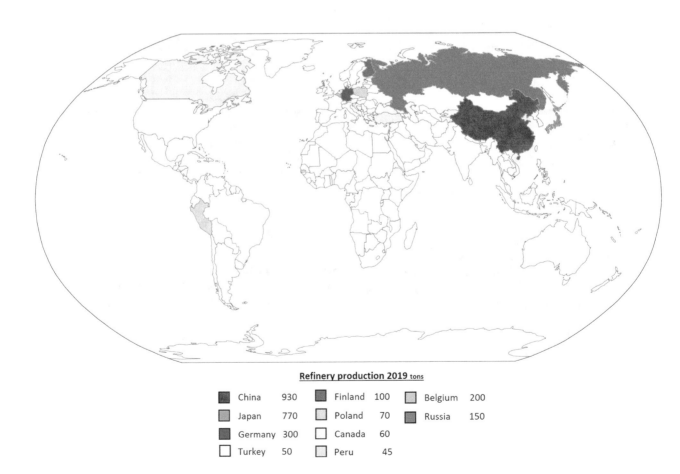

Refinery production 2019 tons

China	930	Finland	100	Belgium	200
Japan	770	Poland	70	Russia	150
Germany	300	Canada	60		
Turkey	50	Peru	45		

Fig. 47.2 List of producing countries based on the US Geological Survey, Mineral Commodity Summaries

Fig. 47.3 Rubber tires. (*Photo* Joaquim Sanz. MGVM)

Added to glass, it is used in construction to block the transmission of solar heat. It gives a reddish color to glass and ceramics.

Metallurgical Industry

In coating materials, selenium enhances both appearance and durability. It is used in the electrolytic production of manganese because it improves production and performance, as well as in the production of certain steels.

Power Generation

Calcium indium copper selenide (CIGS) is a semiconductor used in the manufacture of the conductive thin layer of flexible solar cells, making it an effective alternative to crystalline silicon.

Food

Selenium is used as a dietary supplement for people and livestock as it is an essential element, although it is toxic in excess.

Agriculture

Although not essential, selenium enriches impoverished soils and promotes plant growth and resistance to biotic stress. However, since there is a narrow margin between the beneficial and the toxic concentration of selenium in the human body, it is vital to fine-tune the biofortification strategies of crops to optimize the intake of selenium and thus avoid accumulation of concentrations harmful to health. (Gupta and Gupta) (see link).

Other Fields

Selenium is used as an additive to rubber because it improves its resistance to abrasion (Fig. 47.3), and as pigment in enamels, plastics, paint, and inks that are exposed to heat, UV, and humidity.

Photocopiers and laser printers previously contained a cylinder covered in selenium and tellurium photoreceptors, key elements in making photocopies and prints. Currently, those photoreceptors have been replaced by organic photoreceptors.

Selenium sulfide is a basic ingredient in anti-dandruff shampoo, helping the dry flakes to shed.

47.4 Recycling

A little selenium recycling undertaken, recovered from old photocopiers and laser printers.

Further Reading

Critical Mineral Resources of the United States (2017) Economic and environmental geology and prospects for future supply. https://pubs.usgs.gov/pp/1802/q/pp1802q.pdf. Last accessed May 2021

Gray T, Mann N (2009) The elements. Black Dog & Leventhal Publishers Inc., New York

Gupta M, Gupta S (2017, January 11) An overview of selenium uptake, metabolism, and toxicity in plants. Front Plant Sci. https://www.frontiersin.org/articles/10.3389/fpls.2016.02074/full. Last accessed May 2021

Mata JM, Sanz J (2007) Guia d'identificació de minerals. Manresa, Catalonia. Edicions UPC/Parcir (Catalan 2nd paper edition). 262 p. ISBN 9788483019023. http://hdl.handle.net/2117/90445

Quadbeck-Seeger H-J (2007) World of the elements: elements of the world. Wiley-VCH Verlag GmbH & Co., Germany

Sanz J, Tomasa O (2017) Elements i recursos minerals: aplicacions i reciclatge. Manresa, Catalonia. Zenobita Edicions/Iniciativa Digital Politècnica (Catalan 3rd digital edition). http://hdl.handle.net/2117/105113

Sanz, J.; Tomasa, O. (2018) Elementos y recursos minerales: aplicaciones y reciclaje. Manresa, Catalonia. Zenobita Edicions/Iniciativa Digital Politècnica (Spanish 1st digital edition). http://hdl.handle.net/2117/123674

Stwertka A (2018) A guide to the elements, 4th edn. Oxford University Press, England

USGS (2019) Commodity statistics and information. Selenium. Available at https://www.usgs.gov/centers/nmic/selenium-and-tellurium-statistics-and-information. Last accessed May 2021

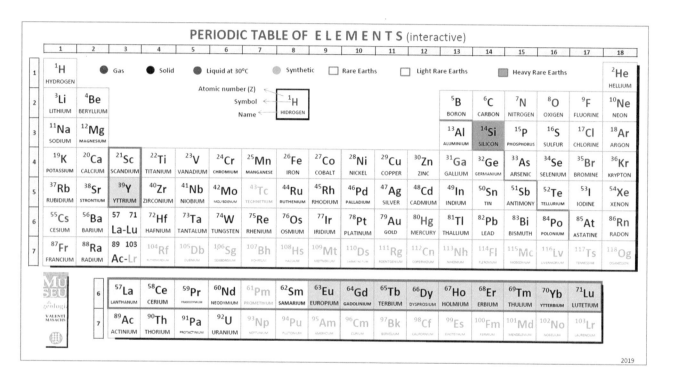

PERIODIC TABLE OF E L E M E N T S (interactive)

2019

- Semi-metal semiconductor
- Second-most abundant element on Earth, after oxygen
- Obtained from white silica sand and pure quartz (Fig. 48.1).
- Assigned the status of a strategic mineral by the EU in 2017.
- A quartz crystal is piezoelectric; a small electric current flows between its ends when under pressure. Conversely, when subjected to slight electrical stress it begins to vibrate precisely at a certain frequency.

48.1 Geology

Silica is present in multiple forms and minerals, but its most common form is quartz (SiO_2). Quartz is the most abundant compound in the Earth's crust, comprising roughly 14%. Quartz crystallization occurs when silica is heated. During the cooling process, silicon and oxygens recombine as molecules formed of one silicon atom and four oxygen atoms. Quartz is found in all rock types, such as granite, gneiss, and sandstone. Silicon is mined both as sand and as vein or lode deposits. However, the high-availability silicon usually extracted from silica sand refers to sands with the composition and grain-size distribution required for

© The Author(s), under exclusive license to Springer Nature Switzerland AG 2022
J. Sanz et al., *Elements and Mineral Resources*, Springer Textbooks in Earth Sciences, Geography and Environment, https://doi.org/10.1007/978-3-030-85889-6_48

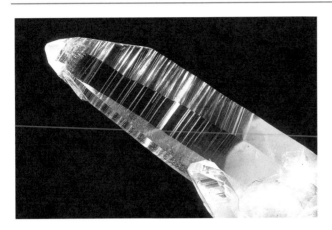

Fig. 48.1 Hyaline quartz (silicon oxide). *Chamonix (France)*. (*Photo Joaquim Sanz. MGVM*)

industrial application. Silicon is produced by heating silica sand to temperatures around 2200 °C.

China is the world's largest silicon producer. The second largest is Russia. In the United States, Arkansas has some world-class deposits of quartz: in 1967, the General Assembly adopted the quartz crystal as its state mineral. One of the world's most noted localities of quartz west of the Mississippi River is Hot Springs, Ouachita, in Arkansas.

48.2   Producing Countries

The world's reserves of silica sand and quartzite for silicon metal are highly abundant and can cover demand for decades with no problem (Fig. 48.2).

48.3   Applications

Electronics Industry

Pure silicon is used in the manufacture of electronic and electrical components such as photoelectric cells, solar panels, microchips, and rectifiers (Fig. 48.3). High performance with silicon instead of graphite anodes is being researched in order to achieve an even higher energy density and better fast-charging capability. Tesla's Model S and Model 3 use silicon carbon anode material. By adding 6 to 10% silicon-based material to synthetic graphite, its battery

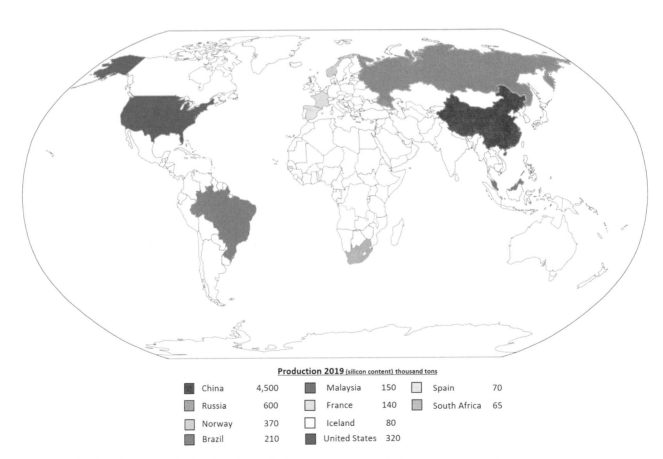

Production 2019 (silicon content) thousand tons					
China	4,500	Malaysia	150	Spain	70
Russia	600	France	140	South Africa	65
Norway	370	Iceland	80		
Brazil	210	United States	320		

Fig. 48.2   List of producing countries based on the US Geological Survey, Mineral Commodity Summaries

Fig. 48.3 Silicon wafer with microchips. (*Photo* Joaquim Sanz. MGVM)

Fig. 48.4 Silicone baking molds. (*Photo* Joaquim Sanz. MGVM)

specific volume rises to over 550 mAh/g and energy density can reach 300 Wh/Kg. Research suggests that in the period to 2030 some anodes silicon in automotive applications will contain up to 30% (Industry News, Backeberg, April 2021: Roskill).

Pure quartz crystal, being piezoelectric and able to maintain a precise vibration frequency, is used in the manufacture of electronic oscillators that create a precise frequency for digital quartz watches, radio and television stations, transmitters, etc.

Metallurgical Industry

Ferrosilicon is used in foundries to improve the properties of steel.

Silicon metal is used to improve the properties of aluminum during the casting process.

Other Fields

The combination of silicon with oxygen results in a polymer called silicone (Fig. 48.4), which is used in lubricants, adhesives, oven containers, electrical cable insulation, valve prostheses, and breast implants.

The combination of silicon and high-temperature carbon forms silicon carbide (also known as carborundum), a material of 9 on the Mohs hardness scale (diamond is 10) that is used as an abrasive and anti-slip; it is a tough material and good electrical insulator (see: silica-quartz).

Hydrated silicon dioxide is an abrasive component of toothpastes. Silicon gel is a porous form of silicon dioxide that is synthetically manufactured from sodium silicate, yielding a useful drying agent.

Crystalline quartz can appear in various colors due to the presence of natural impurities, turning into semi-precious stones such as amethyst (violet color), citrine (light yellow color), and smoked (light brown color), among others.

Cryptocrystalline quartz (formed by microscopic crystals) can appear in beautiful colored forms such as agate or jasper.

48.4 Recycling

Within the European area, solar panels no longer in use or that have stopped functioning are already being collected in order to recover silicon, as well as copper, aluminum, iron, and plastics (see: www.pvcycle.org).

Further Reading

Gray T, Mann N (2009) The elements. Black Dog & Leventhal Publishers Inc., New York

Mata JM, Sanz J (2007) Guia d'identificació de minerals. Manresa, Catalonia. Edicions UPC/Parcir (Catalan 2nd paper edition). 262 p. ISBN 9788483019023. http://hdl.handle.net/2117/90445

Quadbeck-Seeger H-J (2007) World of the elements: elements of the world. Wiley-VCH Verlag GmbH & Co., Germany

Roskill. Market Reports (2020) Silicon and ferrosilicon. Available at https://roskill.com/market-report/silicon-ferrosilicon/. Last accessed May 2021

Sanz J, Tomasa O (2017) Elements i recursos minerals: aplicacions i reciclatge. Manresa, Catalonia. Zenobita Edicions/Iniciativa Digital Politècnica (Catalan 3rd digital edition). http://hdl.handle.net/2117/105113

Sanz J, Tomasa O (2018) Elementos y recursos minerales: aplicaciones y reciclaje. Manresa, Catalonia. Zenobita Edicions/Iniciativa Digital Politècnica (Spanish 1st digital edition). http://hdl.handle.net/2117/123674

Stwertka A (2018) A guide to the elements, 4th edn. Oxford University Press, England

USGS (2021) Commodity statistics and information. Silicon. Available at https://www.usgs.gov/centers/nmic/silicon-statistics-and-information. Last accessed May 2021

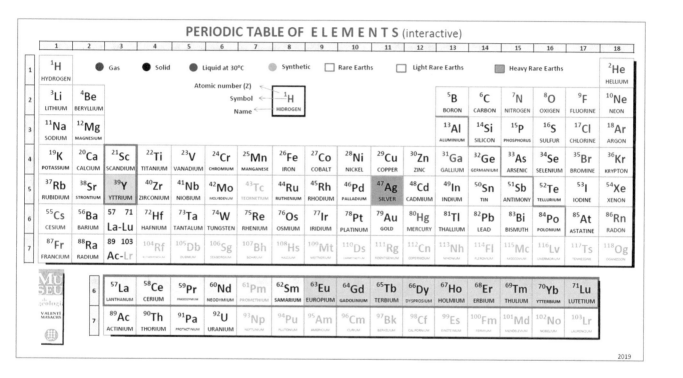

- A noble, soft, highly ductile, and malleable metal (Fig. 49.1)
- The greatest electrical and thermal conductivity of all metals
- Corrosion resistant, but it oxidizes, forming a thin self-protective layer
- The whitest metal and the most reflective
- Many silver salts are sensitive to light
- Since ancient times this metal has been used to make coins
- Obtained from argentite, from argentiferous galena, and as a by-product of copper metallurgy (anodic sludge).

49.1 Geology

Silver is commonly found in hydrothermal veins. Gold and silver are often found together, less so with metals like lead, zinc, copper, and iron, in several kinds of ore deposits. Metals precipitate from ore fluids by various processes depending on specific local conditions. The most common processes are cooling, mixing with other fluids, and pH change. Silver is commonly obtained from argentite (Ag_2S) in assemblages with other sulfides like sphalerite (ZnS), chalcopyrite ($CuFeS_2$), pyrite (FeS_2), and pyrrhotite (FeS). Basinal hydrothermal systems include Mississippi Valley-type (MVT) and sedimentary exhalative (SEDEX) deposits. The MVT deposits are so named because they were

Fig. 49.1 Silver (threads). *Poblet (Catalonia)*. (*Photo* Joaquim Sanz. MGVM)

first found in the mid-west of the United States in the valley of the Mississippi River. These deposits are tectonically related to enrichment of high-salinity metal fluids from sedimentary basins into carbonate platforms. Sedimentary exhalative (SEDEX) deposits consist of layers of sulfides produced and by sedimentary processes and are found within large ancient sedimentary basins. Mexican deposits are the biggest silver mines (e.g. Peñasquito, Fresnillo, Pitarrilla). Bolivia, Peru, Australia, China, and Poland also have important silver deposits.

49.2   Producing Countries

The world's reserves are linked to countries with lead–zinc, copper and gold deposits, and these are Peru (120,000 t), Poland (100,000 t), Australia (90,000 t), Russia (45,000 t), China (41,000 t), Mexico (37,000 t), Chile (26,000 t), the United States (25,000 t), and Bolivia (22,000 t), among others (Fig. 49.2).

49.3   Applications

Electronics Industry

Silver is alloyed with tin to make solder for electronic devices' printed circuit boards. There are cards with a microchip whose contact layer is silver.

The 3D printing technology is evolving to create devices with integrated electronics. In this process, silver inks print the tracks of the conductive circuits to constitute a key element in the manufacture of these electronic devices.

Electrical Industry

Silver is used in alloy with copper in railway catenaries to increase its recrystallization temperature and make it more resistant to the high temperatures reached due to friction with the pantograph (Fig. 49.3).

Battery Industry

Small electronic devices, such as analog watches, often use silver oxide batteries because of the high level of energy that they provide and their long life.

Jewellery

Silver is a precious metal widely used in jewellery (Fig. 49.4). It is also used in gold alloys to give hardness. The 18-carat gold contains 12.5% silver, 12.5% copper, and 75% gold. White gold is comprised of 15% silver, 10% palladium, and 75% gold.

Chemical Industry

Silver salts are used in the manufacture of photographic film, despite current low production due to the introduction of digital photography. They were also used for radiographic (X-ray) plates, which have since become digital (see: europium).

Metallic silver is used as a catalyst for oxidation reactions in the production of formaldehyde from methanol. The oil industry uses silver nitrate as a catalyst.

Medicine

Small amounts of silver are used in certain bandages to prevent infection of wounds and promote healing.

Power Generation

Silver is used in the control rods of pressurized water (PWR) nuclear reactors in the form of an alloy with indium (15%) and cadmium (5%). It plays an important role in the manufacture of solar cells, along with silicon.

Other Fields

Silver is used as a superb reflector of sunlight on the panels of solar power plants, allowing the rays to be concentrated on a liquid or on sodium salts and, with the warmth, to heat water and thus obtain steam, subsequently generating electrical energy.

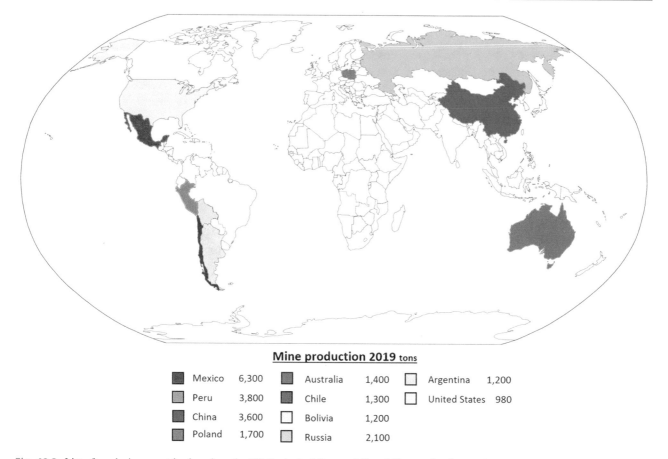

Mine production 2019 tons

◼	Mexico	6,300	◼	Australia	1,400	◻	Argentina	1,200
◻	Peru	3,800	◼	Chile	1,300	◻	United States	980
◼	China	3,600	◻	Bolivia	1,200			
◼	Poland	1,700	◻	Russia	2,100			

Fig. 49.2 List of producing countries based on the US Geological Survey, Mineral Commodity Summaries

Fig. 49.3 Rack railway catenary. (*Photo* Joaquim Sanz. MGVM)

This metal is also used as an electrolytic coating (silvered) to improve the presentation of a metallic object, to increase the electrical conductivity of another metal (in electronic devices), or to make it more resistant to corrosion, improve its weldability, and so on.

Fig. 49.4 A silver ring. (*Photo* Joaquim Sanz. MGVM)

49.4   Recycling

Silver is recycled by recovering it from the residues generated during the manufacture of products with this metal, from alloys, and from industrial residues containing it.

Further Reading

Eurometaux. Introducing the metals. Available at: https://eurometaux.eu/about-our-industry/introducing-metals/ (last accessed May 2021)

Gray T, Mann N (2009) The elements. Black Dog & Leventhal Publishers Inc., New York

Mata JM, Sanz J (2007) Guia d'identificació de minerals. Manresa, Catalonia. Edicions UPC/Parcir (Catalan 2nd paper edition), 262 p. ISBN: 9788483019023. http://hdl.handle.net/2117/90445

Pirajno F (2009) Hydrothermal processes and mineral systems. Springer

Quadbeck-Seeger H-J (2007) World of the elements: elements of the world. Wiley-VCH Verlag GmbH & Co, Germany

Sanz J, Tomasa O (2017) Elements i Recursos minerals: Aplicacions i reciclatge. Manresa, Catalonia. Zenobita Edicions/Iniciativa Digital Politècnica (Catalan 3rd digital edition). http://hdl.handle.net/2117/105113

Sanz J, Tomasa O (2018) Elementos y Recursos minerales: Aplicaciones y reciclaje. Manresa, Catalonia. Zenobita Edicions/Iniciativa Digital Politècnica (Spanish 1st digital edition). http://hdl.handle.net/2117/123674

Stwertka A (2018) A guide to the elements, 4th edn. Oxford University Press, England

USGS. Commodity Statistics and Information. Silver. Available at: https://www.usgs.gov/centers/nmic/silver-statistics-and-information (last accessed May 2021)

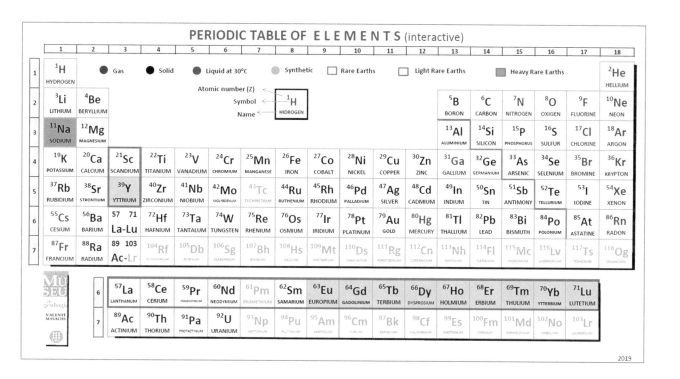

PERIODIC TABLE OF E L E M E N T S (interactive)

- Alkaline mineral
- Reacts very strongly to water, releasing H_2
- Oxidizes quickly with air
- Basic element in human and animal nutrition
- Named after *sodanum*, a medieval Latin word for a sodium-based remedy for migraines
- Its symbol Na, from natrium, is from the ancient Greek word for sodium carbonate, *nitron*
- Fourth most abundant element in the earth's crust
- Obtained from halite (Fig. 50.1) and brine.

50.1 Geology

The main mineral source of sodium is halite. Halite can be found as a mineral (NaCl) or dissolved in water, forming brine. Brine is a solution with a high salt concentration, between 5% and 26–28%. Both halite and brine are closely associated with evaporite deposits, which may have either a lacustrine or marine origin.

The main evaporitic sodium deposits are the saline flats characteristic of arid basins. These basins are shallow and dry apart from when storm flooding turns the pan and surrounding mudflats into a temporary lake. Continental saline

Fig. 50.1 Halite (sodium chloride). *Súria (Catalonia).* (*Photo* Joaquim Sanz. MGVM)

occur in both continental and marginal marine settings, and are known as sabkha. The extent of these deposits vary from one square kilometer to the largest, of thousands of square kilometers, such as Lake Uyuni in Bolivia. Other examples of evaporitic deposits include the sabkhas around the Persian Gulf and the Great Salt Lake of Utah. Another type of halite deposit is a salt dome. These are formed when the salt deposit is buried by layers of sediment. Confined salt has the capacity to deform plastically under pressure and temperature. Therefore, it is forced by the unequal pressures to flow upwards through weaker overlying strata. The resulting formation takes a characteristic elongated mushroom shape and it is composed of relatively pure halite.

50.2 Producing Countries

flats occupy the lowest areas of closed arid basins. The pan is surrounded by a brine-soaked mudflat permeated with evaporite minerals that grow within the sediment. Saline flats

World reserves of sodium are vast and are directly linked to reserves of potassium salts and brine (see: halite and sylvite) (Fig. 50.2).

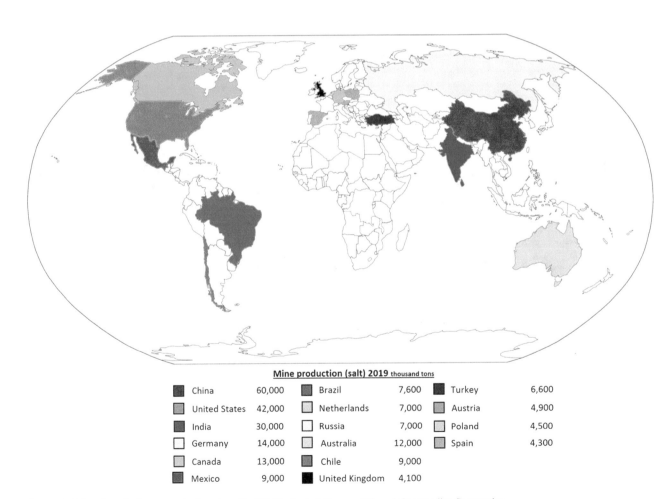

Mine production (salt) 2019 thousand tons					
China	60,000	Brazil	7,600	Turkey	6,600
United States	42,000	Netherlands	7,000	Austria	4,900
India	30,000	Russia	7,000	Poland	4,500
Germany	14,000	Australia	12,000	Spain	4,300
Canada	13,000	Chile	9,000		
Mexico	9,000	United Kingdom	4,100		

Fig. 50.2 List of producing countries based on the US Geological Survey, Mineral Commodity Summaries

Fig. 50.3 Smoked salmon, preserved with sodium chloride. (*Photo* Joaquim Sanz. MGVM)

50.3 Applications

Chemical Industry

The main consumer of sodium chloride is the chemical industry, for the manufacture of chlorine and caustic soda (sodium hydroxide (caustic soda), an ideal chemical agent for unclogging drains and cleaning away deposits).

Sodium chloride is common salt, used in cooking, seasoning, and preserving food (Fig. 50.3) (see: halite).

Sodium sulfate is hygroscopic, so is used as a moisture absorber in laboratories and in the chemical industry (see: thenardite and halite).

Silica gel is a porous form of silicon dioxide that is synthetically manufactured from sodium silicate and becomes a useful drying agent.

Glass Industry

Sodium carbonate (soda ash) and sodium sulfate are vital compounds in the manufacture of glass, as they contribute to lowering the melting point of silica sand.

Metallurgical Industry

Sodium metal is used in metal alloys to improve their structure. It is used as a metal purifier, because it removes scale.

Other Fields

Sodium vapor is used in the manufacture of lamps with good electrical and illumination performance, for lighting city streets and monuments in its typical orange light, although it is already being replaced by LED lamps that both increase the level of light intensity and save electrical energy.

Liquid sodium is used in certain types of nuclear power plants for heat transfer from the reactor core to the electricity-generating turbines. Sodium chloride is used in the manufacture of sodium and nickel batteries (zebra batteries), which work at high temperatures.

50.4 Recycling

Sodium recycling is unknown.

Further Reading

Gray T, Mann N (2009) The elements. Black Dog & Leventhal Publishers Inc., New York

ICL (2021) Iberia Salt. Available at: http://www.icl-ip.com/specialty-minerals/sm-product/salt/ (last accessed May)

Mata JM, Sanz J (2007) Guia d'identificació de minerals. Manresa, Catalonia. Edicions UPC/Parcir (Catalan 2nd paper edition), 262 p. ISBN: 9788483019023. http://hdl.handle.net/2117/90445

Quadbeck-Seeger H-J (2007) World of the elements: elements of the world. Wiley-VCH Verlag GmbH & Co, Germany

Roskill (2021a) Market Reports. Salt. Available at: https://roskill.com/market-report/salt/ (last accessed May 2021)

Roskill (2021b) Market Reports. Soda Ash. Available at: https://roskill.com/market-report/soda-ash/ (last accessed May 2021)

Sanz J, Tomasa O (2017) Elements i Recursos minerals: Aplicacions i reciclatge. Manresa, Catalonia. Zenobita Edicions/Iniciativa Digital Politècnica (Catalan 3rd digital edition). http://hdl.handle.net/2117/105113

Sanz J, Tomasa O (2018) Elementos y Recursos minerales: Aplicaciones y reciclaje. Manresa, Catalonia. Zenobita Edicions/Iniciativa Digital Politècnica (Spanish 1st digital edition). http://hdl.handle.net/2117/123674

Schwab FL (2003) Sedimentary petrology. Encyclopedia of Physical Science and Technology, pp 495–529. https://doi.org/10.1016/b0-12-227410-5/00678-5

Stwertka A (2018) A guide to the elements, 4th edn. Oxford University Press, England

USGS (2021) Commodity Statistics and Information. Salt. Available at: https://www.usgs.gov/centers/nmic/salt-statistics-and-information (last accessed May 2021)

Warren JK (2016) Evaporites: a geological compendium. Springer. https://doi.org/10.1007/978-3-319-13512-0

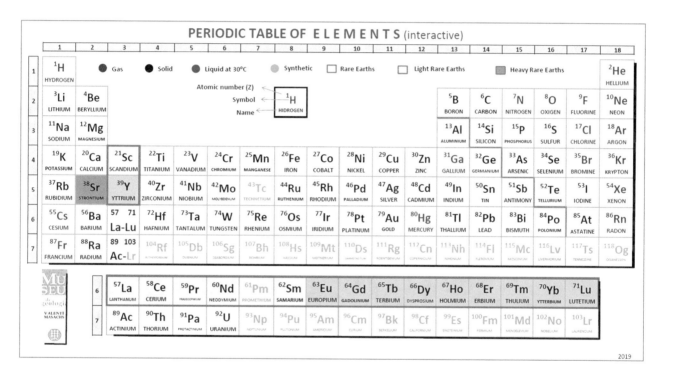

PERIODIC TABLE OF E L E M E N T S (interactive)

- Alkaline earth metal, soft, and somewhat malleable
- Highly reactive in water, from which it releases hydrogen, and also in oxygen
- Identified by English scientist Adair Crawford in 1789 in the Scottish town of Strontian
- Obtained from celestine (Fig. 51.1) and strontianite
- Assigned the status of a strategic mineral by the EU in September 2020.

51.1 Geology

Two strontium-bearing minerals, celestine ($SrSO_4$) and strontianite ($SrCO_3$), contain strontium in quantities sufficient to make their recovery worthwhile. Celestine is more common than strontianite and is the primary source of the world's strontium. Celestine appears as crystals and as massive or fibrous aggregates in sedimentary rocks. It often displays a delicate blue coloration owing to the presence of impurities. Celestine can be found in bedded evaporite

Fig. 51.1 Celestine (strontium sulfate). *La Granja d'Escarp* (*Catalonia*). (*Photo* Joaquim Sanz. MGVM)

deposits in conjunction with gypsum, anhydrite, and halite. It can also occur in cavities within carbonate rocks where it has been precipitated from strontium-bearing groundwater or brines. Strontianite is formed in hydrothermal deposits at low temperatures in limestone and marl or as a secondary mineral in sulfide veins.

51.2 Producing Countries

World reserves are estimated at over 1 billion tons, but there are no estimates available from some countries that may have strontium (Fig. 51.2).

51.3 Applications

Chemical Industry
Strontium nitrate imparts a bright red color to fireworks and emergency signal flares (Fig. 51.3).

Strontium is used in the electrolytic production of zinc, as it removes lead impurities from the ore. Strontium carbonate is used to remove sulfates in wastewater treatments.

Glass and Ceramics Industry
Strontium improves the properties of glass for liquid crystal displays (LCDs). Strontium carbonate is added to the glass to enhance its hardness, scratch resistance, brilliance, and ease of polishing. It is used to glaze tableware ceramics to improve their abrasion resistance and prevent the formation of bubbles during firing.

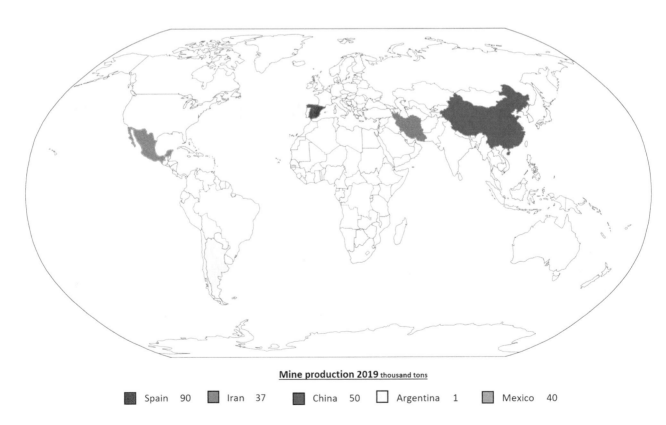

Mine production 2019 thousand tons

Spain 90   Iran 37   China 50   Argentina 1   Mexico 40

Fig. 51.2 List of producing countries based on the US Geological Survey, Mineral Commodity Summaries

Fig. 51.3 Fireworks. (*Image courtesy of* Mireia Arso)

Other Fields

Celestine and barite are used as additives in the muds that facilitate drilling in oil and natural gas wells.

Strontium is used in the manufacture of ceramic ferrite magnets to improve their efficiency.

The radioactive isotope strontium-90, with a half-life of 28.8 years, appears after nuclear explosions that contaminate large areas. It is deposited in the ground and in the grass of meadows where cows may later graze, for instance. The milk from these cows will carry strontium-90, which can cause disorders in people's bones if they consume it.

51.4 Recycling

Strontium recycling is unknown.

Further Reading

Earth Magazine (2021) Mineral Resource of the Month: Strontium. URL: https://www.earthmagazine.org/article/mineral-resource-month-strontium-0 (last accessed May 2021).

Gray T, Mann N (2009) The elements. Black Dog & Leventhal Publishers Inc., New York

Mata JM, Sanz J (2007) Guia d'identificació de minerals. Manresa, Catalonia. Edicions UPC/Parcir (Catalan 2nd paper edition), 262 p. ISBN: 9788483019023. http://hdl.handle.net/2117/90445

Quadbeck-Seeger H-J (2007) World of the elements: elements of the world. Wiley-VCH Verlag GmbH & Co, Germany

Sanz J, Tomasa O (2017) Elements i Recursos minerals: Aplicacions i reciclatge. Manresa, Catalonia. Zenobita Edicions/Iniciativa Digital Politècnica (Catalan 3rd digital edition). http://hdl.handle.net/2117/105113

Sanz J, Tomasa O (2018) Elementos y Recursos minerales: Aplicaciones y reciclaje. Manresa, Catalonia. Zenobita Edicions/Iniciativa Digital Politècnica (Spanish 1st digital edition). http://hdl.handle.net/2117/123674

Stwertka A (2018) A guide to the elements, 4th edn. Oxford University Press, England

USGS (2021) Commodity Statistics and Information. Strontium. Available at: https://ww.usgs.gov/centers/nmic/strontium-statistics-and-information (last accessed May 2021)

PERIODIC TABLE OF E L E M E N T S (interactive)

Fig. 52.1 — Periodic table of elements showing the legend (Gas, Solid, Liquid at 30°C, Synthetic, Rare Earths, Light Rare Earths, Heavy Rare Earths) with Sulfur (16 S) highlighted.

2019

- A yellow non-metal (Fig. 52.1)
- Emits a bad smell in practically all forms
- Found in its native state on many slopes of active volcanoes, forming thick, exploitable layers
- Currently obtained as a by-product of distilling crude oil, in treating natural gas (in the form of H_2S), from types of sulfides, and from evaporite and volcanic deposits.

52.1 Geology

Elemental sulfur constitutes a metastable, intermediate state in the geological conversion of sulfate to sulfide or vice versa. Sulfur sedimentary deposits are formed from hydrogen sulfide resulting from the chemical reaction of sulfate with organic matter. Another possible sulfur origin was formed during or after formation of the enclosing rock. The geological bacterial sulfur cycle is another sulfur source, when oxidizing and reducing bacteria can occur side by side,

© The Author(s), under exclusive license to Springer Nature Switzerland AG 2022
J. Sanz et al., *Elements and Mineral Resources*, Springer Textbooks in Earth Sciences, Geography and Environment, https://doi.org/10.1007/978-3-030-85889-6_52

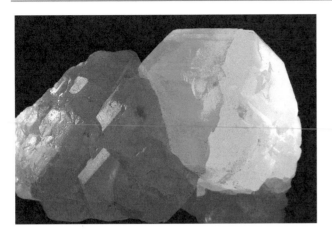

Fig. 52.1 Sulfur. *Sicily* (*Italy*). (*Photo* Joaquim Sanz. MGVM)

and the active stage cannot be determined simply by locating the bacteria. Sulfur historically was usually obtained in pyrite (S_2Fe). The Iberian Pyrite Belt was the most important sulfur source for decades.

The Iberian Pyrite Belt is located in the southwest of the Iberian Peninsula, comprising part of Portugal and Spain. It is one of the world's most important volcanogenic massive sulfide (VMS) districts and has been mined for more than 5000 years. During the nineteenth and twentieth centuries, production was focused on sulfuric acid.

The sequence of the Iberian Pyrite Belt is relatively simple. It begins with a basal sedimentary sequence of Late Devonian age, overlain by the volcano-sedimentary complex (CVS), deposited in an intra-continental basin during the collision of the South Portuguese Zone against the Iberian Massif. Then, in the Early Carboniferous period volcanic activity took place. The whole series is affected by very low-degree metamorphism and a fold and thrust tectonic of the Variscan (Hercynian) orogeny. Most of the mineral deposits in this area consist of massive sulfides over for a wide zone of hydrothermal alteration. In volcanic regions like Sicily, in ancient times sulfur was stacked in brick kilns built on sloping hillsides, with an airspace between them. Next, some sulfur was pulverized, spread over the stacked ore, and ignited, causing the free sulfur to melt and flow downhill. As a mineral, native sulfur under salt domes is thought to be a fossil mineral resource produced by the action of anaerobic bacteria on sulfate deposits. Until the late twentieth century, sulfur was extracted from salt domes.

Sulfur is now obtained as a by-product or is produced from oil, petroleum, natural gas, and related fossil resources, from which it is obtained mainly as hydrogen sulfide in the United States, Russia, and China.

52.2   Producing Countries

The sulfur reserves in petroleum, crude oil, natural gas, and various sulfides are very large. Only crude oil contains naturally high amounts of sulfur (Fig. 52.2).

52.3   Applications

Chemical Industry
Sulfur is mainly used to make sulfuric acid (H_2SO_4) (Fig. 52.3), also in petroleum refinery, for pigments, and for pickling painted, oxidized, and varnished metal surfaces, etc.

Sulfuric acid dissolves phosphorite, an insoluble rock basically formed of calcium phosphate, into a soluble phosphate known as superphosphate. This is one of the most important agricultural fertilizers worldwide, providing the main essential nutrients for plant growth (see: phosphorus).

Sulfur is one of the most widely used fungicides and pesticides for fruit and vegetable crops. By burning a strip of paper impregnated with sulfur inside them, it disinfects wine barrels; there are other chemical means to achieve the same effect.

Other Fields
Sulfur is used in the manufacture of paper and as a bleaching agent. It is one of the components of black powder (gunpowder).

In the production of tires, sulfur is used in vulcanizing rubber, heating it in the presence of sulfur to make it harder and more resistant (Fig. 52.4).

52.4   Recycling

In the manufacture of sulfuric acid and in agriculture, the sulfur from crude oil refinery and natural gas treatment is reused.

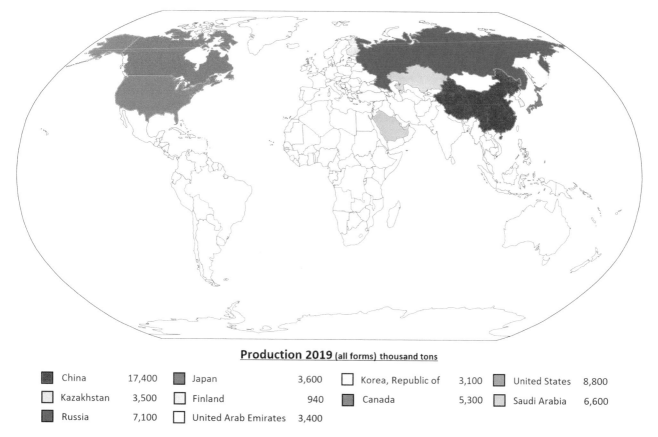

Production 2019 (all forms) thousand tons

■ China	17,400	■ Japan	3,600	☐ Korea, Republic of	3,100	■ United States	8,800
☐ Kazakhstan	3,500	☐ Finland	940	■ Canada	5,300	☐ Saudi Arabia	6,600
■ Russia	7,100	☐ United Arab Emirates	3,400				

Fig. 52.2 List of producing countries based on the US Geological Survey, Mineral Commodity Summaries

Fig. 52.4 Rubber tires. (*Photo* Joaquim Sanz. MGVM)

Fig. 52.3 Sulfuric acid. (*Photo* Joaquim Sanz. MGVM)

Further Reading

Gray T, Mann N (2009) The elements. Black Dog & Leventhal Publishers Inc., New York

Mata JM, Sanz J (2007) Guia d'identificació de minerals. Manresa, Catalonia. Edicions UPC/Parcir (Catalan 2nd paper edition), 262 p. ISBN: 9788483019023. http://hdl.handle.net/2117/90445

Quadbeck-Seeger H-J (2007) World of the elements: elements of the world. Wiley-VCH Verlag GmbH & Co, Germany

Sanz J, Tomasa O (2017) Elements i Recursos minerals: Aplicacions i reciclatge. Manresa, Catalonia. Zenobita Edicions/Iniciativa Digital Politècnica (Catalan 3rd digital edition). http://hdl.handle.net/2117/105113

Sanz J, Tomasa O (2018) Elementos y Recursos minerales: Aplicaciones y reciclaje. Manresa, Catalonia. Zenobita Edicions/Iniciativa Digital Politècnica (Spanish 1st digital edition). http://hdl.handle.net/2117/123674

Stwertka A (2018) A guide to the elements, 4th edn. Oxford University Press, England

Sulfur deposits-An overview. https://www.sciencedirect.com/topics/engineering/sulfur-deposits (last accessed May 2021)

The Iberian Pyritic Belt. http://www.igme.es/patrimonio/GEOSITES/Chapter_04_SGFG.pdf (last accessed May 2021)

USGS (2021) Commodity Statistics and Information. Sulfur. Available at: https://www.usgs.gov/centers/nmic/sulfur-statistics-and-information (last accessed May 2021)

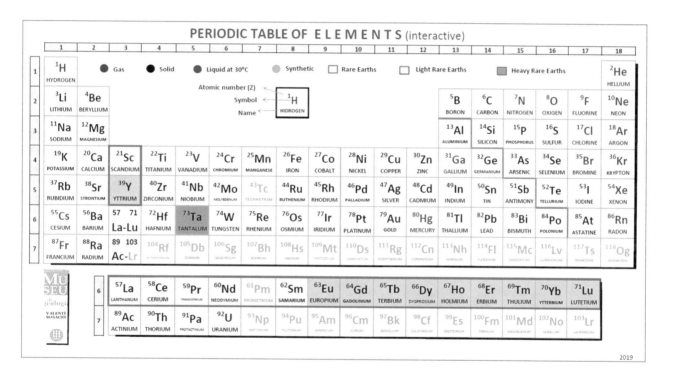

53.1 Geology

The primary ore of tantalum is the columbite-tantalite group. The columbite group is a solid solution with varying proportions of Nb or Ta in its composition. According to the International Mineralogical Association (IMA), 'tantalite' or 'coltan' are not scientific names, yet they are widely used. Many specimens commonly named tantalite are actually minerals of the columbite or tapiolite group.

The main minerals of tantalum-niobium ores from which both are extracted are columbite ($(Mn,Fe)(Ta,Nb)_2O_6$); hatchettolite ($(Ca,U,Ce)_2(Nb,Ti,Ta)_2O_6(OH,F)$; and microlite ($A_{2-m}Ta_2X_{6-w}Z_{-n}$) and ixiolite ($(Nb,Ta,Sn,W,Sc)_3O_6$).

- Heavy, hard, and resistant to corrosion
- Large electricity storage capacity
- Resistant to high temperatures (melts at 2996 °C)
- The chemist A.G. Ekeberg identified it in 1802 in minerals from Ytterby, Sweden, naming it after Tantalos, father of Niobe in Greek mythology
- Assigned the status of a strategic metal by the EU in 2017
- Found in columbite-tantalite (coltan) (Fig. 53.1), tantalite, and cassiterite *(these minerals can have small amounts of thorium and uranium and emit radioactivity)*.

Fig. 53.1 Columbite-tantalite (niobium oxide and tantalum). *Musaca* (*Rwanda*). (*Photo* Joaquim Sanz. MGVM)

The most common genetic types of Nb–Ta tantalum ore are:

1. Rare-metal pegmatites, in which ores are usually represented by zonal vein bodies ranging in size from a few hundred meters to 1–2 km with small amounts of spodumene or petalite: rare-metal Nb–Ta-bearing granites (greisen) are represented by small pipes and granitic domes. A certain amount of Ni-Ta is also extracted from Sn or W greisen deposits. Minas Gerais (Brazil) is an important world producer of tantalum.

2. Placers related to pegmatitic sources and containing cassiterite and minerals of the columbite-tantalite group, and nepheline syenites: the most important deposits are mainly in the DR Congo, whose exploitation is mainly small scale since the mineral is easily separable by gravimetry due to its high specific weight.

53.2 Producing Countries

Note that Rwanda's columbo-tantalite (coltan) production originates in the DR Congo yet counts as Rwandan, because armed groups sell the coltan to both Rwanda and Uganda.

Many open-pit mines in the Kiwu area in the north of the DR Congo turn into mud pits when the rains fall. Children and young people work tirelessly, digging, washing, classifying, and transporting coltan and cassiterite to help their families (IPIS—International Peace Information Services).

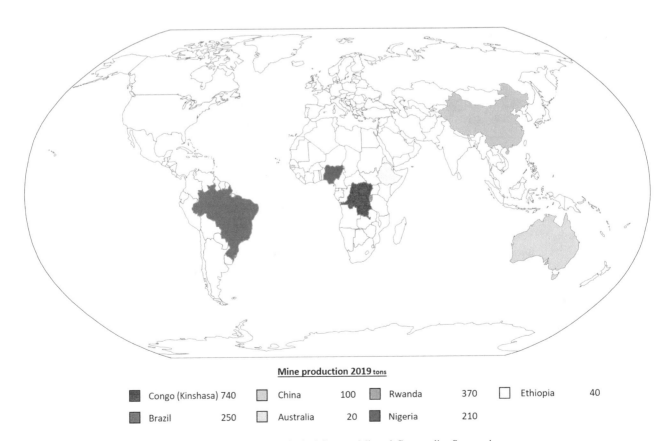

Mine production 2019 tons

■	Congo (Kinshasa) 740	▢	China	100	▨	Rwanda	370	▢	Ethiopia	40
▨	Brazil	250	▢	Australia	20	■	Nigeria	210		

Fig. 53.2 List of producing countries based on the US Geological Survey, Mineral Commodity Summaries

Tantalum carbide is considered stronger than tungsten carbide, and is used in cutting tools. As tantalum is biocompatible and resistant to body fluids, prostheses for joints, cranial plates, screws and surgical instruments, and implants are manufactured.

Chemical Industry

Tantalum can be used in equipment for obtaining aggressive acids such as hydrochloric (HCl) and nitric (HNO_3).

Optical Industry

Tantalum oxide is added to the glass of glasses, cell phones, and cameras to increase its transparency and refractive index, also making it lighter.

53.4 Recycling

The main source of tantalum recycling is scrap generated in the manufacture of capacitors, electronic components, tools, and the alloys based on this metal.

Fig. 53.3 Tiny tantalum capacitor (red) in front of aluminum capacitor (yellow). (*Photo* Joaquim Sanz. MGVM)

The world reserves for which data are available are Australia (55,000 t) and Brazil (34,000 t), and the reserves of other producing countries are unknown (Fig. 53.2).

53.3 Applications

Electronics Industry

The main application of tantalum is in the production of capacitors. The ability to store electricity in small amounts of metal and give a good response to high frequencies has allowed the miniaturization of all electronic devices, such as mobile phones, game consoles, microprocessors, 'smart' weapons, missiles, and so on, by replacing classic aluminum capacitors (larger) by those of tantalum (much smaller) (Fig. 53.3).

Metallurgical Industry

The second application of tantalum is the manufacture of refractory super alloys, due to good resistance to high temperatures and corrosion, for the space industry and in jet engines and gas turbines. It is also used as a thermal spray for surface coatings to protect against wear and the oxidation of parts that operate at high temperatures.

Further Reading

Gray T, Mann N (2009) The elements. Black Dog & Leventhal Publishers Inc., New York

Mata JM, Sanz J (2007) Guia d'identificació de minerals. Manresa, Catalonia. Edicions UPC/Parcir (Catalan 2nd paper edition), 262 p. ISBN: 9788483019023. http://hdl.handle.net/2117/90445

Quadbeck-Seeger H-J (2007) World of the elements: elements of the world. Wiley-VCH Verlag GmbH & Co, Germany

Roskill (2021) Market Reports. Tantalum. Available at: https://roskill.com/market-report/tantalum/ (last accessed May 2021)

Sanz J, Tomasa O (2017) Elements i Recursos minerals: Aplicacions i reciclatge. Manresa, Catalonia. Zenobita Edicions/Iniciativa Digital Politècnica (Catalan 3rd digital edition). http://hdl.handle.net/2117/105113

Sanz J, Tomasa O (2018) Elementos y Recursos minerales: Aplicaciones y reciclaje. Manresa, Catalonia. Zenobita Edicions/Iniciativa Digital Politècnica (Spanish 1st digital edition). http://hdl.handle.net/2117/123674

Stwertka A (2018) A guide to the elements, 4th edn. Oxford University Press, England

Tantalum-Niobium International Study Center. (2021) Available at: https://tanb.org/about-tantalum (last accessed May 2021)

USGS (2021) Commodity Statistics and Information. Tantalum. Available at: https://www.usgs.gov/centers/nmic/niobium-columbium-and-tantalum-statistics-and-information (last accessed May 2021)

PERIODIC TABLE OF E L E M E N T S (interactive)

- The lightest of all chemicals without stable isotopes
- The technetium-99 m isotope comes from the molybdenum-99 isotope, which is prepared in nuclear reactors and in fusion-fission processes (SHINE)
- The metastable isotope technetium-99 m has a half-life of 6 h and emits gamma radiation. It is obtained from a generator containing the molybdenum-99 isotope, which disintegrates into technetium-99 m (Fig. 54.1)

54.1 Applications

Medicine

Technetium-99 m is one of the most widely used radio-pharmaceuticals (about 83%) in diagnostic procedures for organ functioning in the human body (heart, brain, thyroid, lungs, bones, and blood). This metastable isotope, with a half-life of 6 h, is injected into the patient and accumulates in several parts of the body, concentrating in altered parts and emitting 140 keV gamma radiation detectable by gammagraphy (SPECT) devices (Figs. 54.2 and 54.3).

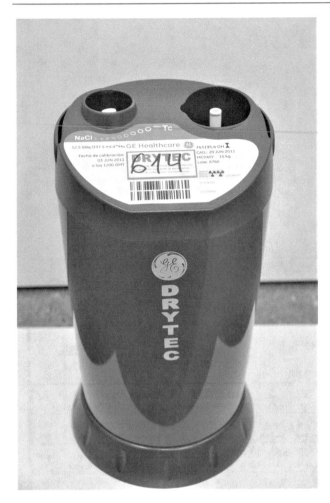

Fig. 54.1 Technetium-99 m generator. (*Photo* Joaquim Sanz. MGVM)

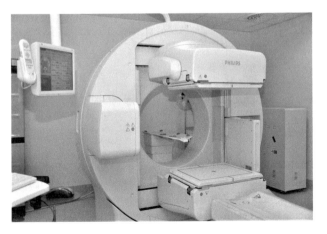

Fig. 54.2 SPECT equipment. (*Photo* Joaquim Sanz. MGVM)

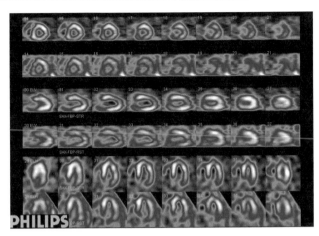

Fig. 54.3 SPECT of the heart with technetium-99 m. (*Image courtesy of* Hospital Sant Pau, Barcelona)

Other Fields

Technetium-99 can be used as a beta radiation standard to calibrate scientific instruments.

54.2 Recycling

Recycling of technetium is unknown.

Further Reading

Flickr. Drytec Technetium Generator (2011). Available at: https://www.flickr.com/photos/iaea_imagebank/sets/72157632748273235/ (last accessed May 2021)

Gray T, Mann N (2009) The elements. Black Dog & Leventhal Publishers Inc., New York

GE Reports. Medical Imaging (2015). Available at: https://www.ge.com/reports/new-production-process-could-help-break-imaging-isotope-shortage-2/ (last accessed May 2021)

Mata JM, Sanz J (2007) Guia d'identificació de minerals. Manresa, Catalonia. Edicions UPC/Parcir (Catalan 2nd paper edition), 262 p. ISBN: 9788483019023. http://hdl.handle.net/2117/90445

O'Driscoll B, González-Jiménez J-M (2016) Petrogenesis of the platinum-group minerals. Rev Miner Geochem 81(1):489–578. https://doi.org/10.2138/rmg.2016.81.09

Sanz J, Tomasa O (2017) Elements i Recursos minerals: Aplicacions i reciclatge. Manresa, Catalonia. Zenobita Edicions/Iniciativa Digital Politècnica (Catalan 3rd digital edition). http://hdl.handle.net/2117/105113

Sanz J, Tomasa O (2018) Elementos y Recursos minerales: Aplicaciones y reciclaje. Manresa, Catalonia. Zenobita Edicions/Iniciativa Digital Politècnica (Spanish 1st digital edition). http://hdl.handle.net/2117/123674

Stwertka A (2018) A guide to the elements, 4th edn. Oxford University Press, England

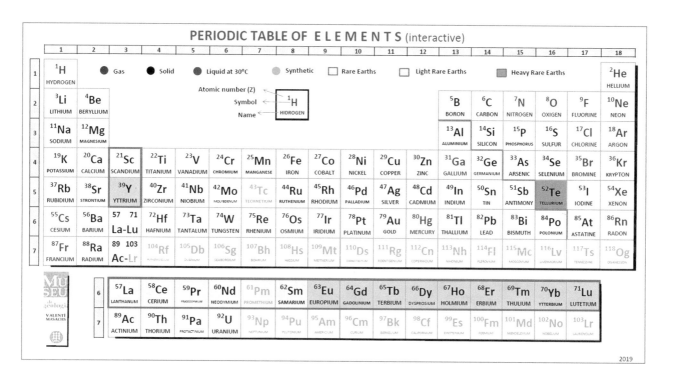

- Semi-metallic semiconductor
- Photoconductive (electrical conductivity increases slightly when exposed to light)
- Named after the Latin word for Earth, *tellus*
- Discovered in 1782 by mine inspector Franz Joseph von Reichenstein, an amateur scientist, experimenting with a mineral found in a Transylvanian gold mine
- Obtained as a by-product of copper refinery and from gold tellurium, such as sylvanite (Fig. 55.1) and calaverite.

55.1 Geology

Tellurium is found mainly as epithermal deposits and occurs in telluride minerals as native tellurium, and as tellurium-bearing sulfosalts in unoxidized ores, as well as in the form of tellurites in secondary ores. Epithermal gold and silver deposits, such as sylvanite $(Au,Ag)_2Te_4$ and calaverite $(AuTe_2)$, can also be a source of tellurium, such as in Colorado, in New Mexico in the United States, and in Japan.

Another kind of deposit with economic interest in its tellurium content is in porphyry. Tellurium is enriched in sulfide and/or telluride phases in most copper porphyry

Fig. 55.1 Silvanite (silver and gold telluride). *Nagiag (Transylvania, Romania).* (*Photo* Joaquim Sanz. MGVM)

deposits, and thus may be a recoverable by-product when milling some of these copper ores. Porphyry deposits are shallowly formed magmatic-hydrothermal deposits typically developed in the cupolas of porphyritic intrusions within active continental margins above subduction zones. The deposits are generally characterized by stockwork veinlets rich in copper- and/or molybdenum-bearing minerals.

The principal source of tellurium globally is as a by-product of copper refinery of ores from large-tonnage, low-grade copper and copper–gold porphyry deposits in the United States.

55.2 Producing Countries

The world's reserves, of known value, are in China (6,600 t), the United States (3,500 t), Canada (800 t), and Sweden (670 t), among other countries (Fig. 55.2).

These values are not high; however, as tellurium is extracted as a by-product of electrolytic refinery of copper and steel from lead, zinc, bismuth, and other metals, it will always be available.

55.3 Applications

Metallurgical Industry

The main use of tellurium is as an addition to steel to improve its mechanical properties, such as its hardness and corrosion resistance. It is also an additive in copper alloys to enhance their machinability without losing electrical conductivity.

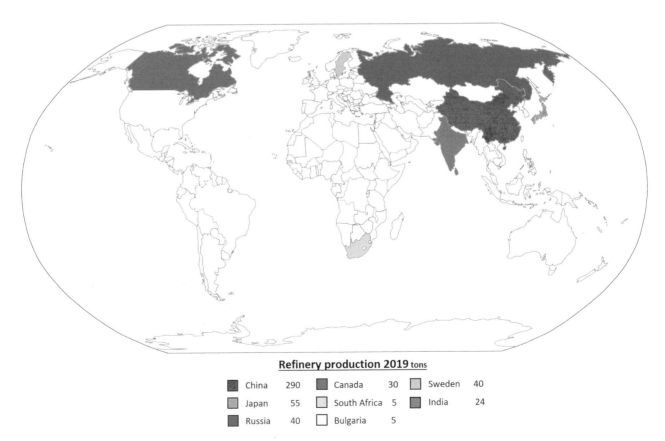

Refinery production 2019 tons

■ China	290	■ Canada	30	■ Sweden	40
■ Japan	55	■ South Africa	5	■ India	24
■ Russia	40	□ Bulgaria	5		

Fig. 55.2 List of producing countries based on the US Geological Survey, Mineral Commodity Summaries

Tellurium, added to lead, improves its hardness and increases its resistance to acids.

Chemical Industry

Tellurium is a catalyst in the production of synthetic fibers. It is used in the production of tires to accelerate the process of rubber vulcanization, where the rubber is heated in the presence of tellurium to replace sulfur or selenium, becoming harder and more resistant.

Power Generation

Tellurium is increasingly used in the manufacture of solar cells alongside cadmium (Fig. 55.3).

Electronics Industry

Cadmium zinc telluride is used in gamma-ray detectors for medicine (in PET—positron emission tomography) and industry. Cadmium manganese telluride is used as an infrared-sensitive semiconductor.

Fig. 55.3 Solar panels. (*Photo* Joaquim Sanz. MGVM)

Tellurium oxide is used in the preparation of the rewritable surface of DVDs and Blu-Ray disks. Bismuth telluride is one of the key compounds in Peltier plates, which produce heat or cold as a result of an electric current passing through the junction of two metals, alloys, or semiconductors (the Peltier effect).

55.4 Recycling

There is very little tellurium recycling, yet its recovery from old solar panels is increasing.

Further Reading

Critical Mineral Resources of the United States (2017) Economic and environmental geology and prospects for future supply. https://pubs.usgs.gov/pp/1802/r/pp1802r.pdf (last accessed February 2021)

Gray T, Mann N (2009) The elements. Black Dog & Leventhal Publishers Inc., New York

Mata JM, Sanz J (2007) Guia d'identificació de minerals. Manresa, Catalonia. Edicions UPC/Parcir (Catalan 2nd paper edition), 262 p. ISBN: 9788483019023. http://hdl.handle.net/2117/90445

O'Driscoll B, González-Jiménez J-M (2016) Petrogenesis of the platinum-group minerals. Rev Miner Geochem 81(1):489–578. https://doi.org/10.2138/rmg.2016.81.09

Pirajno F (2009) Hydrothermal processes and mineral systems. Springer

Quadbeck-Seeger H-J (2007) World of the elements: elements of the world. Wiley-VCH Verlag GmbH & Co, Germany

Sanz J, Tomasa O (2017) Elements i Recursos minerals: Aplicacions i reciclatge. Manresa, Catalonia. Zenobita Edicions/Iniciativa Digital Politècnica (Catalan 3rd digital edition). http://hdl.handle.net/2117/105113

Sanz J, Tomasa O (2018) Elementos y Recursos minerales: Aplicaciones y reciclaje. Manresa, Catalonia. Zenobita Edicions/Iniciativa Digital Politècnica (Spanish 1st digital edition). http://hdl.handle.net/2117/123674

Stwertka A (2018) A guide to the elements, 4th edn. Oxford University Press, England

USGS (2021) Commodity Statistics and Information. Tellurium. Available at: https://www.usgs.gov/centers/nmic/selenium-and-tellurium-statistics-and-information (last accessed May 2021)

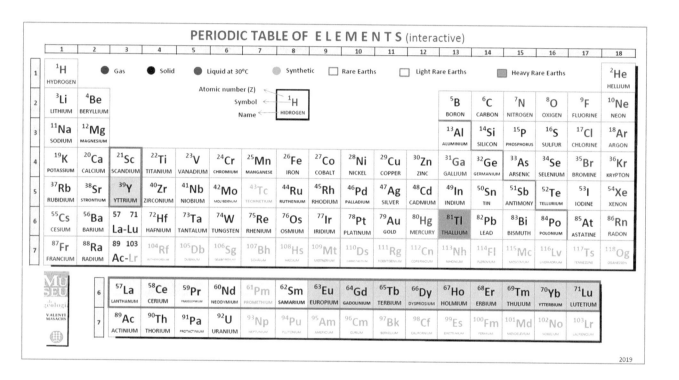

PERIODIC TABLE OF E L E M E N T S (interactive)

(2019)

- Gray, malleable metal
- Soft and heavy
- Looks like tin
- Extremely toxic and carcinogenic
- Discovered in 1861 by Sir William Crookes, who identified it by the bright green emission line of its spectrum
- *Thallos*, a Greek word, means 'green twig'
- Extracted as a by-product of copper, lead, and zinc and copper sulfides (Fig. 56.1)

56.1 Geology

Thallium is mostly found in association with potassium minerals in clays, soils, and granites, but it is not commonly commercially recoverable from those sources. The major source of commercial thallium is trace amounts found in the sulfide ores of copper, lead, zinc, and other metals.

Thallium is a metal contained in pyrites and chalcopyrites and is extracted as a by-product of sulfuric acid production when the ore is roasted. It is also recovered from the

J. Sanz et al., *Elements and Mineral Resources*, Springer Textbooks in Earth Sciences,
Geography and Environment, https://doi.org/10.1007/978-3-030-85889-6_56

Fig. 56.1 Chalcopyrite (copper and iron sulfide). *El Brull* (*Catalonia*). (*Photo* Joaquim Sanz. MGVM)

smelting of lead- and zinc-rich ores. Manganese nodules, which are found on the ocean floor, also contain thallium, but nodule extraction is prohibitively expensive and potentially damaging to the environment.

56.2 Producing Countries

World reserves of thallium in zinc ore deposits are estimated at 17 million kilograms and are in Canada, Europe, and the United States (Fig. 56.2).

56.3 Applications

Electronics Industry

Thallium is used as a sodium iodide crystal dopant in radioactivity detectors (scintillators) (Fig. 56.3).

The electrical conductivity of thallium sulfide changes when exposed to infrared radiation, so it is used in certain photocells and infrared detectors.

Other Fields

Thallium is employed in the manufacture of glass for lenses and prisms for fiber optics. It is used in distress flares to

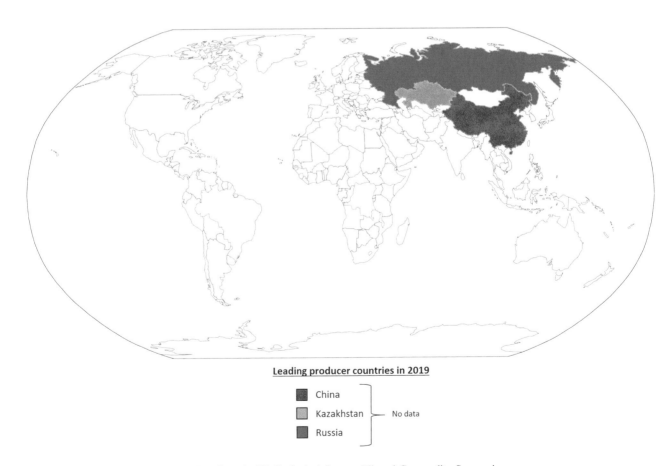

Leading producer countries in 2019

- China
- Kazakhstan
- Russia

No data

Fig. 56.2 List of producing countries based on the US Geological Survey, Mineral Commodity Summaries

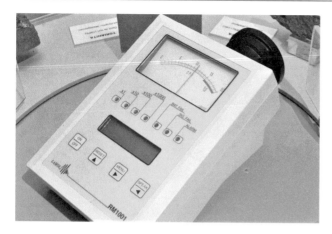

Fig. 56.3 Radioactivity detector. (*Photo* Joaquim Sanz. MGVM)

impart an intense green color. It enhances the density and refractive index of glass.

Thallium salts had been one of the most effective rat poisons known, but its use is currently prohibited due to the high toxicity of this element.

From 1970 to 1980 the thallium-201 isotope (gamma radiation emitter) was used to check the functioning of the myocardium in SPECT scans, but it was replaced by the far more efficient and non-poisonous technetium-99 m (see technetium).

Thallium is a highly potent and carcinogenic poison, even through skin. The best antidote is Prussian Blue (ferric ferrocyanide), which absorbs and neutralizes it. Prussian Blue is also the best antidote to the cesium-137 isotope, one of the radioactive isotopes released into the atmosphere in accidents at nuclear power plants such as Chernobyl and Fukushima.

56.4 Recycling

Thallium recycling is unknown.

Further Reading

Gray T, Mann N (2009) The elements. Black Dog & Leventhal Publishers Inc., New York

Mata JM, Sanz J (2007) Guia d'identificació de minerals. Manresa, Catalonia. Edicions UPC/Parcir (Catalan 2nd paper edition), 262 p. ISBN: 9788483019023. http://hdl.handle.net/2117/90445

New World Encyclopedia. Thallium. https://www.newworldencyclopedia.org/entry/Thallium (last accessed May 2021)

Quadbeck-Seeger H-J (2007) World of the elements: elements of the world. Wiley-VCH Verlag GmbH & Co, Germany

Sanz J, Tomasa O (2017) Elements i Recursos minerals: Aplicacions i reciclatge. Manresa, Catalonia. Zenobita Edicions/Iniciativa Digital Politècnica (Catalan 3rd digital edition). http://hdl.handle.net/2117/105113

Sanz J, Tomasa O (2018) Elementos y Recursos minerales: Aplicaciones y reciclaje. Manresa, Catalonia. Zenobita Edicions/Iniciativa Digital Politècnica (Spanish 1st digital edition). http://hdl.handle.net/2117/123674

Stwertka A (2018) A guide to the elements, 4th edn. Oxford University Press, England

USGS (2007) Mineral resource of the month: Thallium. https://pubs.er.usgs.gov/publication/70045866 (last accessed May 2021)

USGS (2021) Commodity Statistics and Information. Thallium. Available at: https://www.usgs.gov/centers/nmic/thallium-statistics-and-information (last accessed May 2021)

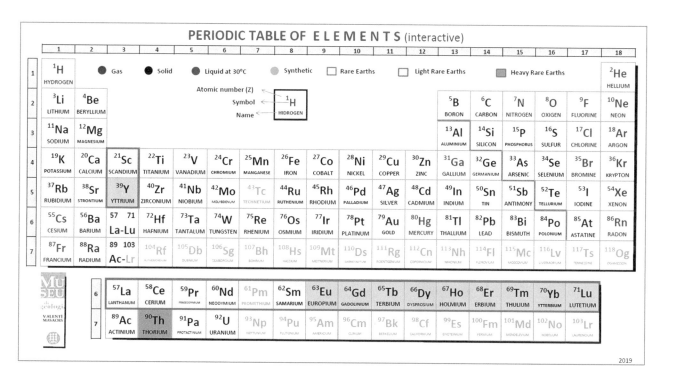

- Radioactive actinide metal
- Soft, ductile, and malleable
- Has a high melting point (2946 °C)
- Found in monazite and thorianite
- Monazite is currently the main ore (Fig. 57.1), superseding thorianite, given the large consumption of that mineral for rare earth extraction.

57.1 Geology

According to the International Atomic Energy Agency (IAEA), thorium does not occur in its metallic form in nature because it is markedly oxyphile, thus occurs as oxide (thorianite (ThO_2)), silicates (thorite ($Th,U)SiO_4$), and phosphates (frequently with rare earth elements).

Deposits can be classified as placers, carbonatites, veins, and alkaline rocks. Placers are formed by weathering, transportation, and wave action. Carbonatite rocks are of

Fig. 57.1 Monacite (cerium, lanthanum, neodymium, and thorium phosphate) *Evje* (*Norway*). (*Photo* Joaquim Sanz. MGVM)

magmatic origin (>50% carbonate minerals), often enriched in magnetite, apatite, fluorite, and accessory Nb–Ta minerals, containing thorium and potentially recovered as a by-product. Vein-type deposits occur close to igneous rocks, often related to carbonatites. They are frequently polymetallic. Alkaline and peralkaline intrusions often have a carbonatite core or plug. Geochemically, these rocks are characterized by an over-saturation of alkali elements and an undersaturation of aluminum.

Nonetheless, the main source of thorium is monazite (Ce, La,Nd,Th)PO$_4$, which is a phosphate of cerium, lanthanum, and neodymium from which it is extracted as a by-product. The thorium content in monazite can reach 26%, but the usual concentration does not exceed 10%. Other common minerals, although with a lower thorium content, are xenotime and zircon.

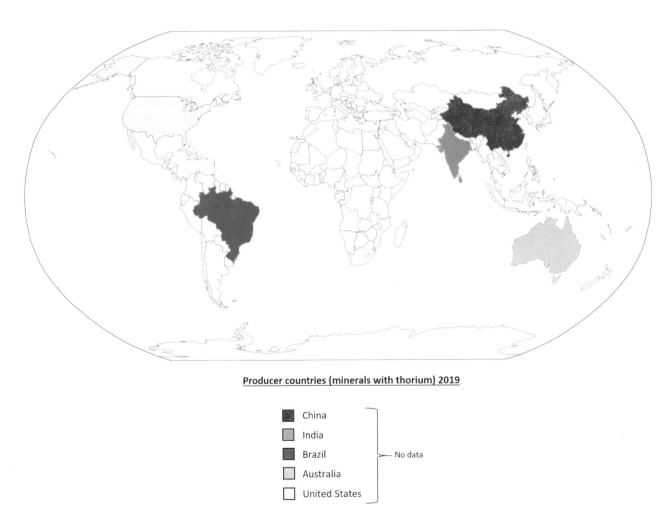

Producer countries (minerals with thorium) 2019

- China
- India
- Brazil
- Australia
- United States

— No data

Fig. 57.2 List of producing countries based on the US Geological Survey, Mineral Commodity Summaries

In India, deposits of placers (sands with heavy minerals) with thorianite may be a great source of thorium.

57.2 Producing Countries

The world's reserves of thorium are associated with the recovery from monazite in placer deposits (sands with heavy minerals) and in rare earth deposits.

The estimated reserves are in India (850,000 t), Brazil (630,000 t), Australia and the United States (600,000 t) (Fig. 57.2).

57.3 Applications

Comment: In general, the use of thorium is being replaced in industry by elements that are less radioactive.

Power Generation

In the Earth's crust, thorium is more abundant than uranium. Thorium-232 can be used as fuel in a suitable nuclear reactor as, although it is not fissile in itself, it absorbs slow neutrons to produce uranium-233, which is indeed fissile. This process reduces the production of long-lived nuclear waste, compared to uranium. This technology is currently being studied in countries such as the United States, Russia, and India, although the economic interest in uranium is slowing its application.

Gas lamp casings now use (non-radioactive) yttrium instead of the (radioactive) thorium that was previously used.

Metallurgical Industry

Thorium is added to magnesium alloys for aircraft engines to increase their resistance to high temperatures (thorium oxide melts at 3300 °C). The TIG (Tungsten Inert Gas) electric welding method for aluminum, magnesium, and stainless-steel alloys use tungsten electrodes with a small amount of thorium oxide (2%) to increase both the current and the life of the electrode and to facilitate the formation and stability of the arc (Fig. 57.3). Cerium (non-radioactive) can substitute thorium with similar results.

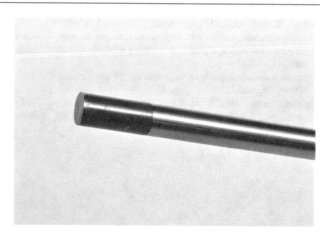

Fig. 57.3 Tungsten-thorium welding electrode. (*Photo* Joaquim Sanz. MGVM)

Optical Industry

Thorium oxide is added to glass to make high-quality lenses in certain scientific cameras and instruments, as it increases the refractive index and decreases the dispersion.

Chemical Industry

Thorium oxide can be used as a catalyst in the conversion of ammonia to nitric acid, in oil refinery and in the production of sulfuric acid.

57.4 Recycling

Thorium recycling is unknown.

Further Reading

Gray T, Mann N (2009) The elements. Black Dog & Leventhal Publishers Inc., New York

Mata JM, Sanz J (2007) Guia d'identificació de minerals. Manresa, Catalonia. Edicions UPC/Parcir (Catalan 2nd paper edition), 262 p. ISBN: 9788483019023. http://hdl.handle.net/2117/90445

Quadbeck-Seeger H-J (2007) World of the elements: elements of the world. Wiley-VCH Verlag GmbH & Co, Germany

Sanz J, Tomasa O (2017) Elements i Recursos minerals: Aplicacions i reciclatge. Manresa, Catalonia. Zenobita Edicions/Iniciativa Digital Politècnica (Catalan 3rd digital edition). http://hdl.handle.net/2117/105113

Sanz J, Tomasa O (2018) Elementos y Recursos minerales: Aplicaciones y reciclaje. Manresa, Catalonia. Zenobita Edicions/Iniciativa Digital Politècnica (Spanish 1st digital edition). http://hdl.handle.net/2117/123674

Stwertka A (2018) A guide to the elements, 4th edn. Oxford University Press, England

Thorium Energy World. Available at: http://www.itheo.org/articles/world%E2%80%99s-first-thorium-reactor-designed (last accessed May 2021)

USGS (2021) Commodity Statistics and Information. Thorium. Available at: https://www.usgs.gov/centers/nmic/thorium-statistics-and-information (last accessed May 2021)

World Nuclear Association. Thorium. Available at: http://www.world-nuclear.org/info/Current-and-Future-Generation/Thorium/ (last accessed May 2021)

PERIODIC TABLE OF ELEMENTS (interactive)

- Malleable, ductile, and heavy metal
- Does not react with oxygen or water, but does react with acids and bases
- Low melting point (232 °C)
- A rare metal
- Obtained from cassiterite (Fig. 58.1)

58.1 Geology

The primary mineral source of tin is cassiterite (SnO_2). Tin's mineralization is commonly associated with igneous intrusion and related ore systems. The classic tin deposit is greisen, an ore deposit characterized by a hydrothermally altered granitic rock. The mineralogical alteration is mostly composed of albite, quartz, and mica, common in European tin deposits. However, there are other tin deposit types, such as vein systems, anorogenic ring complexes, and breccia pipes. In some cases there may be spatial, temporal, and genetic links to classical intrusive ore deposits as porphyry and epithermal systems.

Another important type of cassiterite (SnO_2) deposit is placers formed by the mechanical concentration of resistant minerals of high specific gravity, as in Indonesia. Here these deposits are directly related to local mineralization associated with granite intrusions and represent the world's second largest tin mine.

Fig. 58.1 Cassiterite (tin oxide). *Alt Empordà* (*Catalonia*). (*Photo* Joaquim Sanz. MGVM)

58.2   Producing Countries

Many open-pit mines in the Kiwu area in the north of DR Congo turn into mud pits when the rains fall. To help their families, children and young people work tirelessly digging, washing, classifying, and transporting cassiterite and columbite-tantalite (IPIS—International Peace Information Services).

The world's reserves are in China (1,100 Mt), Indonesia (0.8 Mt), Brazil (0.7 Mt), Bolivia (0.4 Mt), Australia (0.42 Mt), Russia (0.35 Mt), Malaysia (0.25 Mt), the Congo (Kinshasa) (0.15 Mt), and Burma and Peru (0.11 Mt), among others (Fig. 58.2).

58.3   Applications

Metallurgical Industry

Bronze is an alloy of copper and tin (12%). Church and cathedral organ pipes are made of an alloy of 85–75% tin and 15–25% lead (Fig. 58.3). Tin increases the strength of titanium alloys. It is used as a coating for steel sheet (tinned) in the manufacture of cans, which represents 40% of the world's tin consumption; these are mainly used as packaging for food products, paint, oils, etc.

Tin is used in welding, as it adheres to metals at low temperatures. Tin, combined with lead, is used in soldering in the electrical industry, in car manufacture, and in plumbing (Fig. 58.4). Tin and zinc alloys are also used to solder

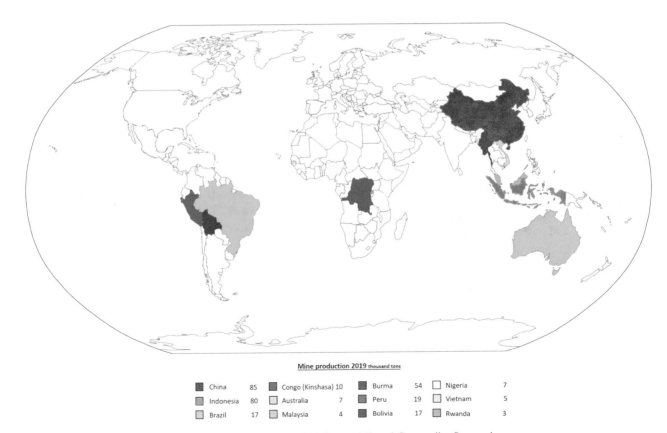

Mine production 2019 thousand tons

China	85	Congo (Kinshasa)	10	Burma	54	Nigeria	7
Indonesia	80	Australia	7	Peru	19	Vietnam	5
Brazil	17	Malaysia	4	Bolivia	17	Rwanda	3

Fig. 58.2 List of producing countries based on the US Geological Survey, Mineral Commodity Summaries

Fig. 58.4 Soldering an integrated electronic circuit. (*Photo* Joaquim Sanz. MGVM)

The use of tin in the anodes of lithium-ion (Li-ion) batteries is also growing, since the tin compounds involved in the manufacture of the anodes impart to these batteries a greater stability (Roskill, 2020).

Tin with indium (ITO) is used in the manufacture of flat screens and, together with gallium, copper, and selenium, is used in the manufacture of solar panels. The allotropic form α-Sn is used in the field of semiconductors.

Medicine

The alloy of gallium, indium, and tin (galinstan) is a substitute for mercury in clinical thermometers.

58.4 Recycling

According to the International Tin Association, in 2018 31% of tin was recovered worldwide from both old, tinned steel cans and electronic circuits, as well as debris generated during industrial production.

Tin can be recycled many times without losing its commercial technical characteristics.

Further Reading

Gray T, Mann N (2009) The elements. Black Dog & Leventhal Publishers Inc., New York

Mata JM, Sanz J (2007) Guia d'identificació de minerals. Manresa, Catalonia. Edicions UPC/Parcir (Catalan 2nd paper edition), 262 p. ISBN: 9788483019023. http://hdl.handle.net/2117/90445

Pirajno F (2009) Hydrothermal processes and mineral systems. Springer

Fig. 58.3 Organ at Montserrat monastery. (*Catalonia*). (*Image by courtesy of* Oleguer Serra)

aluminum; and tin, antimony, and silver alloys are used for applications where a high final mechanical strength is required.

Glass and Ceramics Industry

In the Pilkington glass manufacturing process, the glass floats on a bed of molten tin.

Electronics industry

Nearly 50% of the world's tin consumption is destined for soldering electronic components onto printed circuit boards for conventional mobile phones, smartphones, computers, tablets, smart home devices, and so on. With the introduction of 5G networks, consumption will increase (Roskill, 2020).

Quadbeck-Seeger H-J (2007) World of the elements: elements of the world. Wiley-VCH Verlag GmbH & Co, Germany

Roskill (2020) Market Reports. Tin. Available at: https://roskill.com/market-report/tin/ (last accessed May 2021)

Sanz J, Tomasa O (2017) Elements i Recursos minerals: Aplicacions i reciclatge. Manresa, Catalonia. Zenobita Edicions/Iniciativa Digital Politècnica (Catalan 3rd digital edition). http://hdl.handle.net/2117/105113

Sanz J, Tomasa O (2018) Elementos y Recursos minerales: Aplicaciones y reciclaje. Manresa, Catalonia. Zenobita Edicions/Iniciativa Digital Politècnica (Spanish 1st digital edition). http://hdl.handle.net/2117/123674

Stwertka A (2018) A guide to the elements, 4th edn. Oxford University Press, England

USGS (2021) Commodity Statistics and Information. Tin. Available at: https://www.usgs.gov/centers/nmic/tin-statistics-and-information (last accessed May 2021)

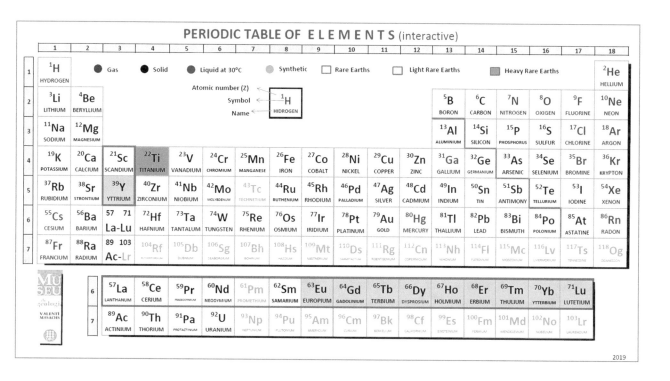

59.1 Geology

- A hard metal and highly resistant to corrosion
- Lighter than steel (40%)
- Stainless and highly biocompatible
- High melting point (1668 °C)
- Discovered by English amateur mineralogist Rev William Gregor in 1791
- Assigned the status of a strategic mineral in September 2020
- Obtained from ilmenite (Fig. 59.1) and rutile.

59.1 Geology

Titanium is the ninth-most abundant element in Earth's crust and, although it is not found as a pure metal in nature, it is found in nearly all rocks and sediments. It has a strong affinity for oxygen, typically forming oxide minerals, mainly ilmenite ($FeTiO_3$) and rutile (TiO_2) but also with other titanium dioxide polymorphs such as anatase and brookite. There are basically types of three titanium-deposit types. The most important are related to coastal shorelines and dunes and older equivalent rocks. Most of the world's titanium supply is from processing ilmenite and rutile in shoreline

Fig. 59.1 Ilmenite (titanium oxide and iron). *Sierra de la Albarrana (Córdoba) Spain.* (*Photo* Joaquim Sanz. MGVM)

(beach) and fluvial (river and stream) deposits of heavy-mineral sand, found along many continental margins.

Other deposit types are related to igneous and metamorphic rocks. Igneous titanium deposits dominated by ilmenite and hemo-ilmenite are hosted by discordant and layered bodies in Proterozoic-age massif anorthosite plutonic complexes. Their primary mineralogy is of anorthosite, charnockite, jotunite, gabbro, norite, or troctolite. Deposits dominated by titaniferous magnetite with minor amounts of ilmenite are hosted in layered and massive intrusions of gabbro, leucogabbro, and norite.

In metamorphic titanium deposits, the recent search for high-grade ore has focused on rutile, because it is the most economically valuable mineral. The only currently active rutile mines in metamorphic rocks are the Daixian rutile deposits in east-central China, consisting of lenses and layers of rutile-bearing high-grade metamorphic rocks that average about double the TiO_2 weight percentage.

59.2 Producing Countries

The world's main reserves of ilmenite are in Australia (250 Mt), China (230 Mt), India (85 Mt),, Brazil (43 Mt), Norway (37 Mt),South Africa (35 Mt), Canada (31 Mt), and Mozambique (14 Mt), among other countries (Fig. 59.2).

For rutile, the countries with significant reserves are Australia (29 Mt), India (7.4 Mt), South Africa (6.1 Mt), and the Ukraine (2.5 Mt), among others (Fig. 59.2).

59.3 Applications

Metallurgical Industry
Titanium is used as an alloying element with iron in steels because it improves grain refinement and acts as a deoxidizing element, and also in stainless steels to reduce the carbon content.

Thanks to its resistance to high temperatures, traction, fatigue, and corrosion, titanium and its alloys are used in the chemical industry, in the military (armor and missiles), in aeronautics (aircraft and spacecraft) (Fig. 59.3), and in the manufacture of ships and submarines.

Chemical Industry
Titanium oxide is widely used as an intensely white pigment and as an opacifier in paper, paint, plastics, food, pharmaceuticals, cosmetics, etc. Since it is used in the manufacture of vehicle paint, its consumption is an indicator of the level of industrial activity in this field. It is also a component of sunscreen.

Construction
Titanium can be used as a building envelope as it is light, resistant, and unalterable. It is a feature of the Guggenheim Museum in Bilbao (Spain) (Fig. 59.4).

Medicine
Titanium is highly biocompatible, non-toxic, and, for the most part, not rejected by the human body (although there have been some cases). It is used in the manufacture of surgical instruments and medical and dental screws and cardiac Pacemakers, in combination with iridium.

Titanium-molybdenum alloy wire (TMA) nickel free, is widely used as an orthodontic archwire. It has good properties between stainless steel and Ni–Ti archwires.

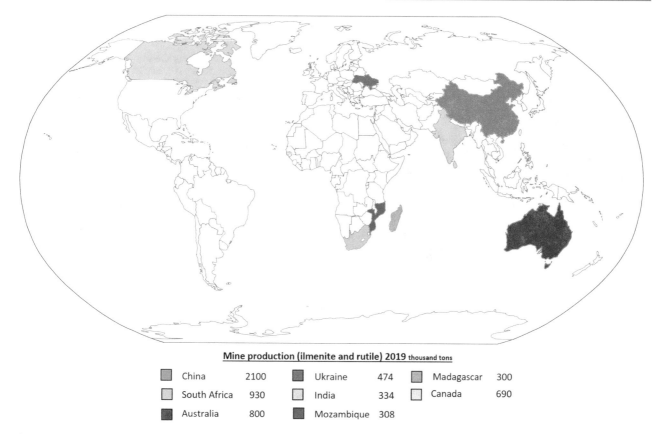

Mine production (ilmenite and rutile) 2019 thousand tons

▫	China	2100	▪	Ukraine	474	▫	Madagascar	300
▫	South Africa	930	▫	India	334	▫	Canada	690
▪	Australia	800	▪	Mozambique	308			

Fig. 59.2 List of producing countries based on the US Geological Survey, Mineral Commodity Summaries

Fig. 59.3 Helicopter engine. (*Photo* Joaquim Sanz. MGVM)

Other Fields

Due to its lightness and resistance it is used in the manufacture of bicycles, wheelchairs, crutches, tennis rackets, glasses, watches, etc. It is also used to build seawater desalination plants and in the manufacture of body piercings.

59.4 Recycling

Titanium is recycled from offcuts from printing, also from remains of the metal already used by various industries.

Fig. 59.4 Cladding on the Guggenheim Museum in Bilbao, Spain. (*Image courtesy of* Joan Closes)

Further Reading

Gray T, Mann N (2009) The elements. Black Dog & Leventhal Publishers Inc., New York

Mata JM, Sanz J (2007) Guia d'identificació de minerals. Manresa, Catalonia. Edicions UPC/Parcir (Catalan 2nd paper edition), 262 p. ISBN: 9788483019023. http://hdl.handle.net/2117/90445

Quadbeck-Seeger H-J (2007) World of the elements: elements of the world. Wiley-VCH Verlag GmbH & Co, Germany

Roskill (2020) Market Reports. Titanium metal. Available at: https://roskill.com/market-report/titanium-metal/ (last accessed May 2021)

Sanz J, Tomasa O (2017) Elements i Recursos minerals: Aplicacions i reciclatge. Manresa, Catalonia. Zenobita Edicions/Iniciativa Digital Politècnica (Catalan 3rd digital edition). http://hdl.handle.net/2117/105113

Sanz J, Tomasa O (2018) Elementos y Recursos minerales: Aplicaciones y reciclaje. Manresa, Catalonia. Zenobita Edicions/Iniciativa Digital Politècnica (Spanish 1st digital edition). http://hdl.handle.net/2117/123674

Stwertka A (2018) A guide to the elements, 4th edn. Oxford University Press, England

Titanium.org. (2021)

USGS (2021) Commodity Statistics and Information. Titanium mineral. Available at: https://www.usgs.gov/centers/nmic/titanium-statistics-and-information (last accessed May 2021)

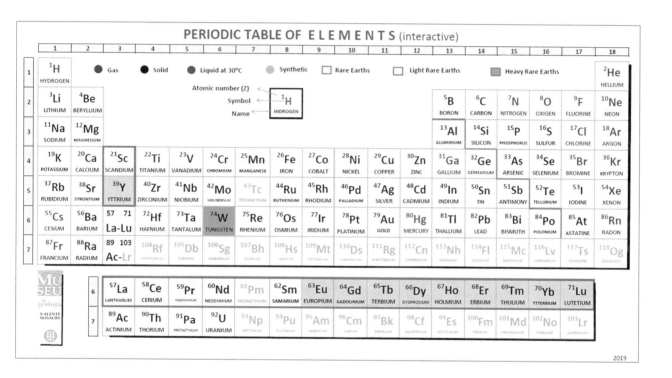

PERIODIC TABLE OF ELEMENTS (interactive)

<!-- periodic table figure -->

2019

- In 1783, brothers Juan José and Fausto de Elhuar discovered this element in wolframite and named it 'wolfram'.
- Currently, the accepted term is tungsten, from the Swedish *tung stem* (heavy stone)
- Very high melting point (3422 °C)
- Can withstand high temperatures
- Very hard, and denser than mercury
- Assigned the status of a strategic metal by the EU in 2017
- Obtained from wolframite, scheelite, and ferberite (Fig. 60.1)

60.1 Geology

The common mineral source of tungsten (W) is wolframite $(Fe^{2+})WO_4$ to $(Mn^{2+})WO_4$. Tungsten's mineralization is commonly associated with igneous intrusion and related ore systems. The classic type of tungsten deposit is greisen, a kind of ore deposit characterized by a hydrothermally altered granitic rock. The mineralogical alteration association is mostly composed of albite, quartz, and mica, very common in European tin-tungsten deposits. However, there are other deposit types, such as vein systems, anorogenic ring complexes, and breccia pipes. In some cases, there may be spatial, temporal, and genetic links with classical intrusive ore deposits as porphyry and epithermal systems.

J. Sanz et al., *Elements and Mineral Resources*, Springer Textbooks in Earth Sciences,
Geography and Environment, https://doi.org/10.1007/978-3-030-85889-6_60

Fig. 60.1 Ferberite (iron tungstate). *Panasqueira (Portugal)*. (Photo Joaquim Sanz. MGVM)

60.2 Producing Countries

The world's most outstanding reserves of tungsten are in China (1900 Mt), followed some way behind by Russia (240,000 t), Vietnam (95,000 t), Spain (54,000 t), the United Kingdom (44,000 t), Austria (10,000 t), and Portugal (3100 t), among others (Fig. 60.2).

60.3 Applications

Electrical Industry
Tungsten, being resistant to high temperatures and having a very low vapor pressure, is practically the only material employed for filaments in projection lamps, halogen, fluorescent, high-intensity discharge lamps, etc., and as an electron-emitting cathode in X-ray tubes.

Metallurgical Industry
Tungsten, together with carbon and cobalt, tungsten carbide (widia), forms a high-strength alloy used for making cutting tools such as milling cutters, drills, saws, drilling bits for oil and gas wells, etc.

The addition of tungsten (10%) increases the total hardness of high-speed steels (HSS) and makes it possible to work them at high temperatures (Fig. 60.3).

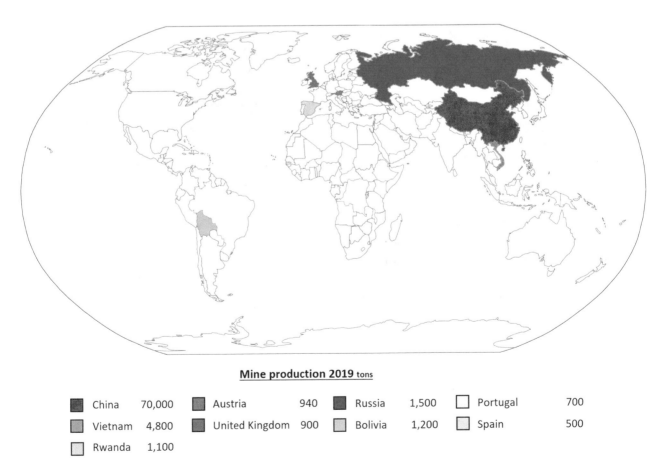

Mine production 2019 tons

China 70,000	Austria 940	Russia 1,500	Portugal 700
Vietnam 4,800	United Kingdom 900	Bolivia 1,200	Spain 500
Rwanda 1,100			

Fig. 60.2 List of producing countries based on the US Geological Survey, Mineral Commodity Summaries

Fig. 60.3 Tungsten steel drill bit. (*Photo* Joaquim Sanz. MGVM)

Fig. 60.4 Tip of a widia ballpoint pen. (*Photo* Joaquim Sanz. MGVM)

The ball at the tip of a ballpoint pen is made of tungsten carbide (widia) (Fig. 60.4).

When tungsten is added to steel, it forms a very strong superalloy used in aerospace, the automotive industry, and power generation turbines.

Tungsten steel is used in the manufacture of diamond saw blades, as it increases their mechanical strength.

Tungsten electrodes are used in TIG (tungsten inert gas) welding in the absence of oxygen due to the high melting point of the element (3422 °C), since the electrode is not consumed during use.

Several types of alloys are used to obtain optimal results in electric arc ignition, stability, and reduction of erosion at the tip. There are pure tungsten and other types alloyed with cerium, lanthanum, zirconium, and thorium, as needed. These last types (with thorium) are being replaced by non-radioactive cerium (see: thorium).

Other Fields

Tungsten is used for fishing weights and shotgun pellets, replacing lead, but it is more expensive and less is now consumed.

Tungsten is a good protector against gamma ionizing radiation, as a substitute for lead, with less thickness and, consequently, less weight. It is used in shielding against radioactive sources for treatments and in collimators for diagnostic and imaging equipment in nuclear medicine.

60.4 Recycling

The level of recycling of metal and tungsten carbides (widia) is very high.

Further Reading

Gray T, Mann N (2009) The elements. Black Dog & Leventhal Publishers Inc., New York

Mata JM, Sanz J (2007) Guia d'identificació de minerals. Manresa, Catalonia. Edicions UPC/Parcir (Catalan 2nd paper edition), 262 p. ISBN: 9788483019023. http://hdl.handle.net/2117/90445

Pirajno F (2009) Hydrothermal processes and mineral systems. Springer

Quadbeck-Seeger H-J (2007) World of the elements: elements of the world. Wiley-VCH Verlag GmbH & Co, Germany

Roskill (2021) Market Reports. Tungsten. Available at: https://roskill.com/market-report/tungsten/ (last accessed May 2021)

Sanz J, Tomasa O (2017) Elements i Recursos minerals: Aplicacions i reciclatge. Manresa, Catalonia. Zenobita Edicions/Iniciativa Digital Politècnica (Catalan 3rd digital edition). http://hdl.handle.net/2117/105113

Sanz J, Tomasa O (2018) Elementos y Recursos minerales: Aplicaciones y reciclaje. Manresa, Catalonia. Zenobita Edicions/Iniciativa Digital Politècnica (Spanish 1st digital edition). http://hdl.handle.net/2117/123674

Stwertka A (2018) A guide to the elements, 4th edn. Oxford University Press, England

USGS (2021) Commodity Statistics and Information. Tungsten. Available at: https://www.usgs.gov/centers/nmic/tungsten-statistics-and-information (last accessed May 2021)

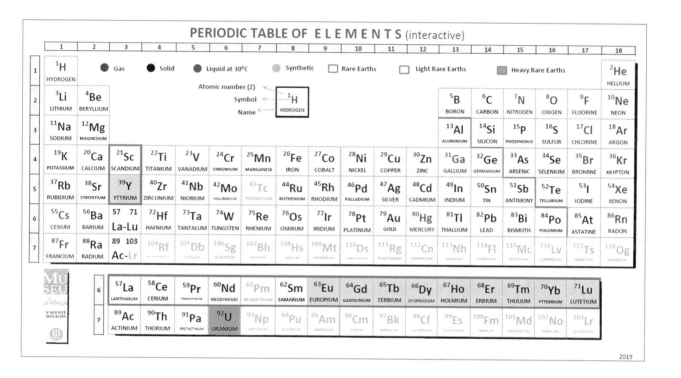

- An actinide metal that emits radioactivity naturally
- Very dense
- In 1789, the German chemist Martin Klaproth found an unknown element in pitchblende
- In 1841, the French chemist Eugène-Melchior Réligot first achieved its isolation
- In 1896, the French physicist Henri Becquerel discovered that uranium was radioactive
- The first radioactive element in the periodic table to be discovered
- Consists of three isotopes: uranium-238; uranium-235; and uranium-234: uranium-235 is the useful fissile isotope

- Much depleted uranium (U-238) (hard and very dense) is obtained as a by-product of uranium-235 enrichment
- Obtained mainly from uraninite, autunite, torbernite (Fig. 61.1), and carnotite.

61.1 Geology

Uranium occurs in a number of geological environments. The major uranium primaries ore minerals are uraninite (UO_2) or pitchblende (U_3O_8). However, a range of other uranium

Fig. 61.1 Torbernite (uranium-copper phosphate). *Don Benito (Badajoz) Spain.* (*Photo* Joaquim Sanz. MGVM)

minerals, such as carnotite ($K_2(UO_2)_2(VO_4)_2 \cdot 3H_2O$) or brannerite ($(U,Ca,Y,Ce)(Ti,Fe)_2O_6$ and secondary like gummite (secondary uranium oxides (yellow-orange)), autunite ($Ca(UO_2)_2(PO_4)_2.10\text{-}12H_2O$), torbernite ($Cu(UO_2)_2(PO_4)_2.8\text{-}12H_2O$), or saleeite ($Mg(UO_2)_2(PO_4)_2.10H_2O$) are found in particular deposits (IAEA 2021).

Uranium deposits can be grouped into 15 types. However, we are focused on the economic types: polymetallic iron-oxide; sandstone; Paleoproterozoic quartz-pebble conglomerate; and surficial uranium deposits.

The world's largest uranium ore deposit, Olympic Dam, is of the polymetallic iron-oxide variety and is recovered as a by-product with copper. The mine is located in a hematite-rich granite breccia complex overlain by approximately 300 m of flat-lying sedimentary rocks of the Stuart Shelf geological province in Australia.

Sandstone uranium deposits represent about 28% of the worldwide assured uranium resources. They are the main deposits of economic importance in Kazakhstan, Uzbekistan, the United States and Niger. They are found in sandstones deposited in a continental fluvial or marginal marine sedimentary environment. Uranium is precipitated under reducing conditions. Impermeable shale or mudstone units usually occur immediately above and below the mineralized area. The most well-known examples are the Kazakh (e.g. Budenovskoye, Tortkuduk, and Moynkum, among others) or Nigerian (e.g. Akouta, Arlit, and Imourare, among others) ore deposits.

Detrital uranium may occur in some Archean–Early Paleoproterozoic quartz-pebble conglomerate deposits. These deposits originated in fluvial transport of detrital uraninite and can be related to rare earth elements (REE) as well as thorium. They may occur unconformably, overlying a granitic and metamorphic basement. An example of this type of ore deposit is Elliot Lake in Canada.

As suggested by their name, surficial uranium deposits have the mineralization nearly at the surface, and they are shallow ore deposits. They were formed by weathering processes enhanced by an arid or semi-arid climate. The typical mineralization in this type is carnotite ($K_2(UO_2)_2(VO_4)_2 \cdot 3H_2O$)); however, other uranium minerals can be found together with calcrete, calcite, gypsum, and dolomite, among others. The Langer Heinrich in Namibia is the largest surficial uranium ore deposit.

61.2 Producing Countries

As they represent strategic information, there is little information on world reserves. However, countries such as Kazakhstan, Canada, and Australia all have uranium, and it is understood that countries such as Niger, Russia, and Namibia have good reserves (Fig. 61.2).

61.3 Applications

Enriched uranium (with a uranium-235 content of between 3 and 10%).

Power Generation

The main application of enriched uranium is as a nuclear fuel to generate heat to make electrical energy. One kilogram of uranium-235 can produce 80 terajoules of energy, equivalent to 3000 tons of coal (Fig. 61.3).

Enriched uranium (with a uranium-235 content of between 20 and 90%).

Military Industry

Uranium is basically used to make atomic bombs and nuclear warheads. Also, it serves as the propellant for atomic submarines.

Depleted uranium (with a uranium-235 content of about 0.7%; very low-level radioactivity).

Military Industry

Depleted uranium is mainly used in the manufacture of anti-armor projectiles because, upon impact, it punctures the armored vehicle, ignites it, and becomes volatile inside. It is used as a counterweight for military missiles and aircraft. In addition, it is used to manufacture armor for military vehicles (tanks).

Depleted uranium, by virtue of its high density, is a suitable material for shielding against gamma radiation; this is why it is used to make containers for storing and transporting radioactive materials.

Mine production 2019 tons (world nuclear association)

■ Kazakhstan 22,808	■ Russia 2,911	■ Australia 6,613	□ China 1,885
▨ Canada 6,938	▨ Uzbekistan 2,404	▨ Namibia 5,476	▨ Ukraine 801
■ Niger 2,983	□ United States 582		

Fig. 61.2 List of producing countries based on the IAEA: International Atomic Energy Agency

Fig. 61.3 Nuclear power station. (*Image courtesy of* Javier Castelo)

Medicine

Depleted uranium is used as a protective container for the radioactive sources employed in the treatment of cancers.

Other Fields

Radium ($RaCl_2$) can be obtained by electrolysis from uranium production.

61.4 Recycling

Spent uranium and plutonium pellets from nuclear power plants are reprocessed for use as new fuel (MOX).

Reference

IAEA (2021) World Uranium Geology, Exploration, Resources and Production. https://www.iaea.org/publications/14687/world-uranium-geology-exploration-resources-and-production

Further Reading

Dahlkamp FJ (2013) Uranium ore deposits. Springer Science & Business Media. https://doi.org/10.1007/978-3-662-02892-6

Gray T, Mann N (2009) The elements. Black Dog & Leventhal Publishers Inc., New York

IAEA: International Atomic Energy Agency (2021) http://www.iaea.org (last accessed May 2021)

Mata JM, Sanz J (2007) Guia d'identificació de minerals. Manresa, Catalonia. Edicions UPC/Parcir (Catalan 2nd paper edition), 262 p. ISBN: 9788483019023. http://hdl.handle.net/2117/90445

Pirajno F (2009) Hydrothermal processes and mineral systems. Springer

Quadbeck-Seeger H-J (2007) World of the elements: elements of the world. Wiley-VCH Verlag GmbH & Co, Germany

Sanz J, Tomasa O (2017) Elements i Recursos minerals: Aplicacions i reciclatge. Manresa, Catalonia. Zenobita Edicions/Iniciativa Digital Politècnica (Catalan 3rd digital edition). http://hdl.handle.net/2117/105113

Sanz J, Tomasa O (2018) Elementos y Recursos minerales: Aplicaciones y reciclaje. Manresa, Catalonia. Zenobita Edicions/Iniciativa Digital Politècnica (Spanish 1st digital edition). http://hdl.handle.net/2117/123674

Stwertka A (2018) A guide to the elements, 4th edn. Oxford University Press, England

United States Nuclear Regulatory Commission. Uranium Enrichment (2020). Available at: http://www.nrc.gov/materials/fuel-cycle-fac/ur-enrichment.html (last accessed May 2021)

World Nuclear Association (2021) https://www.world-nuclear.org/ (last accessed May 2021)

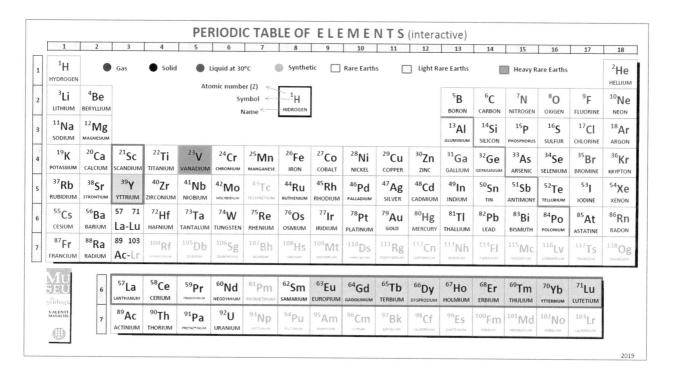

- Ductile and highly corrosion-resistant metal
- Protected against further oxidation by a thin layer of oxide
- High resistance to both acids and bases
- Discovered in 1801 by Mexican professor of mineralogy Andrés Manuel del Rio
- Assigned the status of a strategic metal by the EU in 2017
- Obtained from magnetites rich in titanium and vanadium (Fig. 62.1)
- Vanadium is also present in bauxite deposits, crude oil, coal, and oil shale.

62.1 Geology

Vanadium occurs in nature in a wide variety of minerals. The following four main types of mineral deposits are recognized: vanadiferous titanomagnetite (VTM); sandstone-hosted vanadium (SSV); shale-hosted; and vanadates.

VTM deposits are found throughout the world and are the principal source of vanadium. The most economically significant are the Bushveld Complex in South Africa, the Panzhihua layered intrusion in China, the the Kachkanar massif in the Ural Mountains in Russia, Windimurra Complex in Western Australia, and the Bell River Complex

Fig. 62.1 Magnetite with titanium and vanadium. (*South Africa*). (*Photo* Joaquim Sanz. MGVM)

(Matagami deposit) and Lac Doré Complex in Quebec, Canada.

These deposits consist of magmatic accumulations of magnetite and ilmenite, being more than about 1% rutile. They commonly contain 0.2–1% V_2O_5 but some zones, for example the Bushveld Complex, contain more than 1.5%. VTM deposits are hosted mainly within mafic and ultramafic igneous rocks, most commonly anorthosite and gabbro. Some vanadiferous deposits are hosted in zoned mafic to ultramafic complexes with high levels of chromium and platinum-group elements.

SSV have been identified on all continents, and many are known to have vanadium enrichment. These deposits of vanadium- and uranium-bearing sandstone have average reserves and ore grades ranging from 0.1 to 1 weight percentage of vanadium. The United States has been and is currently the main producer of vanadium from SSV deposits, particularly from the Colorado Plateau.

Shale-hosted vanadium deposits are metalliferous black shales from primarily Late Proterozoic and Phanerozoic marine successions. The term 'shale' covers a range of carbonaceous rocks that include marls and mudstones. They typically contain high concentrations of organic matter (greater than 5%) and reduced sulfur (greater than 1%, mainly as pyrite), as well as a suite of metals such as copper, molybdenum, nickel, PGMs, silver, uranium, vanadium, and zinc. Concentrations regularly exceed 0.18% V_2O_5 and can be as high as 1.7% V_2O_5. Vanadium concentrations correlate to organic carbon in black shales, suggesting that the vanadium is incorporated into organic matter upon its burial. Vanadiferous black shales are commonly found in North America.

Vanadates of lead, zinc, and copper form in the oxidized zones of base-metal deposits, especially in areas of

arid climate and deep oxidation. The copper-lead–zinc vanadate ores in the Otavi Mountain of northern Namibia were once considered to be among the largest vanadium deposits in the world. Other areas with known vanadate deposits include Angola, South Africa, Zambia, and Zimbabwe. Small deposits occur in Argentina, Mexico, and the United States (Arizona, California, Nevada, and New Mexico), but are unlikely ever to be economically significant resources.

62.2 Producing Countries

The world reserves are in the following countries: China (9.5 Mt), Russia (5 Mt), Australia (4 Mt), South Africa (3.5 Mt), Brazil (0.1 Mt), among others (Fig. 62.2).

62.3 Applications

Metallurgical Industry

The main application of this metal is for the ferrovanadium alloy used as an additive in the manufacture of steels to improve the mechanical properties of iron: tools, vehicle axels, knives, gears, vehicle armor, etc.(Fig. 62.3).

Ferrovanadium is being used more and more in the manufacture of the slab foundations of reinforced concrete for buildings in countries with a high risk of earthquakes, so that they can better withstand the movements and any fires caused by seismic activity (Fig. 62.4).

Roskill (2019) estimates that the use of ferrovanadium is being superseded by ferroniobium due to a differential in prices.

Vanadium, added to titanium and aluminum alloys, enhances their mechanical strength and provides stability at high temperatures. Such alloys are used in aircraft engine manufacture.

Battery Industry

Research continues on lithium-vanadium phosphate batteries, which have a higher energy density, more power, greater safety, and shorter recharge time than lithium-cobalt batteries for both electric vehicles and portable electronics.

Chemical Industry

The second main application of vanadium is in vanadium pentoxide, which is a chemical catalyst (without being consumed); i.e. it speeds up the chemical reaction in obtaining sulfuric acid in both the contact and the wet process.

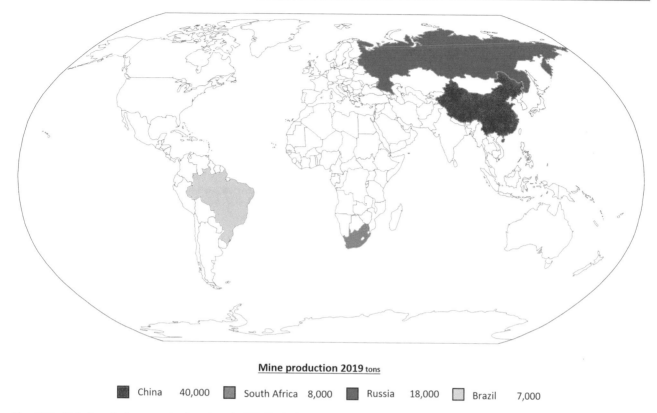

Mine production 2019 tons

■ China 40,000 ■ South Africa 8,000 ■ Russia 18,000 □ Brazil 7,000

Fig. 62.2 List of producing countries based on the US Geological Survey, Mineral Commodity Summaries

Fig. 62.3 Chrome-vanadium steel wrench. (*Photo* Joaquim Sanz. MGVM)

Fig. 62.4 Vanadium steel forging. (*Photo* Joaquim Sanz. MGVM)

62.4 Recycling

The recycling of vanadium is carried out by recovery from steels containing it and from catalysts used in chemical processes, these being the main source.

According to Roskill (2019), the recovery of vanadium from the spent catalysts produced by oil refineries will increase, as it will as a by-product of burning coal and oil shales.

Further Reading

Critical Mineral Resources of the United States (2017) Economic and Environmental Geology and Prospects for Future Supply Vanadium. https://pubs.usgs.gov/pp/1802/u/pp1802u.pdf

Gray T, Mann N (2009) The elements. Black Dog & Leventhal Publishers Inc., New York

Largo Resources. Operations. Maracás Menchen Mine (2020). https://largoresources.com/ (last accessed May 2021)

Mata JM, Sanz J (2007) Guia d'identificació de minerals. Manresa, Catalonia. Edicions UPC/Parcir (Catalan 2nd paper edition), 262 p. ISBN: 9788483019023. http://hdl.handle.net/2117/90445

Quadbeck-Seeger H-J (2007) World of the elements: elements of the world. Wiley-VCH Verlag GmbH & Co, Germany

Roskill (2019) Market Reports. Vanadium. Available at: https://roskill.com/market-report/vanadium/ (last accessed May 2021)

Sanz J, Tomasa O (2017) Elements i Recursos minerals: Aplicacions i reciclatge. Manresa, Catalonia. Zenobita Edicions/Iniciativa Digital Politècnica (Catalan 3rd digital edition). http://hdl.handle.net/2117/105113

Sanz J, Tomasa O (2018) Elementos y Recursos minerales: Aplicaciones y reciclaje. Manresa, Catalonia. Zenobita Edicions/Iniciativa Digital Politècnica (Spanish 1st digital edition). http://hdl.handle.net/2117/123674

Stwertka A (2018) A guide to the elements, 4th edn. Oxford University Press, England

USGS (2021) Commodity Statistics and Information. Vanadium. Available at: https://www.usgs.gov/centers/nmic/vanadium-statistics-and-information (last accessed May 2021)

PERIODIC TABLE OF E L E M E N T S (interactive)

- A fragile metal
- Low melting point (420 °C)
- A thin layer of rust protects it from further oxidation
- Essential element for the human body
- Obtained mainly from sphalerite (Fig. 63.1) and smithsonite.

63.1 Geology

Zinc is most commonly found in hydrothermal veins. Zinc and lead are often found together, less so with copper and iron, in several kinds of ore deposits. Metals precipitate from ore fluids by various processes, depending on the specific local conditions. The most common are cooling, mixing with other fluids, and changes in pH. Zinc is most commonly obtained from sphalerite (ZnS) in assemblages with other sulfides like galena (PbS), chalcopyrite ($CuFeS_2$), pyrite

J. Sanz et al., *Elements and Mineral Resources*, Springer Textbooks in Earth Sciences,
Geography and Environment, https://doi.org/10.1007/978-3-030-85889-6_63

Fig. 63.1 Sphalerite (zinc sulfide). *Vilaller (Catalonia)*. (*Photo* Joaquim Sanz. MGVM)

(FeS$_2$), and pyrrhotite (FeS). Basinal hydrothermal systems include Mississippi Valley-type (MVT) and sedimentary exhalative (SEDEX) deposits. The MVT deposits are so named because they were first found in the mid-west United States in the valley of the Mississippi River. These deposits are tectonically related to enrichment of high-salinity metal fluids from sedimentary basins into carbonate platforms. Sedimentary exhalative (SEDEX) deposits consist of layers of lead–zinc-iron sulfides from sedimentary processes, and are found within large ancient sedimentary basins. Famous SEDEX deposits include Broken Hill, Mount Isa, and McArthur River in Australia and Sullivan in British Columbia. Zinc is also found in volcanogenic massive sulfide (VMS) deposits and skarn deposits. Typical zinc-lead VMS and skarn deposits are found at Francisco I. Madero and Velardeña mine in Mexico.

63.2   Producing Countries

The world's reserves are in Australia (68,000 t), China (44,000 t), Mexico and Russia (22,000 t), Peru (19,000 t), Kazakhstan (12,000 t), the United States (11,000 t), India (7500 t), Bolivia (4800 t), Sweden (3600 t), and Canada (2200 t), among others (Fig. 63.2).

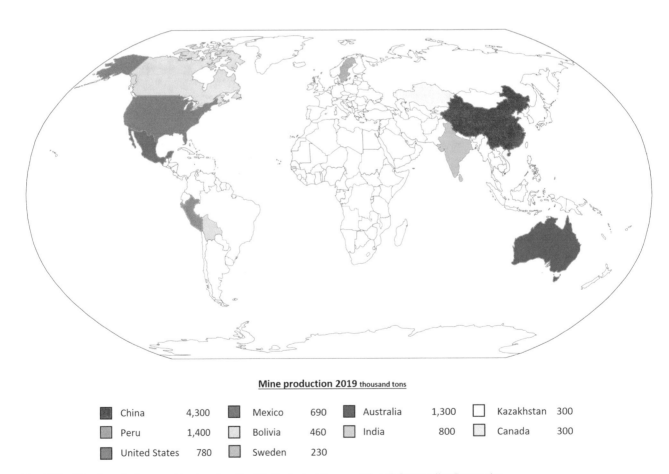

Mine production 2019 thousand tons

■ China	4,300	■ Mexico	690	■ Australia	1,300	□ Kazakhstan	300
■ Peru	1,400	□ Bolivia	460	■ India	800	□ Canada	300
■ United States	780	□ Sweden	230				

Fig. 63.2 List of producing countries based on the US Geological Survey, Mineral Commodity Summaries

Fig. 63.4 The *fiscorn*, the Catalan bass-flugelhorn, a brass musical instrument. (*Photo* Joaquim Sanz. MGVM)

Fig. 63.3 Galvanized steel post for traffic lights. (*Photo* Joaquim Sanz. MGVM)

63.3 Applications

Metallurgical Industry

The main application of zinc, accounting for approximately 55% of consumption, is as an anti-corrosion agent in galvanizing iron and steel (Fig. 63.3).

The second main application of zinc is in the manufacture of zamak alloy, which contains zinc as a base metal alloyed with aluminum, magnesium, and copper. This is used in the manufacture of molds for die casting. Another important application is the manufacture of brass (an alloy of copper and zinc) for making musical instruments, faucets, and decorative accessories (Fig. 63.4).

Zinc is used in the manufacture of sacrificial anodes, which are pieces of this metal that are electrically connected to steel structures such as bridges, railways, and ships' hulls. The zinc dissolves away until it eventually disappears; meanwhile, the steel is protected from oxidation.

Battery Industry

Zinc chloride, together with carbon and manganese dioxide, forms the classic dry cell batteries that are economical but do not allow for very high energy consumption, which are now being displaced by alkaline or Ni-MH batteries. Tiny zinc-air batteries are the power source for hearing aids (oxygen reacts with the zinc to generate energy).

Other Fields

Zinc oxide is used in the manufacture of some deodorants and also white paint, rubber, and suncream.

Sulfates, oxides, and chelates of zinc are used in agriculture as fertilizers, in both curative and preventive veterinary medicine, and in cereal crops and fruit trees to treat nutrient deficiencies.

63.4 Recycling

The main source is chemical residue from galvanization, galvanized steel scrap, and brass offcuts.

Further Reading

Gray T, Mann N (2009) The elements. Black Dog & Leventhal Publishers Inc., New York

Mata JM, Sanz J (2007) Guia d'identificació de minerals. Manresa, Catalonia. Edicions UPC/Parcir (Catalan 2nd paper edition). 262 p. ISBN: 9788483019023. http://hdl.handle.net/2117/90445

Pirajno F (2009) Hydrothermal processes and mineral systems. Springer

Quadbeck-Seeger H-J (2007) World of the elements: elements of the world. Wiley-VCH Verlag GmbH & Co., Germany

Sanz J, Tomasa O (2017) Elements i Recursos minerals: Aplicacions i reciclatge. Manresa, Catalonia. Zenobita Edicions/Iniciativa Digital Politècnica (Catalan 3rd digital edition). http://hdl.handle.net/2117/105113

Sanz J, Tomasa O (2018) Elementos y Recursos minerales: Aplicaciones y reciclaje. Manresa, Catalonia. Zenobita Edicions/Iniciativa Digital Politècnica (Spanish 1st digital edition). http://hdl.handle.net/2117/123674

Stwertka A (2018) A guide to the elements, 4th edn. Oxford University Press, England

USGS (2021) Commodity statistics and information. Zinc. Available at https://www.usgs.gov/centers/nmic/zinc-statistics-and-information. Last accessed May 2021

Zinc.org (2021) http://www.zinc.org. Last accessed May 2021

PERIODIC TABLE OF ELEMENTS (interactive)

Fig. 64.1 Periodic table

2019

- Hard, resistant, and abrasive metal
- Highly resistant to corrosion
- Refractory and withstands high temperatures (1855 °C)
- It is very difficult to extract pure zirconium, as hafnium is always present as an impurity
- Obtained from zircon (Fig. 64.1) between the minerals of heavy sands and as a by-product of treatment of minerals such as ilmenite and rutile in the titanium extraction process.

64.1 Geology

Zircon ($ZrSiO_4$) is the most common naturally occurring zirconium-bearing mineral. Most zircon forms as a product of primary crystallization in igneous rocks, and it is always associated with hafnium. The world's largest primary deposit of zirconium associated with alkaline igneous rocks is in a single locality on the Kola Peninsula of Murmanskaya Oblast, Russia, where baddeleyite (zirconium oxide) is recovered as a by-product of mining apatite and magnetite. However, it is a special case because at present there are few primary igneous deposits of zirconium-bearing minerals of economic value.

Fig. 64.1 Zircon (zirconium silicate). *Madagascar* . (*Photo* Joaquim Sanz. MGVM)

minerals such as ilmenite and rutile (for titanium), chromite (for chromium), and monazite (for rare earth elements) in sedimentary systems, particularly in coastal environments. In coastal deposits, heavy-mineral enrichment occurs when sediment is repeatedly reworked by wind, waves, currents, and tidal processes. The resulting heavy-mineral sand deposits, called placers or paleoplacers, form preferentially at relatively low latitudes on passive continental margins and supply the entirety of the world's zircon.

Zircon makes up a relatively small percentage of economic heavy minerals in most deposits, and is produced primarily as a by-product of mining heavy-mineral sand for titanium minerals.

64.2   Producing Countries

Due to its economic viability, the main zirconium ore deposits worldwide are heavy-mineral sands produced by the weathering and erosion of pre-existing rocks and its concentration with other economically important heavy

The world's most notable reserves are in Australia (42,000 t), and some way behind come South Africa (6500 t) and Mozambique (1800 t), among other countries (Fig. 64.2).

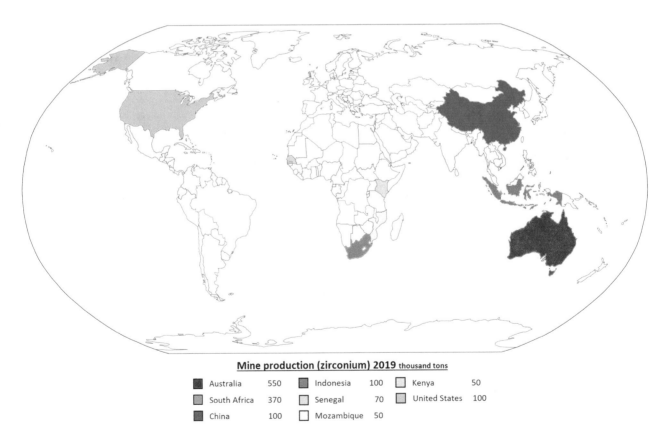

Mine production (zirconium) 2019 thousand tons

■ Australia	550	■ Indonesia	100	☐ Kenya	50
☐ South Africa	370	☐ Senegal	70	☐ United States	100
■ China	100	☐ Mozambique	50		

Fig. 64.2 List of producing countries based on the US Geological Survey, Mineral Commodity Summaries

64.3 Applications

Power Generation

There is a high level of consumption of zirconium in the manufacture of nuclear reactor fuel-element rods in the form of zircaloy (an alloy of zirconium and other metals), due to its high corrosion resistance and low neutron capture.

Ceramic Industry

The main consumption of zirconium is in the manufacture of glazed ceramic tiles, which are highly resistant to heat while acting as an opacifier.

The parts of gas turbine engines that must withstand high temperatures in order to produce more electricity and less CO_2 are protected by a thin, highly refractory ceramic layer consisting of zirconium and yttrium oxide.

Zirconium oxide, together with cerium oxide, forms part of the ceramic support for vehicle catalytic converters, enhancing their resistance to high temperatures (Fig. 64.3).

Jewelry

When it has a good color and transparency, zirconium is used in jewelry as a gemstone. Synthetic cubic zirconium oxide (zirconite) also is also as an imitation gem. It has a hardness of 8.5 on the Mohs scale and a high refractive index, and its appearance is similar to that of a diamond.

Medicine

Zirconium oxide (zirconia) is used in the manufacture of dental implants and in ceramic prostheses to correct deformations of the big toe joint (hallux rigidus) (Fig. 64.4), as it is biocompatible.

Other Fields

When zirconite and alumina are melted, fused zirconia-alumina is obtained. This is used as a

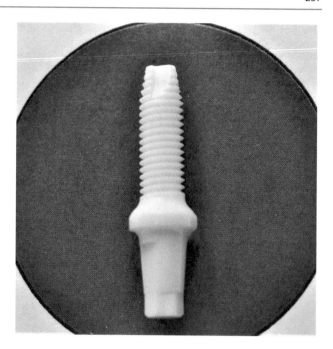

Fig. 64.4 Zirconium oxide ceramic prostheses. (*Photo* Joaquim Sanz. MGVM)

high-performance refractory, with chemical stability and resistance to mechanical and thermal shock, widely used in linings of spacecraft exposed to the high temperatures encountered upon re-entry to the Earth's atmosphere.

Zirconium is also employed in the manufacture of abrasive disks, mechanical bearings, kitchen knives, etc.

Zirconium acetate is used to make waterproof textiles and fibers.

64.4 Recycling

Recycled zirconium is recovered from scrap produced during the treatment and manufacture of elements with this metal and from used or worn-out refractory molds.

Further Reading

Alkane.com.au. Zirconium (2021) http://www.alkane.com.au/products/zirconium/. Last accessed May 2021

Critical Mineral Resources of the United States (2017) Economic and environmental geology and prospects for future supply. Available at https://pubs.usgs.gov/pp/1802/v/pp1802v.pdf

Gray T, Mann N (2009) The elements. Black Dog & Leventhal Publishers Inc., New York

Mata JM, Sanz J (2007) Guia d'identificació de minerals. Manresa, Catalonia. Edicions UPC/Parcir (Catalan 2nd paper edition). 262 p. ISBN: 9788483019023. http://hdl.handle.net/2117/90445

Pirajno F (2009) Hydrothermal processes and mineral systems. Springer

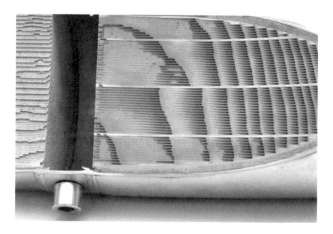

Fig. 64.3 Ceramic support of a vehicle's catalytic converter. (*Photo* Joaquim Sanz. MGVM)

Quadbeck-Seeger H-J (2007) World of the elements: elements of the world. Wiley-VCH Verlag GmbH & Co., Germany

Sanz J, Tomasa O (2017) Elements i Recursos minerals: Aplicacions i reciclatge. Manresa, Catalonia. Zenobita Edicions/Iniciativa Digital Politècnica (Catalan 3rd digital edition). http://hdl.handle.net/2117/105113

Sanz J, Tomasa O (2018) Elementos y Recursos minerales: Aplicaciones y reciclaje. Manresa, Catalonia. Zenobita Edicions/Iniciativa Digital Politècnica (Spanish 1st digital edition). http://hdl.handle.net/2117/123674

Stwertka A (2018) A guide to the elements, 4th edn. Oxford University Press, England

USGS (2019) Commodity statistics and information. Zirconium (2021). Available at https://www.usgs.gov/centers/nmic/zirconium-and-hafnium-statistics-and-information. Last accessed May 2021

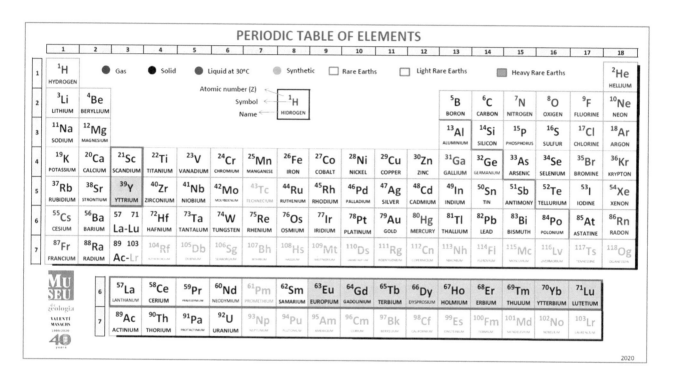

PERIODIC TABLE OF ELEMENTS

- Elements in the lanthanide series, plus scandium and yttrium, are known as the rare earths and are either light and heavy
- Found mainly in monazite (Fig. 65.1), bastnäsite, xenotime, and heavy rare earth-impregnated laterite clays from southeast China (Fig. 65.2)

At the end of 2019, China remained the world leader in rare earth production and refinery, responsible for a proportion in the order of 77% (Roskill 2021). Australia is some way behind in terms of production, then come countries such as the United States, Myanmar, Russia, India, and Thailand, among others.

Increasingly, countries are investigating their potential deposits of rare earths in an attempt to become less dependent on China.

Lightweight Rare Earths (LREE)
Lanthanum (La), cerium (Ce), praseodymium (Pr), neodymium (Nd), and samarium (Sm) were assigned the status of strategic metals by the EU in 2020.

Heavy Rare Earths (HREE)
Europium (Eu), gadolinium (Gd), terbium (Tb), dysprosium (Dy), holmium (Ho), erbium (Er), tullium (Tm), ytterbium (Yb) and lutetium (Lu) were also assigned the status of strategic metals by the EU in 2020.

© The Author(s), under exclusive license to Springer Nature Switzerland AG 2022
J. Sanz et al., *Elements and Mineral Resources*, Springer Textbooks in Earth Sciences, Geography and Environment, https://doi.org/10.1007/978-3-030-85889-6_65

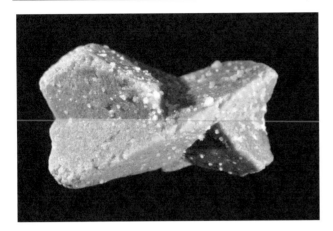

Fig. 65.1 Monazite (cerium, lanthanum, yttrium, and neodymium phosphate). *Minas Geraes* (*Brazil*). (*Photo* Joaquim Sanz. MGVM)

Fig. 65.2 Clay with rare earths. (*China*). (*Photo* Joaquim Sanz. MGVM)

Other Elements (Found in Other Minerals)

Yttrium (Y) and scandium (Sc), also assigned the status of strategic metals by the EU in 2020, strictly speaking are not rare earths; however, their physicochemical properties are very similar to those of rare earths, therefore they are included in this group.

65.1 Geology

The variations are substantial within the REE class of ore deposits, and the formation of an REE ore deposit gives little information about its classification. Besides, classifying REE ore deposits on the basis of only their genetic evolution would quickly induce misinterpretations. For instance, the class of copper-gold-uranium-REE-iron (IOCG) includes the well-known Olympic Dam deposit (South Australia), the iron deposits of Kiruna (Sweden), the iron-REE deposits of

Box Bixby and Pea Ridge (Missouri, United States) and possibly the REE-rich Bayan Obo (Mongolia), the Palabora carbonatite-hosted copper, and the Vergenoug iron-fluorine deposit (South Africa). When examined in detail, these deposits are remarkably different.

There are several types of natural (primary) REE resources, including those formed by high-temperature geological processes (carbonatites, alkaline rocks, vein, and skarn deposits) and those formed by low-temperature processes (placers, laterites, bauxites, and ion-adsorption clays) (Goodenough et al. 2018).

Clay-fixing REE by adsorption in clays (Aagaard 1974) in the south of China comprises the main source of world-wide resources (ion-adsorption clays) (Bao and Zhao 2008). The various deposits produce different types of REE: weathering of two mica granites produces laterites rich in Y-HREE, whilst weathering of biotitic granites generates laterites rich in LREE (Chi and Tian 2008).

65.2 Applications

Magnet Manufacturing

Rare earth magnets (such as Nd-Fe-B neodymium, combined with praseodymium and dysprosium) are the most powerful and smallest permanent magnets available. They are used in the electric motors of hybrid cars and in the generators of wind turbines, achieving improved performance in power generation from wind. They are also used in magnetic resonance equipment, air-conditioning systems, elevator motors, and in electronic devices with a small volume that need strong magnets, such as small headphones (for media players, smartphones), speakers, microphones, etc.

Battery Industry

Rare earth minerals are the key component of the nickel-metal hydride (lanthanum) (Ni-MH) batteries used in hybrid cars and some electric vehicles, and in laptops, etc.

Automotive Industry

Rare earth minerals play a vital role in the manufacture of many vehicles' exhaust catalytic converters.

Electric and Electronic Industry

Rare earth minerals are used in the manufacture of LEDs and as phosphorophores in plasma screens, LCDs, energy-saving bulbs, and fluorescent tubes (phosphorus: compound that presents luminescence when excited by electrons) (with properties of the element of phosphorus).

Glass and Ceramics Industry

Rare earth minerals are used for high-quality polishing of crystals, and are also absorbent of ultraviolet light. They are used as colorants and opacifiers in classic ceramics and in technical ceramics to give them heat resistance, low weight, and great hardness. They are also used in the preparation of glazes.

Medicine

Rare earth minerals are used in YAG lasers for medical purposes (ophthalmology, dermatology, gynecology, oral surgery, and urology).

Other Uses

Rare earth minerals are involved in crude oil distillation processes.

65.3 Recycling

Neodymium and dysprosium from wind turbine magnets, electric motors, and hard disks are already being recycled (Santoku Corporation), as are lanthanum and nickel from Ni-MH batteries (Umicore).

At its La Rochelle factory (France), the Solvay company started recycling fluorescent tubes and energy-saving lamps in 2011 due to their rare earth content, such as cerium, lanthanum, europium, terbium, gadolinium, and yttrium. However, in 2016, this type of recycling ceased due to the fall in rare-earth prices (compared to 2011), the reduction in the supply of used fluorescent tubes (due to the progressive introduction of LED-based lighting), and the stabilization of rare earth exports from China.

Research is continuing into methods of recovering these valuable items of old electronic equipment ('urban mining') as, although the content is very low, expensive, and difficult to extract, growing demand for new equipment and the risks to the supply from the world's leading producer, China, justify the recovery processes and costs, the more so in view of the expected increase in electric vehicles.

At several universities, research is being conducted into using microorganisms to 'digest' metals that are present in small amounts on many electronic circuit boards (e.g. in smartphones, tablets, and computers). The Universitat Politècnica de Catalunya (UPC) (Polytechnic University of Catalonia) (EPSEM) is working on this issue (Biometallum) in the chemistry laboratories at Manresa (Barcelona), Catalonia.

65.4 Price Changes in Rare Earths, 2009–2021

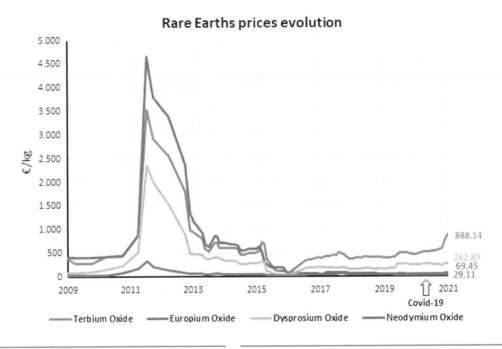

References

Aagaard P (1974) Rare earth elements adsorption on clay minerals. Bulletin Du Groupe Français Des Argiles 26(2):193–199. https://doi.org/10.3406/argil.1974.1217

Bao Z, Zhao Z (2008) Geochemistry of mineralization with exchangeable REY in the weathering crusts of granitic rocks in South China. Ore Geol Rev 33(3–4):519–535. https://doi.org/10.1016/j.oregeorev.2007.03.005

Chi R, Tian J (2008) Weathered crust elution-deposited rare earth ores. Nova Science Publishers

Goodenough KM, Wall F, Merriman D (2018) The rare earth elements: demand, global resources, and challenges for resourcing future generations. Nat Resour Res 27(2):201–216. https://doi.org/10.1007/s11053-017-9336-5

Further Reading

Alkane Resources (2021) Rare earths. http://www.alkane.com.au/products/rare-earths-overview/. Last accessed May 2021

CIT UPC (2021) Biometallum project: recovery of metals from electric and electronic waste. Biogap Group. https://cit.upc.edu/en/portfolio-item/biometallum_project_recovery_of_metals_from_electric_and_electronic_waste_biogap_group/. Last accessed May 2021

Roskill (2021) Market reports. Rare earths. https://roskill.com/market-report/rare-earths/. Last accessed May 2021

USGS (2021) Commodity statistics and information. Rare earths. https://www.usgs.gov/centers/nmic/rare-earths-statistics-and-information. Last accessed May 2021

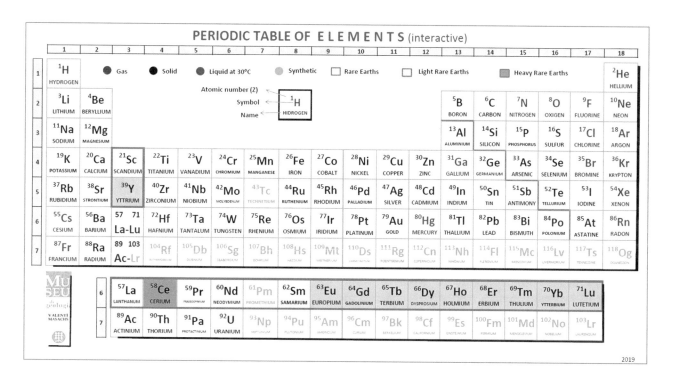

- The most abundant REE metal
- Member of the light rare earths (LREE)
- Discovered in 1803, and named after the asteroid Ceres
- Ductile and malleable
- Oxidizes readily in contact with air
- Pyrophoric, and easily ignited by scratching or grinding
- Obtained from bastnäsite (Fig. 66.1), monazite, and xenotime.

66.1 Geology

The variations are substantial within the REE class of ore deposits, and the formation of an REE ore deposit gives little information about its classification. Besides, classifying REE ore deposits on the basis of only their genetic evolution would quickly induce misinterpretations. For instance, the class of copper-gold-uranium-REE-iron (IOCG) includes the well-known Olympic Dam deposit (South Australia), the iron deposits of Kiruna (Sweden), the iron-REE deposits of Box Bixby and Pea Ridge (Missouri, United States) and

Fig. 66.1 Bastnäsite-(Ce) (cerium, lanthanum carbonate). *Peshawar (Pakistan)*. (*Photo* Joaquim Sanz. MGVM)

possibly the REE-rich Bayan Obo (Mongolia), the Palabora carbonatite-hosted copper, and the Vergenoug iron-fluorine deposit (South Africa). When examined in detail, these deposits are remarkably different.

There are several types of natural (primary) REE resources, including those formed by high-temperature geological processes (carbonatites, alkaline rocks, vein, and skarn deposits) and those formed by low-temperature processes (placers, laterites, bauxites, and ion-adsorption clays) (Goodenough et al. 2018).

Clay-fixing REE by adsorption in clays (Aagaard 1974) in the south of China comprises the main source of worldwide resources (ion-adsorption clays) (Bao and Zhao 2008). The various deposits produce different types of REE: weathering of two mica granites produces laterites rich in Y-HREE, whilst weathering of biotitic granites generates laterites rich in LREE (Chi and Tian 2008).

Cerium is obtained from bastnäsite-(Ce), (Ce, La) (CO_3) F, monazite-(Ce), (Ce, La, Nd, Th)PO_4, and xenotime-(Y), YPO_4. The most important deposits in terms of the exploitation of bastnäsite and xenotime are the Bayan Obo and Maoniuping carbonatites in China and at Mount Weld in Australia, among others. Monazite, too, is found in carbonatite deposits in Mount Weld and Bayan Obo.

Monazite and xenotime are found in placer deposits. Both are highly resistant to transport and are common in detritic sediments, accumulating in alluvial and beach placers with zircon, ilmenite, and rutile. Important placer deposits are found in India, Malaysia, Myanmar, and Brazil, among others. Monazite is a widely distributed mineral, appearing as an accessory in granitic igneous rocks, gneissic metamorphic rocks, and detrital sands.

66.2 Producing Countries

The main world reserves of rare earths are in China (44 Mt), Brazil and Vietnam (22 Mt), Russia (12 Mt), India (6.9 Mt), Australia (3.4 Mt), Greenland (1.5 Mt), and the United States (1.4 Mt), also in Myanmar, Canada, Madagascar, and Thailand (Fig. 66.2).

66.3 Applications

Automotive Industry
Cerium oxide is part of the ceramic support for the catalytic converters that change vehicle exhaust emissions into CO_2 and H_2O to reduce environmental pollution (Fig. 66.3).

Electronics and Lamp Industry
Combined with yttrium, cerium is used in the manufacture of white-light LEDs. It is also used in the manufacture of fluorescent tubes to improve energy and light efficiency.

Glass and Ceramics Industry
Cerium oxide, combined with tin oxide, is used in the manufacture of car windscreens and solar panels to absorb ultraviolet radiation.

Cerium oxide is used to polish the surface of windscreens, LCDs and plasma screens, x-ray tube cathodes, and computer hard drives. It serves to polish precious stones and glass lenses. In addition, it is used as a dye, as the combination of cerium and titanium imparts a golden color to glass.

Cerium is used in the manufacture of dental prostheses.

Chemical Industry
Cerium, added to diesel fuel, allows the fuel to burn more cleanly and with less contamination.

Other Fields
Ferrocerium, a material based on iron, cerium, lanthanum, neodymium, praseodymium, and magnesium, is used to make sparking lighter 'flints', as it sparks when scratched. It is also used in the manufacture of steels (HSLA) to eliminate free oxygen and sulfur and in magnesium and aluminum alloys.

Yttrium oxide and cerium oxide are substitutes for thorium in gas-light casings because they are heat resistant, give very intense light when heated, and are not radioactive.

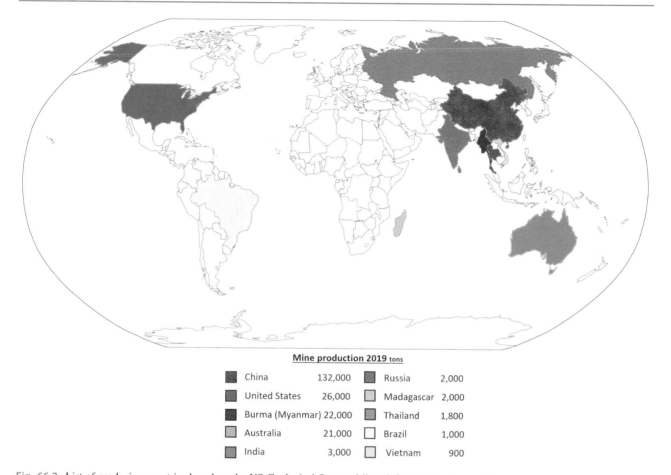

Mine production 2019 tons

■	China	132,000	■	Russia	2,000
■	United States	26,000	▨	Madagascar	2,000
■	Burma (Myanmar)	22,000	■	Thailand	1,800
▨	Australia	21,000	☐	Brazil	1,000
■	India	3,000	▨	Vietnam	900

Fig. 66.2 List of producing countries based on the US Geological Survey, Mineral Commodity Summaries

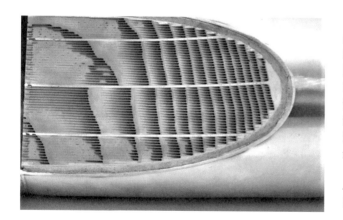

Fig. 66.3 Ceramic holder for a catalytic converter. (*Photo* Joaquim Sanz. MGVM)

66.4 Recycling

The cerium used in glass polishing can be recycled after separation of the silica and alumina that remain attached at the end.

At its La Rochelle factory (France), the Solvay company started recycling fluorescent tubes and energy-saving lamps in 2011 due to their rare earth content, such as cerium, lanthanum, europium, terbium, gadolinium, and yttrium. However, in 2016, this type of recycling ceased due to the fall in rare-earth prices (compared to 2011), the reduction in the supply of used fluorescent tubes (due to the progressive introduction of LED-based lighting), and the stabilization of rare earth exports from China.

References

Aagaard P (1974) Rare earth elements adsorption on clay minerals. Bulletin Du Groupe Français Des Argiles 26(2):193–199. https://doi.org/10.3406/argil.1974.1217

Bao Z, Zhao Z (2008) Geochemistry of mineralization with exchangeable REY in the weathering crusts of granitic rocks in South China. Ore Geol Rev 33(3–4):519–535. https://doi.org/10.1016/j.oregeorev.2007.03.005

Chi R, Tian J (2008) Weathered crust elution-deposited rare earth ores. Nova Science Publishers

Goodenough KM, Wall F, Merriman D (2018) The rare earth elements: demand, global resources, and challenges for resourcing future generations. Nat Resour Res 27(2):201–216. https://doi.org/10.1007/s11053-017-9336-5

Further Reading

Alkane Resources (2021) Rare earths http://www.alkane.com.au/products/rare-earths-overview/. Last accessed May 2021

Roskill (2021) Market reports. Rare earths. https://roskill.com/market-report/rare-earths/. Last accessed May 2021

USGS (2021) Commodity statistics and information. Rare earths. https://www.usgs.gov/centers/nmic/rare-earths-statistics-and-information. Last accessed May 2021

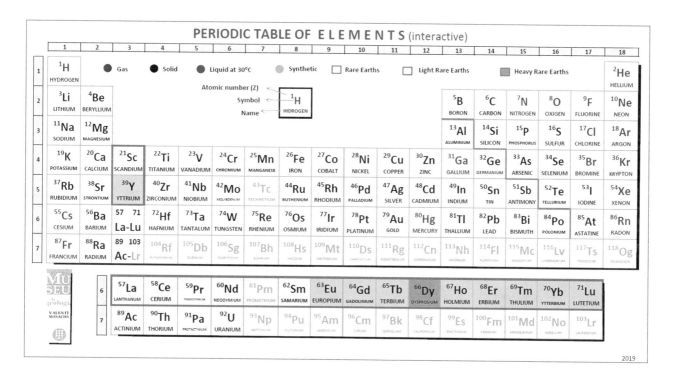

PERIODIC TABLE OF E L E M E N T S (interactive)

- A member of the heavy rare earths (HREE)
- Discovered in a sample of erbium oxide in 1886 by the French chemist, Lecoq de Boisbaudran
- Available since 1950, when modern chemical techniques such as ion-exchange separation were developed
- Ductile and malleable
- Has a silvery metallic sheen
- Found mainly in xenotime, in the heavy rare earth clays of China (Fig. 67.1), and in smaller quantities in monazite and bastnäsite.

67.1 Geology

The variations are substantial within the REE class of ore deposits, and the formation of an REE ore deposit gives little information about its classification. Besides, classifying REE ore deposits on the basis of only their genetic evolution would quickly induce misinterpretations. For instance, the class of copper-gold-uranium-REE-iron (IOCG) includes the well-known Olympic Dam deposit (South Australia), the iron deposits of Kiruna (Sweden), the iron-REE deposits of Box Bixby and Pea Ridge (Missouri, United States) and possibly the REE-rich Bayan Obo (Mongolia), the Palabora carbonatite-hosted copper, and the Vergenoug iron-fluorine

Fig. 67.1 Clay with rare earths and dysprosium. (*China*). (*Photo Joaquim Sanz. MGVM*)

deposit (South Africa). When examined in detail, these deposits are remarkably different.

There are several types of natural (primary) REE resources, including those formed by high-temperature geological processes (carbonatites, alkaline rocks, vein, and skarn deposits) and those formed by low-temperature processes (placers, laterites, bauxites, and ion-adsorption clays) (Goodenough et al. 2018).

Clay-fixing REE by adsorption in clays (Aagaard 1974) in the south of China comprises the main source of world-wide resources (ion-adsorption clays) (Bao and Zhao 2008). The various deposits produce different types of REE: weathering of two mica granites produces laterites rich in Y-HREE, whilst weathering of biotitic granites generates laterites rich in LREE (Chi and Tian 2008).

Dysprosium is basically obtained from ion-adsorption clays and from xenotime-(Y), YPO_4, but it can also be extracted from monazite-(Ce), $(Ce,La,Nd,Th)PO_4$, and from bastnäsite-(Ce), (Ce,La) $(CO_3)F$. The most important deposits for the exploitation of bastnäsite and xenotime are the Bayan Obo and Maoniuping carbonatites in China and in Mount Weld in Australia, among others. Monazite is also found in carbonatite deposits in Mount Weld and Bayan Obo.

Monazite and xenotime are found in placer deposits. Both are highly resistant to transport and are common in detritic sediments, accumulating in alluvial and beach placers with zircon, ilmenite, and rutile. Important placer deposits are found in India, Malaysia, Myanmar, and Brazil, among others. Monazite is a widely distributed mineral, appearing as an accessory in granitic igneous rocks, gneissic metamorphic rocks, and detrital sands.

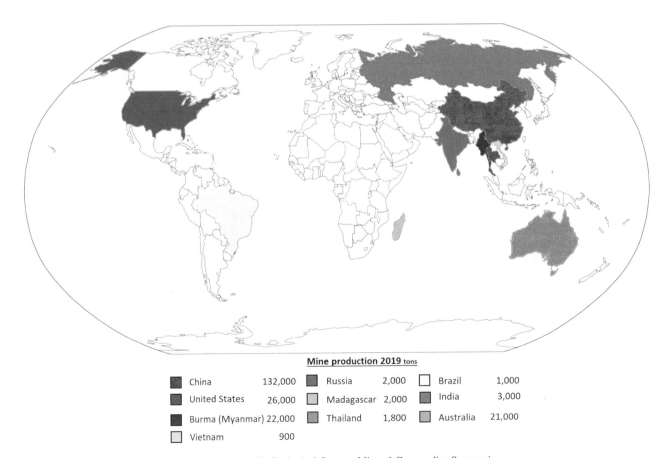

Mine production 2019 tons

China	132,000	Russia	2,000	Brazil	1,000
United States	26,000	Madagascar	2,000	India	3,000
Burma (Myanmar)	22,000	Thailand	1,800	Australia	21,000
Vietnam	900				

Fig. 67.2 List of producing countries based on the US Geological Survey, Mineral Commodity Summaries

Fig. 67.3 Charging an electric vehicle. (*Photo* Joaquim Sanz. MGVM)

67.2 Producing Countries

The main world reserves of rare earths are in China (44 Mt), Brazil and Vietnam (22 Mt), Russia (12 Mt), India (6.9 Mt), Australia (3.4 Mt), Greenland (1.5 Mt), and the United States (1.4 Mt), and other countries such as Myanmar, Canada, Madagascar, and Thailand (Fig. 67.2).

67.3 Applications

Magnet Manufacturing

Dysprosium, along with neodymium and terbium, is used in the manufacture of permanent magnets. It helps to maintain and increase the strength of the magnetic field despite high temperatures. These magnets have applications in electric vehicle engines (Fig. 67.3), hybrid vehicles, and wind turbines, enhancing performance. Dysprosium is also one of the main components of computer hard-disk magnets.

Metallurgical Industry

Dysprosium tends to improve the resistance of steels to corrosion. Dysprosium, iron, and terbium form an alloy with the most powerful known magnetoresistance at room temperature, terphenol-D. This is used in naval sonar systems and magnetomechanical sensors (see rare earths: terbium).

Power Generation

Dysprosium has a high neutron absorption capacity, which is why it can be used in the manufacture of control rods for nuclear reactors, whose function is to regulate the nuclear chain reaction that produces heat.

Electrical Industry

Dysprosium iodide is used in the manufacture of high-intensity halogen lamps for lighting and projection because it improves the quality of the spectrum, especially in the red band.

67.4 Recycling

Processes are being developed for recycling recovered magnets that contain dysprosium, from electric motors (cars, bicycles, air-conditioning compressors, etc.), wind turbines, computer hard drives, loudspeakers, and other electronic devices, using ISR technology (Innord's separation of REE).

References

Aagaard P (1974) Rare earth elements adsorption on clay minerals. Bulletin Du Groupe Français Des Argiles 26(2):193–199. https://doi.org/10.3406/argil.1974.1217

Bao Z, Zhao Z (2008) Geochemistry of mineralization with exchangeable REY in the weathering crusts of granitic rocks in South China. Ore Geol Rev 33(3–4):519–535. https://doi.org/10.1016/j.oregeorev.2007.03.005

Chi R, Tian J (2008) Weathered crust elution-deposited rare earth ores. Nova Science Publishers

Goodenough KM, Wall F, Merriman D (2018) The rare earth elements: demand, global resources, and challenges for resourcing future generations. Nat Resour Res 27(2):201–216. https://doi.org/10.1007/s11053-017-9336-5

Further Reading

Alkane Resources (2021) Rare earths. http://www.alkane.com.au/products/rare-earths-overview/. Last accessed May 2021

Geomega. Innord Inc. (2021) Rare earths separation. Dysprosium. https://ressourcesgeomega.ca. Last accessed May 2021

Roskill (2021) Market reports. Rare earths. https://roskill.com/market-report/rare-earths/. Last accessed May 2021

USGS (2021) Commodity statistics and information. Rare earths. https://www.usgs.gov/centers/nmic/rare-earths-statistics-and-information. Last accessed May 2021

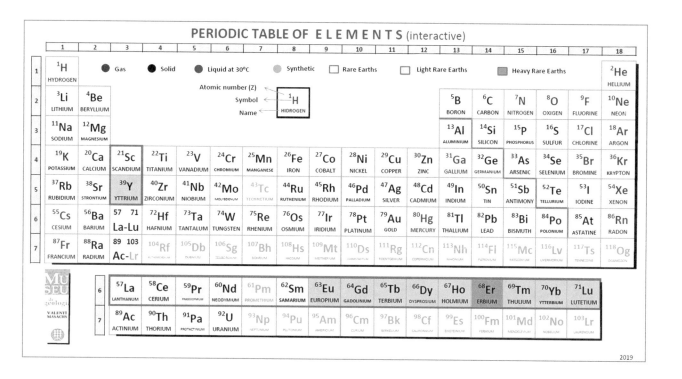

PERIODIC TABLE OF **E L E M E N T S** (interactive)

- A member of the heavy rare earths (HREE)
- A fairly scarce rare earth metal
- Discovered in 1843 by Carl Gustaf Mosander
- Malleable and ductile
- Stable in contact with air, and does not rust as easily as other rare earth metals
- Found in the heavy rare earth-impregnated clays of China, and also in xenotime (Fig. 68.1)

68.1 Geology

The variations are substantial within the REE class of ore deposits, and the formation of an REE ore deposit gives little information about its classification. Besides, classifying REE ore deposits on the basis of only their genetic evolution would quickly induce misinterpretations. For instance, the class of copper-gold-uranium-REE-iron (IOCG) includes the well-known Olympic Dam deposit (South Australia), the iron deposits of Kiruna (Sweden), the iron-REE deposits of

Fig. 68.1 Xenotime-Y (yttrium phosphate with cerium and erbium). *Novo Horizonte (Bahia, Brazil). (Photo Joaquim Sanz. MGVM)*

Box Bixby and Pea Ridge (Missouri, United States) and possibly the REE-rich Bayan Obo (Mongolia), the Palabora carbonatite-hosted copper, and the Vergenoug iron-fluorine deposit (South Africa). When examined in detail, these deposits are remarkably different.

There are several types of natural (primary) REE resources, including those formed by high-temperature

geological processes (carbonatites, alkaline rocks, vein, and skarn deposits) and those formed by low-temperature processes (placers, laterites, bauxites, and ion-adsorption clays) (Goodenough et al. 2018).

Clay-fixing REE by adsorption in clays (Aagaard 1974) in the south of China comprises the main source of worldwide resources (ion-adsorption clays) (Bao and Zhao 2008). The various deposits produce different types of REE: weathering of two mica granites produces laterites rich in Y-HREE, whilst weathering of biotitic granites generates laterites rich in LREE (Chi and Tian 2008).

Erbium is basically obtained from ion-adsorption clays and from xenotime-(Y), YPO$_4$.The most important deposits of xenotime in terms of exploitation are in the Bayan Obo and Maoniuping carbonatites in China and in Mount Weld in Australia, among others.

Xenotime is found in placer deposits. It is highly resistant to transport and common in detritic sediments, accumulating in alluvial and beach placers with zircon, ilmenite, and rutile. Important placer deposits are found in India, Malaysia, Myanmar, and Brazil, among others.

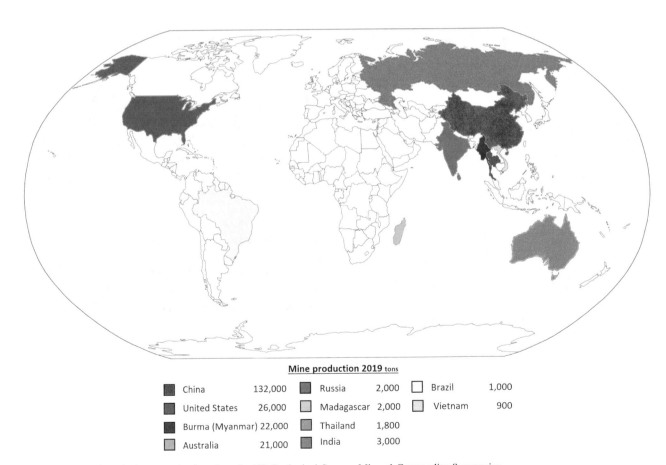

Fig. 68.2 List of producing countries based on the US Geological Survey, Mineral Commodity Summaries

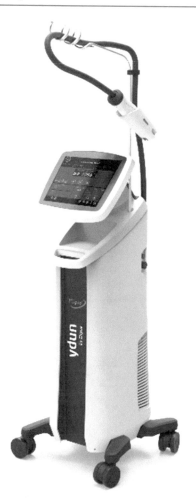

Fig. 68.3 Erbium-YAG laser equipment. (*Image courtesy of* Candel Medical)

68.2 Producing Countries

The main world reserves of rare earths are in China (44 Mt), Brazil and Vietnam (22 Mt), Russia (12 Mt), India (6.9 Mt), Australia (3.4 Mt), Greenland (1.5 Mt), and the United States (1.4 Mt), and other countries such as Myanmar, Canada, Madagascar, and Thailand (Fig. 68.2).

68.3 Applications

Glass and Ceramics Industry

Erbium is used as a pink dye in the manufacture of glass and ceramics, as well as for making zirconia for jewelry.

Electronics Industry

Erbium is used in optical fiber communications, since it allows direct amplification of a light impulse inside the cable without converting it into an electrical signal, so that it leaves the erbium-doped glass fiber with much more intensity than when it entered.

Power Generation

Erbium has a high neutron absorption capacity, which is why it can be used in the manufacture of nuclear reactor control rods to regulate the nuclear chain reaction that produces heat.

Medicine

Erbium is a component of the Er-YAG laser used in dermatology to eliminate skin imperfections. Its technical characteristics mean that its effect is restricted to the skin's superficial layers and does not damage the deeper layers (Fig. 68.3).

Other Fields

Erbium is used as a component of the glass of certain photographic filters.

68.4 Recycling

The recycling of erbium is unknown.

References

Aagaard P (1974) Rare earth elements adsorption on clay minerals. Bulletin Du Groupe Français Des Argiles 26(2):193–199. https://doi.org/10.3406/argil.1974.1217

Bao Z, Zhao Z (2008) Geochemistry of mineralization with exchangeable REY in the weathering crusts of granitic rocks in South China. Ore Geol Rev 33(3–4):519–535. https://doi.org/10.1016/j.oregeorev.2007.03.005

Chi R, Tian J (2008) Weathered crust elution-deposited rare earth ores. Nova Science Publishers

Goodenough KM, Wall F, Merriman D (2018) The rare earth elements: demand, global resources, and challenges for resourcing future generations. Nat Resour Res 27(2):201–216. https://doi.org/10.1007/s11053-017-9336-5

Further Reading

Alkane Resources (2021) Rare earths. http://www.alkane.com.au/products/rare-earths-overview/. Last accessed May 2021

Candela Medical (2021) https://candelamedical.com/na/provider/product/nordlys. Last accessed May 2021

Roskill (2021) Market reports. Rare earths. https://roskill.com/market-report/rare-earths/. Last accessed May 2021

USGS (2021) Commodity statistics and information. Rare earths. https://www.usgs.gov/centers/nmic/rare-earths-statistics-and-information. Last accessed May 2021

PERIODIC TABLE OF ELEMENTS (interactive)

2019

- Most reactive of the heavy rare earths (HREE)
- Discovered in 1901
- Named after the continent of Europe
- As soft as lead and quite ductile
- Oxidizes quickly in contact with air
- Mainly found in the heavy rare earth-impregnated clays of China, in bastnäsite (Fig. 69.1), and in xenotime.

69.1 Geology

The variations are substantial within the REE class of ore deposits, and the formation of an REE ore deposit gives little information about its classification. Besides, classifying REE ore deposits on the basis of only their genetic evolution would quickly induce misinterpretations. For instance, the class of copper-gold-uranium-REE-iron (IOCG) includes the well-known Olympic Dam deposit (South Australia), the

Fig. 69.1 Bastnäsite-Ce (cerium, lanthanum carbonate with europium). *Peshawar (Pakistan)*. (*Photo* Joaquim Sanz. MGVM)

iron deposits of Kiruna (Sweden), the iron-REE deposits of Box Bixby and Pea Ridge (Missouri, United States) and possibly the REE-rich Bayan Obo (Mongolia), the Palabora carbonatite-hosted copper, and the Vergenoug iron-fluorine deposit (South Africa). When examined in detail, these deposits are remarkably different.

There are several types of natural (primary) REE resources, including those formed by high-temperature geological processes (carbonatites, alkaline rocks, vein, and skarn deposits) and those formed by low-temperature processes (placers, laterites, bauxites, and ion-adsorption clays) (Goodenough et al. 2018).

Clay-fixing REE by adsorption in clays (Aagaard 1974) in the south of China comprises the main source of worldwide resources (ion-adsorption clays) (Bao and Zhao 2008). The various deposits produce different types of REE: weathering of two mica granites produces laterites rich in Y-HREE, whilst weathering of biotitic granites generates laterites rich in LREE (Chi and Tian 2008).

Europium is basically obtained from ion-adsorption clays and from bastnäsite-(Ce), (Ce, La) (CO$_3$) F, yet it can also be extracted from xenotime-(Y), YPO$_4$. The most important deposits for the exploitation of bastnäsite and xenotime are the Bayan Obo and Maoniuping carbonatites in China and Mount Weld in Australia, with some others. Xenotime is also found in the placer fields of India, Malaysia, Myanmar, Brazil, among others. Xenotime is found in placer deposits, as it is highly resistant to transport and common in detritic

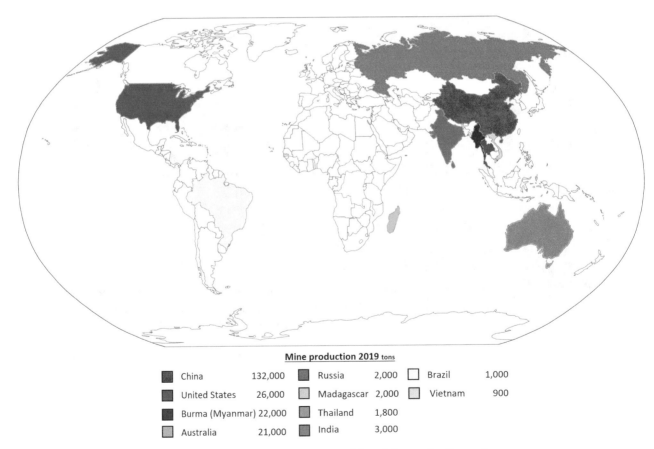

Mine production 2019 tons

China	132,000	Russia	2,000	Brazil	1,000
United States	26,000	Madagascar	2,000	Vietnam	900
Burma (Myanmar)	22,000	Thailand	1,800		
Australia	21,000	India	3,000		

Fig. 69.2 List of producing countries based on the US Geological Survey, Mineral Commodity Summaries

sediments, accumulating in alluvial placers and beach placers, with zircon, ilmenite, and rutile.

69.2 Producing Countries

The main world reserves of rare earths are China (44 Mt), Brazil and Vietnam (22 Mt), Russia (12 Mt), India (6.9 Mt), Australia (3.4 Mt), Greenland (1.5 Mt), and the United States (1.4 Mt), then countries such as Myanmar, Madagascar, Canada, and Thailand (Fig. 69.2).

69.3 Applications

Electronics Industry

Europium (III) oxide is used as a red phosphorous (phosphorophore) material, along with terbium and yttrium, in plasma and LCD screens, and in military technology (phosphorophore: a compound that exhibits luminescence when excited by electrons) (has the properties of elemental phosphorus).

Lamp Industry

Europium is used in the manufacture of compact fluorescent (energy-saving) tubes to improve energy efficiency and obtain a warmer shade than that of traditional fluorescent lighting.

Medicine

Europium is an ingredient in the medical preparation for the detection of Down syndrome and other genetic diseases. Europium is used as a phosphorus, along with bromine, barium, and fluoride, in digital x-ray images, which have superseded earlier film images (Fig. 69.3).

Other Fields

Europium is used in the manufacture of phosphorescent paint: once exposed to a source of intense light, it maintains its luminosity for hours. It is also used in the printing ink for euro banknotes that fluoresce strongly with a red color under ultraviolet light.

69.4 Recycling

Research continues into the recycling of REE elements recovered from old consumer electronics equipment ('urban mining'), since the current chemical separation methods are expensive and unprofitable for recycling companies.

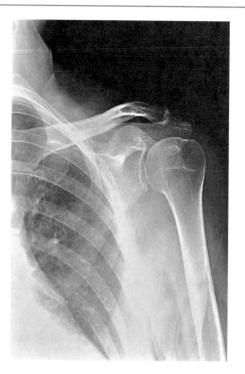

Fig. 69.3 Digital x-ray. (*Image courtesy of* Hospital Sant Pau, Barcelona)

References

Aagaard P (1974) Rare earth elements adsorption on clay minerals. Bulletin Du Groupe Français Des Argiles 26(2):193–199. https://doi.org/10.3406/argil.1974.1217

Bao Z, Zhao Z (2008) Geochemistry of mineralization with exchangeable REY in the weathering crusts of granitic rocks in South China. Ore Geol Rev 33(3–4):519–535. https://doi.org/10.1016/j.oregeorev.2007.03.005

Chi R, Tian J (2008) Weathered crust elution-deposited rare earth ores. Nova Science Publishers

Goodenough KM, Wall F, Merriman D (2018) The rare earth elements: demand, global resources, and challenges for resourcing future generations. Nat Resour Res 27(2):201–216. https://doi.org/10.1007/s11053-017-9336-5

Further Reading

Alkane Resources (2021) Rare earths. http://www.alkane.com.au/products/rare-earths-overview/. Last accessed May 2021

Roskill (2021) Market reports. Rare earths. https://roskill.com/market-report/rare-earths/. Last accessed May 2021

USGS (2021) Commodity statistics and information. Rare earths. https://www.usgs.gov/centers/nmic/rare-earths-statistics-and-information. Last accessed May 2021

PERIODIC TABLE OF E L E M E N T S (interactive)

- A member of the heavy rare earths (HREE)
- Discovered in 1886
- Named after the mineral gadolinite and the distinguished Finnish chemist who examined it, Johan Gadolin
- Malleable and ductile
- Relatively stable in contact with air
- Its compounds are paramagnetic
- Found in clays impregnated with rare earths (Fig. 70.1), in xenotime, and in monazite.

70.1 Geology

The variations are substantial within the REE class of ore deposits, and the formation of an REE ore deposit gives little information about its classification. Besides, classifying REE ore deposits on the basis of only their genetic evolution would quickly induce misinterpretations. For instance, the class of copper-gold-uranium-REE-iron (IOCG) includes the well-known Olympic Dam deposit (South Australia), the iron deposits of Kiruna (Sweden), the iron-REE deposits of Box Bixby and Pea Ridge (Missouri, United States) and

Fig. 70.1 Clay with rare earths and gadolinium. (*China*). (*Photo* Joaquim Sanz. MGVM)

possibly the REE-rich Bayan Obo (Mongolia), the Palabora carbonatite-hosted copper, and the Vergenoug iron-fluorine deposit (South Africa). When examined in detail, these deposits are remarkably different.

There are several types of natural (primary) REE resources, including those formed by high-temperature geological processes (carbonatites, alkaline rocks, vein, and skarn deposits) and those formed by low-temperature processes (placers, laterites, bauxites, and ion-adsorption clays) (Goodenough et al. 2018).

Clay-fixing REE by adsorption in clays (Aagaard 1974) in the south of China comprises the main source of world-wide resources (ion-adsorption clays) (Bao and Zhao 2008). The various deposits produce different types of REE: weathering of two mica granites produces laterites rich in Y-HREE, whilst weathering of biotitic granites generates laterites rich in LREE (Chi and Tian 2008).

Gadolinium is basically obtained from ion-adsorption clays and from xenotime-(Y), YPO_4, but it can also be extracted from monazite $(Ce,La,Nd,Th)PO_4$. The most important deposits for the exploitation of xenotime are in the Bayan Obo and Maoniuping carbonatites in China and in Mount Weld in Australia, among others. Xenotime and monazite are also found in the placer deposits of India, Malaysia, Myanmar, Brazil, among others. Monazite and xenotime are both are very resistant to transport and are common in detritic sediments, accumulating in alluvial placers and beach placers with zircon, ilmenite, and rutile.

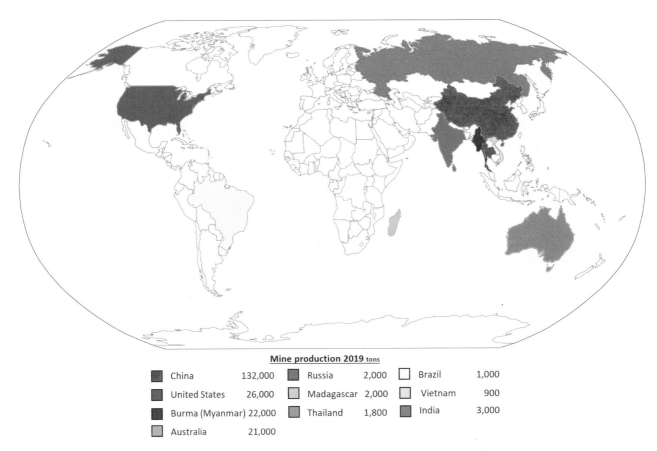

Mine production 2019 tons

China	132,000	Russia	2,000	Brazil	1,000
United States	26,000	Madagascar	2,000	Vietnam	900
Burma (Myanmar)	22,000	Thailand	1,800	India	3,000
Australia	21,000				

Fig. 70.2 List of producing countries based on the US Geological Survey, Mineral Commodity Summaries

70.2 Producing Countries

The main world reserves of rare earths are in China (44 Mt), Brazil and Vietnam (22 Mt), Russia (12 Mt), India (6.9 Mt), Australia (3.4 Mt), Greenland (1.5 Mt), and the United States (1.4 Mt), with other countries such as Myanmar, Madagascar, Canada, and Thailand (Fig. 70.2).

70.3 Applications

Manufacture of Magnets
Gadolinium is used to improve the remanence and strengthen the corrosion resistance of samarium-cobalt permanent magnets.

Electronic and Lighting Industries
Terbium-doped gadolinium is used as a green phosphorous material (*phosphorophore*) in plasma and LCD screens (phosphorophore: a compound that exhibits luminescence when excited by electrons) (carrier of phosphorus element properties).

Gadolinium is used in the manufacture of fluorescent tubes to improve energy and light efficiency.

Metallurgical Industry
Gadolinium is added to iron and chromium alloys to increase their corrosion resistance and improve steel machining.

Medicine
Gadolinium compounds are paramagnetic (weakly magnetic), but respond to the strong magnetic fields of MRIs. When injected into the patient, they enhance the contrast of the blood vessel images of certain parts of the body,

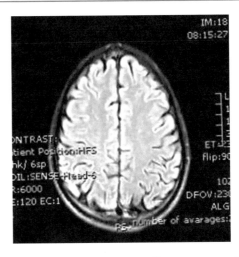

Fig. 70.4 MRI with gadolinium contrast. (*Images courtesy of* Hospital Sant Pau, Barcelona)

thus facilitating the work of the diagnostic specialist (Figs. 70.3, 70.4).

In positron emission tomography (PET), gadolinium oxyortosilicate (GSO) crystals have been used as radiation detectors, but today yttria-doped lutetium oxyortosilicate (LYSO) crystals are already incorporated in the latest so-called digital PETs (see: lutetium).

Power Generation
Gadolinium has a high neutron absorption capacity, so it can be used in the manufacture of the control rods for nuclear reactors, whose role is to regulate the nuclear chain reaction that produces heat.

Other Fields
The use of gadolinium alloy with germanium and silicon in magnetic cooling is being explored, as it could reduce electrical heating and cooling costs by 80%.

70.4 Recycling

The Solvay company started in 2011, at the La Rochelle (France) factory, recycling fluorescent lamps and energy-saving lamps due to their content of rare earths such as: cerium, lanthanum, europium, terbium, gadolinium, and yttrium.

However, in 2016, this type of recycling ceased due to the decrease in rare earth prices (compared to 2011), the decrease in used fluorescent tubes (due to the progressive introduction of LED-based lamps) and the stabilization of rare earth exports from China.

Research continues into the recycling of REE elements obtained from old consumer electronics equipment ('urban mining'), since current chemical separation methods are expensive and unprofitable for recycling companies.

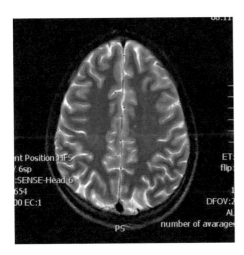

Fig. 70.3 MRI without gadolinium contrast. (*Images courtesy of* Hospital Sant Pau, Barcelona)

References

Aagaard P (1974) Rare earth elements adsorption on clay minerals. Bulletin Du Groupe Français Des Argiles 26(2):193–199. https://doi.org/10.3406/argil.1974.1217

Bao Z, Zhao Z (2008) Geochemistry of mineralization with exchangeable REY in the weathering crusts of granitic rocks in South China. Ore Geol Rev 33(3–4):519–535. https://doi.org/10.1016/j.oregeorev.2007.03.005

Chi R, Tian J (2008) Weathered crust elution-deposited rare earth ores. Nova Science Publishers

Goodenough KM, Wall F, Merriman D (2018) The rare earth elements: demand, global resources, and challenges for resourcing future generations. Nat Resour Res 27(2):201–216. https://doi.org/10.1007/s11053-017-9336-5

Further Reading

Alkane Resources (2021) Rare earths. http://www.alkane.com.au/products/rare-earths-overview/. Last accessed May 2021

Roskill (2021) Market reports. Rare earths. https://roskill.com/market-report/rare-earths/. Last accessed May 2021

USGS (2021) Commodity statistics and information. Rare earths. https://www.usgs.gov/centers/nmic/rare-earths-statistics-and-information. Last accessed May 2021

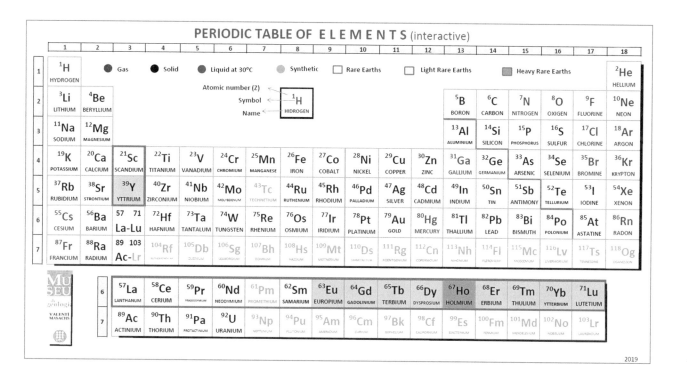

71.1 Geology

The variations are substantial within the REE class of ore deposits, and the formation of an REE ore deposit gives little information about its classification. Besides, classifying REE ore deposits on the basis of only their genetic evolution would quickly induce misinterpretations. For instance, the class of copper-gold-uranium-REE-iron (IOCG) includes the well-known Olympic Dam deposit (South Australia), the iron deposits of Kiruna (Sweden), the iron-REE deposits of Box Bixby and Pea Ridge (Missouri, United States) and possibly the REE-rich Bayan Obo (Mongolia), the Palabora

- A member of the heavy rare earth group (HREE)
- Isolated in 1879 by the Swedish chemist Per Teodor Cleve, who named it after his native city, Stockholm—in Latin, Holmia
- Extraordinary magnetic properties
- The element with the greatest magnetic intensity
- Highly absorbent of neutrons generated in nuclear fission
- Found mainly in xenotime (Fig. 71.1) and the heavy rare earth-impregnated clays of China.

Fig. 71.1 Xenotime-Y (yttrium phosphate with cerium and holmium). *Novo Horizonte (Bahia, Brazil)*. (*Photo* Joaquim Sanz. MGVM)

carbonatite-hosted copper, and the Vergenoug iron-fluorine deposit (South Africa). When examined in detail, these deposits are remarkably different.

There are several types of natural (primary) REE resources, including those formed by high-temperature geological processes (carbonatites, alkaline rocks, vein, and skarn deposits) and those formed by low-temperature processes (placers, laterites, bauxites, and ion-adsorption clays) (Goodenough et al. 2018).

Clay-fixing REE by adsorption in clays (Aagaard 1974) in the south of China comprises the main source of worldwide resources (ion-adsorption clays) (Bao and Zhao 2008). The various deposits produce different types of REE: weathering of two mica granites produces laterites rich in Y-HREE, whilst weathering of biotitic granites generates laterites rich in LREE (Chi and Tian 2008).

Holmium is basically obtained from ion-adsorption clays and from xenotime-(Y), YPO_4. The most important deposits for exploitation of xenotime are the Bayan Obo and Maoniuping carbonatites in China and in Mount Weld in Australia, among others. Xenotime is also found in placer fields of India, Malaysia, Myanmar, and Brazil, as it is highly resistant to transport and common in detritic sediments, accumulating in alluvial placers and beach placers with zircon, ilmenite, and rutile.

71.2 Producing Countries

The main world reserves of rare earths are in China (44 Mt), Brazil and Vietnam (22 Mt), Russia (12 Mt), India (6.9 Mt), Australia (3.4 Mt), Greenland (1.5 Mt), and the United States (1.4 Mt), and other countries such as Myanmar, Canada, Madagascar, and Thailand (Fig. 71.2).

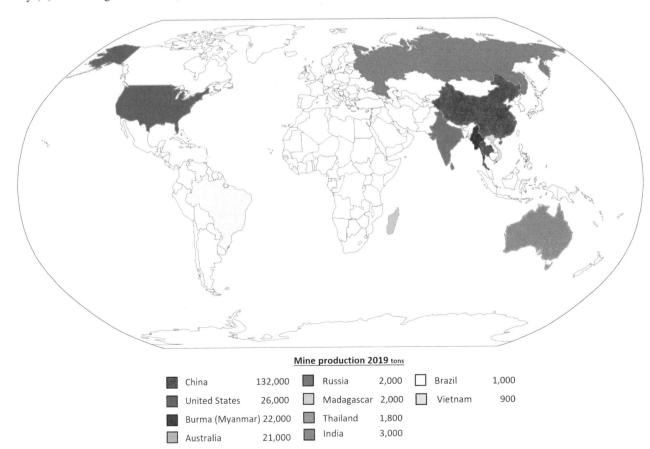

Mine production 2019 tons

China	132,000	Russia	2,000	Brazil	1,000
United States	26,000	Madagascar	2,000	Vietnam	900
Burma (Myanmar)	22,000	Thailand	1,800		
Australia	21,000	India	3,000		

Fig. 71.2 List of producing countries based on the US Geological Survey, Mineral Commodity Summaries

71.3 Applications

Glass and Ceramics Industry

Holmium is an excellent dye for yellow and red glass and for zirconite in jewelry, giving a peach or yellow color depending on the type of incident light.

Magnet Manufacturing

Holmium can be used to create strong magnetic fields by placing it between strong magnets, as a 'magnetic flux concentrator'.

Medicine

Holmium is used in the manufacture of solid-state lasers, such as Ho-YAG (holmium-yttrium-aluminum-garnet).

The holmium laser is very useful in day surgery, for instance for procedures for benign prostate enlargement (BPE), bladder or kidney stones, eyes, teeth, skin, knees, etc., because of its great safety and efficiency, low penetration, and the fact that it is strongly absorbed by the water in the tissues, causing vaporization and subsequent cauterization of the blood vessels. It is the laser with the best performance in the enucleation of the prostate (Fig. 71.3) (see: phosphorus or potassium KTP laser; see: thulium laser).

Power Generation

Holmium has a high neutron absorption capacity, which is why it can be used to make nuclear reactors' control rods, whose role is to regulate the nuclear chain reaction that produces heat.

71.4 Recycling

Holmium recycling is unknown.

References

Aagaard P (1974) Rare earth elements adsorption on clay minerals. Bulletin Du Groupe Français Des Argiles 26(2):193–199. https://doi.org/10.3406/argil.1974.1217

Bao Z, Zhao Z (2008) Geochemistry of mineralization with exchangeable REY in the weathering crusts of granitic rocks in South China. Ore Geol Rev 33(3–4):519–535. https://doi.org/10.1016/j.oregeorev.2007.03.005

Chi R, Tian J (2008) Weathered crust elution-deposited rare earth ores. Nova Science Publishers

Goodenough KM, Wall F, Merriman D (2018) The rare earth elements: demand, global resources, and challenges for resourcing future generations. Nat Resour Res 27(2):201–216. https://doi.org/10.1007/s11053-017-9336-5

Further Reading

Alkane Resources (2021) Rare earths. http://www.alkane.com.au/products/rare-earths-overview/. Last accessed May 2021

Roskill (2021) Market reports. Rare earths. https://roskill.com/market-report/rare-earths/. Last accessed May 2021

USGS (2021) Commodity statistics and information. Rare earths. https://www.usgs.gov/centers/nmic/rare-earths-statistics-and-information. Last accessed May 2021

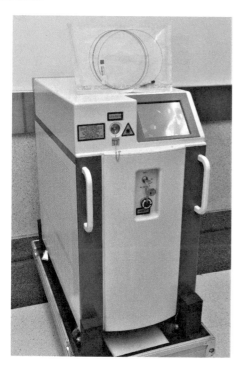

Fig. 71.3 Holmium laser equipment. (*Photo* Joaquim Sanz. MGVM)

PERIODIC TABLE OF ELEMENTS (interactive)

Fig. 72.1 Periodic table of elements

- The first of the lanthanide series elements
- Discovered in 1839 by the Swedish chemist Carl Gustaf Mosander
- A member of the light rare earths (LREE)
- Malleable, sectile, and soft
- Oxidizes rapidly upon contact with air
- Found in clays impregnated with rare earths (Fig. 72.1), in monazite, and in bastnäsite.

72.1 Geology

The variations are substantial within the REE class of ore deposits, and the formation of an REE ore deposit gives little information about its classification. Besides, classifying REE ore deposits on the basis of only their genetic evolution would quickly induce misinterpretations. For instance, the class of copper-gold-uranium-REE-iron (IOCG) includes the well-known Olympic Dam deposit (South Australia), the iron deposits of Kiruna (Sweden), the iron-REE deposits of Box Bixby and Pea Ridge (Missouri, United States) and

Fig. 72.1 Clay with rare earths and lanthanum. (*China*). (*Photo* Joaquim Sanz. MGVM)

possibly the REE-rich Bayan Obo (Mongolia), the Palabora carbonatite-hosted copper, and the Vergenoug iron-fluorine deposit (South Africa). When examined in detail, these deposits are remarkably different.

There are several types of natural (primary) REE resources, including those formed by high-temperature geological processes (carbonatites, alkaline rocks, vein, and skarn deposits) and those formed by low-temperature processes (placers, laterites, bauxites, and ion-adsorption clays) (Goodenough et al. 2018).

Clay-fixing REE by adsorption in clays (Aagaard 1974) in the south of China comprises the main source of world-wide resources (ion-adsorption clays) (Bao and Zhao 2008). The various deposits produce different types of REE: weathering of two mica granites produces laterites rich in Y-HREE, whilst weathering of biotitic granites generates laterites rich in LREE (Chi and Tian 2008).

Lanthanum is basically obtained from ion-adsorption clays and from bastnäsite-(La), (La, Ce) (CO$_3$)F, but it can also be extracted from monazite-(La), (La, Ce, Nd, Th) PO$_4$. The most important deposits in the exploitation of bastnäsite are the Bayan Obo and Maoniuping carbonatites in China and in Mount Weld in Australia, among others. Monazite is found in placer deposits, as it is very resistant to transport and common in detritic sediments, accumulating in alluvial placers and beach placers with zircon, ilmenite, and rutile in India, Malaysia, Myanmar, and Brazil, among others.

72.2   Producing Countries

The main world reserves of rare earths are in China (44 Mt), Brazil and Vietnam (22 Mt), Russia (12 Mt), India (6.9 Mt), Australia (3.4 Mt), Greenland (1.5 Mt), and the United States (1.4 Mt), with other countries such as Myanmar, Madagascar, Canada, and Thailand (Fig. 72.2)

72.3   Applications

Optical Industry
Lanthanum oxide improves the alkaline resistance of glass, so it is used in the production of special products such as infrared absorption glasses (night-vision equipment) and the lenses of cameras and microscopes, improving visual clarity.

Battery Industry
Lanthanum is a component of the rechargeable nickel metal hydride (Ni-MH) batteries used in many hybrid cars (Fig. 72.3), in computers, electronic equipment, and portable power tools, although nowadays Li-ion batteries are displacing them.

Lanthanum is used as a catalyst in hydrogen fuel cells in new-generation hydrogen vehicles.

Electronics and Lamp Industry
Lanthanum activated by terbium and tullium is used as a phosphorescent material (phosphorophore) that gives the green and blue colors on plasma and LCD screens (phosphorophore: a compound that presents luminescence when excited by electrons) (has the properties of phosphorus element).

Lanthanum is used in the manufacture of fluorescent tubes to improve energy and light efficiency.

Medicine
Lanthanum carbonate is used in the treatment of kidney failure and hyperphosphatemia, a disease caused by excess phosphorus in the blood.

Chemical Industry
In the oil industry, lanthanum is used in the cracking process, which is the breaking down of complex molecules to obtain simpler ones that convert dense crude oil into gas-oils, gasolines, and gases (butane, propane, etc.)

Other Fields
Ferrocerium, a material of a mixture of iron, cerium, lanthanum, neodymium, praseodymium, and magnesium, is used to make sparking lighter 'flints'.

Another use of lanthanum is in the manufacture of high-light intensity electrodes for projectors with arc lamps, for theaters, cinemas, TV studios, etc.

72.4   Recycling

Toyota Motor Europe recycles 90% of the batteries from its hybrid cars, and from 2011 SNAM (France) is recovering lanthanum, nickel, and other metals from these sources.

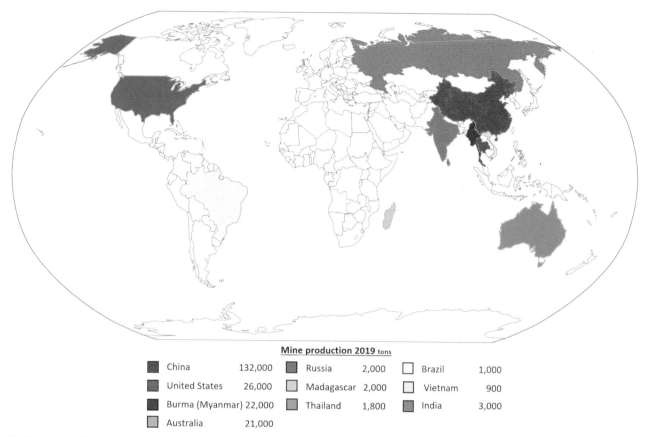

Fig. 72.2 List of producing countries based on the US Geological Survey, Mineral Commodity Summaries

Mine production 2019 tons					
China	132,000	Russia	2,000	Brazil	1,000
United States	26,000	Madagascar	2,000	Vietnam	900
Burma (Myanmar)	22,000	Thailand	1,800	India	3,000
Australia	21,000				

Fig. 72.3 Hybrid taxi with Ni-MH battery. (*Photo* Joaquim Sanz. MGVM)

At its La Rochelle factory (France), the Solvay company started recycling fluorescent tubes and energy-saving lamps in 2011 due to their rare earth content, such as cerium, lanthanum, europium, terbium, gadolinium, and yttrium. However, in 2016, this type of recycling ceased due to the fall in rare-earth prices (compared to 2011), the reduction in the supply of used fluorescent tubes (due to the progressive introduction of LED-based lighting), and the stabilization of rare earth exports from China.

Lanthanum could be recovered from spent catalytic converters using the FCC (fluidized catalytic cracking) process.

Research continues into the recycling of REE elements recovered from old consumer electronics equipment ('urban mining'), since current chemical separation methods are expensive and unprofitable for recycling companies.

References

Aagaard P (1974) Rare earth elements adsorption on clay minerals. Bulletin du Groupe français des Argiles 26(2):193–199. https://doi.org/10.3406/argil.1974.1217

Bao Z, Zhao Z (2008) Geochemistry of mineralization with exchangeable REY in the weathering crusts of granitic rocks in South China. Ore Geol Rev 33(3–4):519–535. https://doi.org/10.1016/j.oregeorev.2007.03.005

Chi R, Tian J (2008) Weathered crust elution-deposited rare earth ores. Nova Science Publishers

Goodenough KM, Wall F, Merriman D (2018) The rare earth elements: demand, global resources, and challenges for resourcing future generations. Nat Resour Res 27(2):201–216. https://doi.org/10.1007/s11053-017-9336-5

Further Reading

Alkane Resources (2021) Rare earths. http://www.alkane.com.au/products/rare-earths-overview/. Last accessed May 2021

Roskill (2021) Market reports. Rare earths. https://roskill.com/market-report/rare-earths/. Last accessed May 2021

Toyota Europe (2021) Vehicle recycling. https://www.toyota-europe.com/world-of-toyota/feel/environment/better-earth/recycle. Last accessed May 2021

USGS (2021) Commodity statistics and information. Rare earths. https://www.usgs.gov/centers/nmic/rare-earths-statistics-and-information. Last accessed May 2021

PERIODIC TABLE OF E L E M E N T S (interactive)

- A member of the heavy rare earths (HREE)
- Discovered in 1907 and named after the ancient name for Paris, 'Lutetia'
- The densest and toughest of all the rare earth elements
- Extremely rare
- Found in xenotime (Fig. 73.1) and in the rare-earth clays of China.

73.1 Geology

The variations are substantial within the REE class of ore deposits, and the formation of an REE ore deposit gives little information about its classification. Besides, classifying REE ore deposits on the basis of only their genetic evolution would quickly induce misinterpretations. For instance, the class of copper-gold-uranium-REE-iron (IOCG) includes the well-known Olympic Dam deposit (South Australia), the iron deposits of Kiruna (Sweden), the iron-REE deposits of Box Bixby and Pea Ridge (Missouri, United States) and

Fig. 73.1 Xenotime-Y (yttrium phosphate with cerium and lutetium). *Novo Horizonte (Bahia, Brazil)*. (*Photo* Joaquim Sanz. MGVM)

possibly the REE-rich Bayan Obo (Mongolia), the Palabora carbonatite-hosted copper, and the Vergenoug iron-fluorine deposit (South Africa). When examined in detail, these deposits are remarkably different.

There are several types of natural (primary) REE resources, including those formed by high-temperature geological processes (carbonatites, alkaline rocks, vein, and skarn deposits) and those formed by low-temperature processes (placers, laterites, bauxites, and ion-adsorption clays) (Goodenough et al. 2018).

Clay-fixing REE by adsorption in clays (Aagaard 1974) in the south of China comprises the main source of worldwide resources (ion-adsorption clays) (Bao and Zhao 2008). The various deposits produce different types of REE: weathering of two mica granites produces laterites rich in Y-HREE, whilst weathering of biotitic granites generates laterites rich in LREE (Chi and Tian 2008).

Lutetium is obtained basically from ion-adsorption clays and from xenotime-(Y), YPO_4. The most important deposits in terms of the exploitation of xenotime are the Bayan Obo and Maoniuping carbonatites in China and at Mount Weld in Australia. Xenotime is also found in India, Malaysia, Myanmar, and Brazil, in placer deposits. It is highly resistant to transport and common in detritic sediments, accumulating in alluvial and beach placers with zircon, ilmenite, and rutile.

73.2 Producing Countries

The world's main reserves of rare earths are in China (44 Mt), Brazil and Vietnam (22 Mt), Russia (12 Mt), India (6.9 Mt), Australia (3.4 Mt), Greenland (1.5 Mt), and the United States (1.4 Mt), and also in Myanmar, Canada, Madagascar, and Thailand (Fig. 73.2).

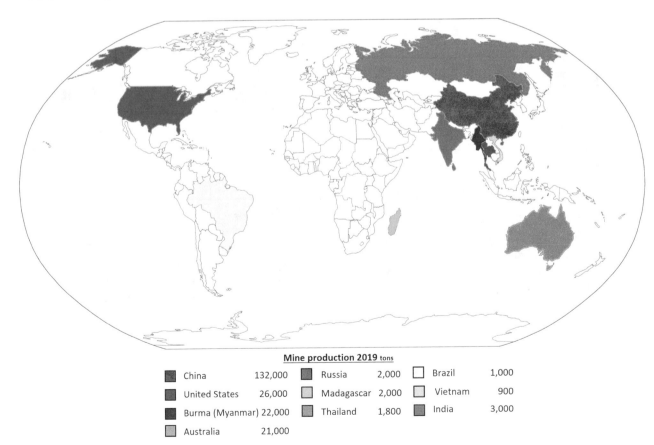

Mine production 2019 tons

China	132,000	Russia	2,000	Brazil	1,000
United States	26,000	Madagascar	2,000	Vietnam	900
Burma (Myanmar)	22,000	Thailand	1,800	India	3,000
Australia	21,000				

Fig. 73.2 List of producing countries based on the US Geological Survey, Mineral Commodity Summaries

73.3 Applications

Electronics Industry

Lutetium is used as a dopant in gallium gadolinium garnet (GGG) to manufacture magnetic memory devices for computers (bubble memory). It is also used to obtain organic LEDs (OLEDs) (diodes with a layer of organic compounds that react and produce light under electrical stimulation), and are already used in the screens of mobile phones, tablets, TVs, and portable electronic equipment, allowing them to curve, if the designer wishes (Fig. 73.3).

Chemical Industry

Lutetium is used as a catalyst in the oil industry in the cracking process, in hydrogenation (adding one or more hydrogen molecules to a compound), and in polymerization (obtaining a polymer by repeating a base monomer).

Medicine

Lutetium tantalate is the densest white material known. It is used as a phosphorescent material (phosphorophore) for x-ray equipment and high-intensity discharge lamps. (phosphorophore: a compound that exhibits luminescence when excited by electrons) (has the properties of elemental phosphor).

Lutetium orthosilicate is used in nuclear medicine in PET (positron emission tomography) equipment, in just 20 min obtaining an image of the human body with the healthy areas are differentiated from altered ones.

Lutetium orthosilicate acts as a sensor of photons emitted by the protons of the radioisotope injected into a patient (usually fluorine-18).

The isotope Lu-177 (dotatate) is a radiopharmaceutical used in patients with inoperable metastases from neuroendocrine tumors in the gastrointestinal tract (stomach, intestine, and pancreas). It emits both beta radiation, which has a therapeutic effect, and gamma radiation, which allows imaging of the progress of the treatment.

73.4 Recycling

Lutetium recycling is unknown.

References

Aagaard P (1974) Rare earth elements adsorption on clay minerals. Bulletin Du Groupe Français Des Argiles 26(2):193–199. https://doi.org/10.3406/argil.1974.1217

Bao Z, Zhao Z (2008) Geochemistry of mineralization with exchangeable REY in the weathering crusts of granitic rocks in South China. Ore Geol Rev 33(3–4):519–535. https://doi.org/10.1016/j.oregeorev.2007.03.005

Chi R, Tian J (2008) Weathered crust elution-deposited rare earth ores. Nova Science Publishers

Goodenough KM, Wall F, Merriman D (2018) The rare earth elements: demand, global resources, and challenges for resourcing future generations. Nat Resour Res 27(2):201–216. https://doi.org/10.1007/s11053-017-9336-5

Further Reading

Alkane Resources (2021) Rare earths. http://www.alkane.com.au/products/rare-earths-overview/. Last accessed May 2021

Roskill (2021) Market reports. Rare earths. https://roskill.com/market-report/rare-earths/. Last accessed May 2021

USGS (2021) Commodity statistics and information. Rare earths. https://www.usgs.gov/centers/nmic/rare-earths-statistics-and-information. Last accessed May 2021

Fig. 73.3 Mobile phone with OLED display. (*Image courtesy of Samsung*)

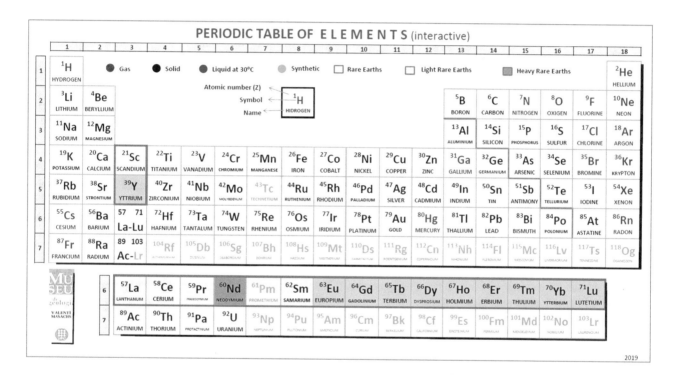

- A member of the group of light rare earths (LREE)
- Discovered in 1885 by the Austrian mineralogist, Welsbach
- Named from two Greek words, *neos* (new) and *didymos* (twin)
- One of the most reactive elements
- Excellent magnetic properties
- Has a metallic silver sheen
- Rapidly oxidizes upon contact with air
- Found in clays impregnated with the rare earths monazite (Fig. 74.1) and bastnäsite.

74.1 Geology

The variations are substantial within the REE class of ore deposits, and the formation of an REE ore deposit gives little information about its classification. Besides, classifying REE ore deposits on the basis of only their genetic evolution would quickly induce misinterpretations. For instance, the class of copper-gold-uranium-REE-iron (IOCG) includes the well-known Olympic Dam deposit (South Australia), the iron deposits of Kiruna (Sweden), the iron-REE deposits of Box Bixby and Pea Ridge (Missouri, United States) and possibly the REE-rich Bayan Obo (Mongolia), the Palabora carbonatite-hosted copper, and the Vergenoug iron-fluorine

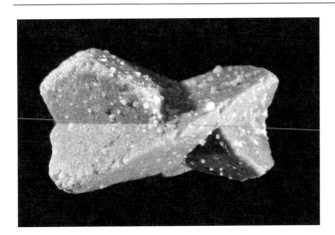

Fig. 74.1 Monazite-(La) (twinned crystals) (cerium, lanthanum, neodymium phosphate). *Minas Geraes* (*Brazil*). (*Photo* Joaquim Sanz. MGVM)

deposit (South Africa). When examined in detail, these deposits are remarkably different.

There are several types of natural (primary) REE resources, including those formed by high-temperature geological processes (carbonatites, alkaline rocks, vein, and skarn deposits) and those formed by low-temperature processes (placers, laterites, bauxites, and ion-adsorption clays) (Goodenough et al. 2018).

Clay-fixing REE by adsorption in clays (Aagaard 1974) in the south of China comprises the main source of worldwide resources (ion-adsorption clays) (Bao and Zhao 2008). The various deposits produce different types of REE: weathering of two mica granites produces laterites rich in Y-HREE, whilst weathering of biotitic granites generates laterites rich in LREE (Chi and Tian 2008).

Neodymium is basically obtained from ion-adsorption clays, from monazite-(Nd) (Nd, Ce, La)PO_4, and bastnäsite (Ce, La, Y, Nd) (CO_3)F. The most important deposits in the exploitation of bastnäsite are in the Bayan Obo and Maoniuping carbonatites in China and in Mount Weld in Australia, among others. Monazite is found in carbonatite deposits such as Mount Weld and Bayan Obo, as well as in the placer deposits of India, Malaysia, and Brazil, as it is highly resistant to transport and common in detritic sediments, accumulating in alluvial and beach placers with zircon, ilmenite, and rutile. Monazite is a widely distributed mineral, appearing as an accessory in granitic igneous rocks, gneissic metamorphic rocks, and detrital sands. Neodymium is the Nd-dominant member of the monazite series.

74.2 Producing Countries

The main world reserves of rare earths are located in China (44 Mt), Brazil and Vietnam (22 Mt), Russia (12 Mt), India (6.9 Mt), Australia (3.4 Mt), Greenland (1.5 Mt), and the United States (1.4 Mt), and also in Myanmar, Canada, Madagascar, and Thailand (Fig. 74.2).

74.3 Applications

Automotive, Electrical and Electronic Industries

Neodymium (Nd-Fe-B) magnets are the most powerful and most compact permanent magnets available. Although they are fragile, they are cheaper, lighter, and more powerful than cobalt-samarium magnets, but need dysprosium to function at moderate temperatures. They are used in the 90% electric motors of electric and hybrid cars (Fig. 74.3) and in the generators of electric turbines, where they enhance the generation of energy from wind (Fig. 74.4). They are also used in air-conditioning equipment, in lifts, in computer hard disks, in small-volume electronic devices such as headphones, loudspeakers, and microphones, and also in military vehicles and aircraft (see: dysprosium).

The presence of praseodymium together with neodymium (NdPr) contributes to maintaining the power of the magnet even at high working temperatures.

Metallurgical Industry

The synthetic yttrium-aluminum garnet (YAG), neodymium doped, is the basic component of the powerful Nd-YAG lasers used for welding (aluminum, titanium, and stainless metal) and cutting metals and other materials.

Medicine

Neodymium, boron, and iron are components of the large permanent magnets in open magnetic resonance imaging (MRI) equipment. $NdYVO_4$-YAG lasers (neodymium-yttrium-vanadate/yttrium-aluminum-garnet) are vital in ophthalmology, dermatology, and otolaryngology (Fig. 74.5).

Other Applications

Ferrocerium, a material made of iron, cerium, lanthanum, neodymium, praseodymium, and magnesium, is used to make sparking lighter 'flints'.

Neodymium is used as a catalyst in the vulcanization of rubber for Formula 1 car tires; it makes them stick to the asphalt and last longer.

74.4 Recycling

Commercial recycling of rare earths is still very scarce. Prices have fallen, so recycling is not competitive; however, some companies are recycling neodymium from magnets recovered from the electric motors that contain it, such as from hybrid cars and electric vehicles, wind turbines, and electronic devices (hard disks, speakers, etc.).

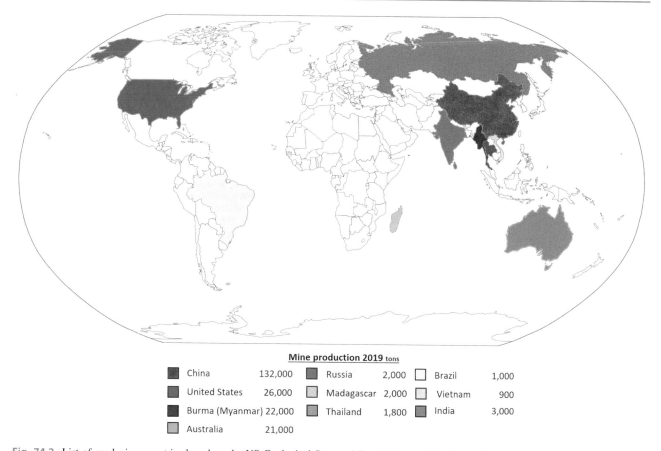

Fig. 74.2 List of producing countries based on the US Geological Survey, Mineral Commodity Summaries

Mine production 2019 tons

■ China	132,000	■ Russia	2,000	□ Brazil	1,000
■ United States	26,000	■ Madagascar	2,000	■ Vietnam	900
■ Burma (Myanmar)	22,000	■ Thailand	1,800	■ India	3,000
■ Australia	21,000				

Fig. 74.3 Electric vehicle being charged. (*Photo* Joaquim Sanz. MGVM)

Fig. 74.4 Wind turbine with neodymium magnets. (*Photo* Joaquim Sanz. MGVM)

For the time being, recycling is carried out using organic solvents (Solvent Extraction—SX). The company Géoméga-Innord is successfully testing (2019) a new technology (ISR) without organic solvents as a clean technology for separation of rare earth elements. It is a sustainable method and a competitive alternative to the SX system, achieving recycling with a very high purity level for neodymium, praseodymium, terbium, and dysprosium. However, as the market for hybrid electric vehicles grows and our current wind turbines need to be changed, the volume of permanent magnets with neodymium for recycling will increase and promote interest in the recovery of these elements.

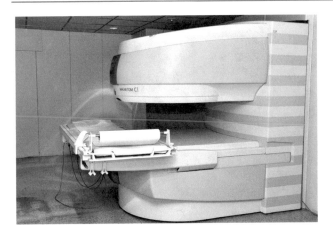

Fig. 74.5 MRI equipment. (*Photo* Joaquim Sanz. MGVM)

References

Aagaard P (1974) Rare earth elements adsorption on clay minerals. Bulletin Du Groupe Français Des Argiles 26(2):193–199. https://doi.org/10.3406/argil.1974.1217

Bao Z, Zhao Z (2008) Geochemistry of mineralization with exchangeable REY in the weathering crusts of granitic rocks in South China.

Ore Geol Rev 33(3–4):519–535. https://doi.org/10.1016/j.oregeorev.2007.03.005

Chi R, Tian J (2008) Weathered crust elution-deposited rare earth ores. Nova Science Publishers

Goodenough KM, Wall F, Merriman D (2018) The rare earth elements: demand, global resources, and challenges for resourcing future generations. Nat Resour Res 27(2):201–216. https://doi.org/10.1007/s11053-017-9336-5

Further Reading

Alkane Resources (2021) Rare earths. http://www.alkane.com.au/products/rare-earths-overview/. Last accessed May 2021

European Training Network for the Design and Recycling of Rare Earth Permanent Magnet Motors and Generators in Hybrid and Full Electric Vehicles (DEMETER) (2020) https://etn-demeter.eu. Last accessed May 2021

Geomega. Innord Inc (2021) Rare earths separation. Neodymium. https://ressourcesgeomega.ca. Last accessed May 2021

Roskill (2021) Market reports. Rare earths. https://roskill.com/market-report/rare-earths/. Last accessed May 2021

USGS (2021) Commodity statistics and information. Rare earths. https://www.usgs.gov/centers/nmic/rare-earths-statistics-and-information. Last accessed May 2021

PERIODIC TABLE OF ELEMENTS (interactive)

- A member of the light rare earths (LREE)
- Identified in 1885 by Carl Auer von Welsbach, an Austrian mineralogist
- Reactive, and in air it develops a green-colored layer
- Ductile and more resistant to corrosion than lanthanum, europium, cerium, or neodymium
- Found in monazite, bastnäsite, and in clays impregnated with rare earths (Fig. 75.1)

75.1 Geology

The variations are substantial within the REE class of ore deposits, and the formation of an REE ore deposit gives little information about its classification. Besides, classifying REE ore deposits on the basis of only their genetic evolution would quickly induce misinterpretations. For instance, the class of copper-gold-uranium-REE-iron (IOCG) includes the well-known Olympic Dam deposit (South Australia), the iron deposits of Kiruna (Sweden), the iron-REE deposits of Box Bixby and Pea Ridge (Missouri, United States) and

Fig. 75.1 Clay with rare earths and praseodymium. (*China*). (*Photo* Joaquim Sanz. MGVM)

possibly the REE-rich Bayan Obo (Mongolia), the Palabora carbonatite-hosted copper, and the Vergenoug iron-fluorine deposit (South Africa). When examined in detail, these deposits are remarkably different.

There are several types of natural (primary) REE resources, including those formed by high-temperature geological processes (carbonatites, alkaline rocks, vein, and skarn deposits) and those formed by low-temperature processes (placers, laterites, bauxites, and ion-adsorption clays) (Goodenough et al. 2018).

Clay-fixing REE by adsorption in clays (Aagaard 1974) in the south of China comprises the main source of worldwide resources (ion-adsorption clays) (Bao and Zhao 2008). The various deposits produce different types of REE: weathering of two mica granites produces laterites rich in Y-HREE, whilst weathering of biotitic granites generates laterites rich in LREE (Chi and Tian 2008).

Praseodymium is basically obtained from monazite-(Ce), (Nd, Ce, La, Th)PO_4, from bastnäsite (Ce, La, Y,) $(CO_3)F$, and from ion-adsorption clays. The most important deposits in the exploitation of bastnäsite are the Bayan Obo and Maoniuping carbonatites in China and in Mount Weld in Australia, among others.

Monazite is found in carbonatite deposits such as Mount Weld and Bayan Obo, as well as in the placer deposits of India, Malaysia, and Brazil. It is highly resistant to transport and common in detritic sediments, accumulating in alluvial and beach placers with zircon, ilmenite, and rutile. Monazite is a widely distributed mineral, appearing as an accessory in granitic igneous rocks, gneissic metamorphic rocks, and detrital sands.

75.2 Producing Countries

The main world reserves of rare earths are in China (44 Mt), Brazil and Vietnam (22 Mt), Russia (12 Mt), India (6.9 Mt), Australia (3.4 Mt), Greenland (1.5 Mt), and the United States (1.4 Mt), and also in Myanmar, Canada, Madagascar, and Thailand (Fig. 75.2).

75.3 Applications

Permanent-Magnet Manufacture
Praseodymium helps to maintain the strength and performance of the magnetic field in neodymium magnets (NdFeB), even when subjected to high temperatures in electric vehicles (Fig. 75.3) and wind turbines, besides making them more resistant to corrosion.

Glass and Ceramics Industry
Neodymium praseodymium salts are used in the manufacture of glass for protective eyewear for electric welders, as well as to produce glass that protects against ultraviolet light. In ceramics, praseodymium oxide is used together with zirconium oxide as a bright yellow pigment.

Power Generation
Praseodymium is used together with europium in the manufacture of energy-saving lamps, to improve their energy efficiency.

Metallurgical Industry
The alloy of praseodymium with magnesium forms a metal of high mechanical resistance that is used in aircraft engines, preventing their corrosion.

Other Fields
Ferrocerium, a material made of iron, cerium, lanthanum, neodymium, praseodymium, and magnesium, is used to make sparking lighter 'flints'. In electric arc lamps, used to illuminate scenes when making movies, the carbon electrodes are doped with praseodymium to impart a tone similar to daylight.

75.4 Recycling

Recycling processes are being developed for the praseodymium in the magnets containing it recovered from electric motors, wind turbines, and electronic equipment using ISR technology (Innord's separation of REE).

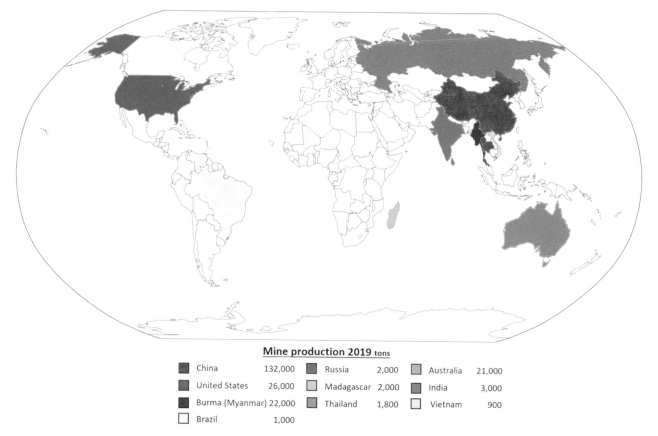

Mine production 2019 tons

■ China	132,000	■ Russia	2,000	■ Australia	21,000
■ United States	26,000	□ Madagascar	2,000	■ India	3,000
■ Burma (Myanmar)	22,000	■ Thailand	1,800	□ Vietnam	900
□ Brazil	1,000				

Fig. 75.2 List of producing countries based on the US Geological Survey, Mineral Commodity Summaries

Fig. 75.3 Electric vehicle being charged. (*Photo* Joaquim Sanz. MGVM)

References

Aagaard P (1974) Rare earth elements adsorption on clay minerals. Bulletin Du Groupe Français Des Argiles 26(2):193–199. https://doi.org/10.3406/argil.1974.1217

Bao Z, Zhao Z (2008) Geochemistry of mineralization with exchangeable REY in the weathering crusts of granitic rocks in South China. Ore Geol Rev 33(3–4):519–535. https://doi.org/10.1016/j.oregeorev.2007.03.005

Chi R, Tian J (2008) Weathered crust elution-deposited rare earth ores. Nova Science Publishers

Goodenough KM, Wall F, Merriman D (2018) The rare earth elements: demand, global resources, and challenges for resourcing future generations. Nat Resour Res 27(2):201–216. https://doi.org/10.1007/s11053-017-9336-5

Further Reading

Alkane Resources (2021) Rare earths. http://www.alkane.com.au/products/rare-earths-overview/. Last accessed May 2021

Geomega. Innord Inc. Rare Earths Separation (2020) https://ressourcesgeomega.ca. Last accessed May 2021

Roskill (2021) Market reports. Rare earths. https://roskill.com/market-report/rare-earths/. Last accessed May 2021

USGS (2021) Commodity statistics and information. Rare earths. https://www.usgs.gov/centers/nmic/rare-earths-statistics-and-information. Last accessed May 2021

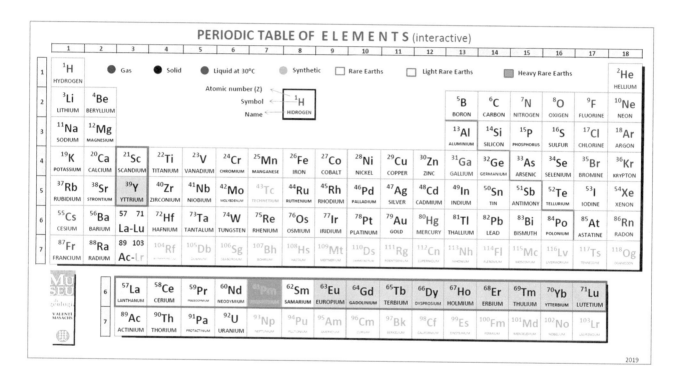

- Not found in nature
- Named after the Greek god Prometheus
- A radioactive element that emits beta particles
- Obtained as a by-product of uranium fission (Fig. 76.1)

76.1 Applications

Radiation and Energy Generation

Promethium is used as a source of beta radiation to measure thickness. It is deployed as a light source for signals that require reliable and independent operation, along with phosphorus, which absorbs beta radiation and produces phosphorescence light.

The isotope promethium-147 can be used in the manufacture of nuclear batteries, in which cells convert the beta emissions into electrical current over a five-year lifetime. Promethium chloride was used for many years, along with zinc sulfide, in the manufacture of luminous watch paint after radio salts were discouraged because they are radioactive. When neodymium is subjected to intense neutron radiation in a nuclear reactor, it transforms into promethium.

76.2 Recycling

Recycling of promethium is unknown.

Further Readings

American Elements (2021) Promethium. https://www.americanelements. com/pm.html. Last accessed May 2021

Lenntech (2021) Promethium. http://www.lenntech.com/periodic/elements/ pm.htm. Last accessed May 2021

Fig. 76.1 Nuclear fuel element: the blue color is due to promethium. (*Image courtesy of* Javier Castelo)

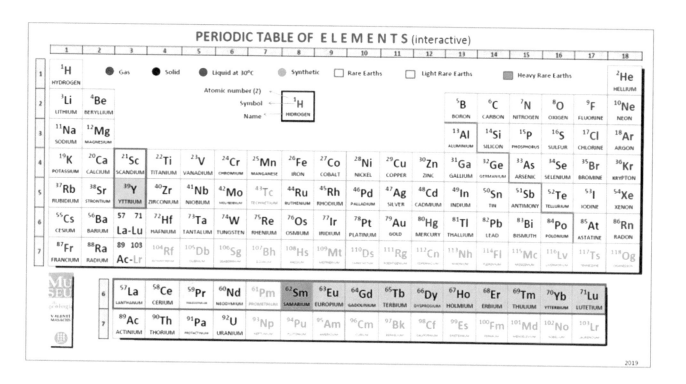

- A member of the light rare earths (LREE)
- Discovered in 1879 in samarskite by the French chemist Paul-Émile Lecoq, Who named it after this mineral
- Good magnetic properties
- Has a high melting point
- Found in the monazite (Fig. 77.1), Rare-earth clays, and xenotime

77.1 Geology

The variations are substantial within the REE class of ore deposits, and the formation of an REE ore deposit gives little information about its classification. Besides, classifying REE ore deposits on the basis of only their genetic evolution would quickly induce misinterpretations. For instance, the class of copper–gold–uranium–REE–iron (IOCG) includes the well-known Olympic Dam deposit (South Australia), the

Fig. 77.1 Monazite-(Sm) (cerium, Samarium, gadolinium, thorium phosphate). Evje (Norway) (*Photo* Joaquim Sanz. MGVM)

iron deposits of Kiruna (Sweden), the iron-REE deposits of Box Bixby and Pea Ridge (Missouri, United States) and possibly the REE-rich Bayan Obo (Mongolia), the Palabora carbonatite-hosted copper, and the Vergenoug iron-fluorine

deposit (South Africa). When examined in detail, these deposits are remarkably different.

There are several types of natural (primary) REE resources, including those formed by high-temperature geological processes (carbonatites, alkaline rocks, vein, and skarn deposits) and those formed by low-temperature processes (placers, laterites, bauxites, and ion-adsorption clays) (Goodenough et al. 2018).

Clay-fixing REE by adsorption in clays (Aagaard 1974) in the south of China comprises the main source of world-wide resources (ion-adsorption clays) (Bao and Zhao 2008). The various deposits produce different types of REE: weathering of two mica granites produces laterites rich in Y-HREE, whilst weathering of biotitic granites generates laterites rich in LREE (Chi and Tian 2008).

Samarium is basically obtained from monazite-(Sm), (Sm, Gd, Ce, Th) PO_4, from ion-adsorption clays and from xenotime-(Y), YPO_4. Monazite is found in carbonatite deposits such as Mount Weld and Bayan Obo, as well as in the placer deposits of India, Malaysia, and Brazil, among others. Monazite and xenotime are found in placer deposits. Both are highly resistant to transport and are common in

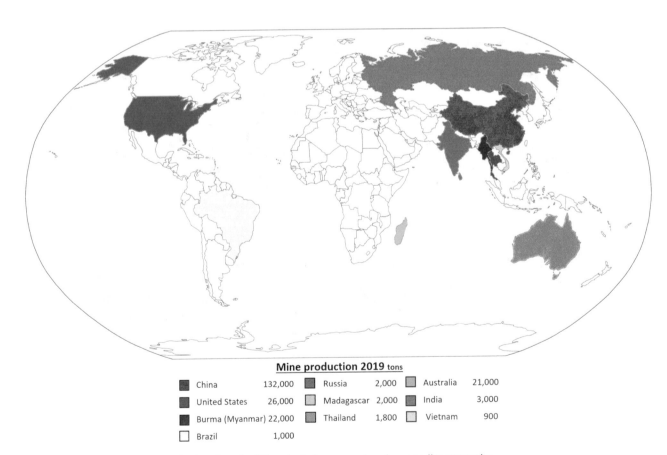

Mine production 2019 tons

China	132,000	Russia	2,000	Australia	21,000
United States	26,000	Madagascar	2,000	India	3,000
Burma (Myanmar)	22,000	Thailand	1,800	Vietnam	900
Brazil	1,000				

Fig. 77.2 List of producing countries based on the US geological survey, mineral commodity summaries

detritic sediments, accumulating in alluvial and beach placers with zircon, ilmenite, and rutile. Monazite is a widely distributed mineral, appearing as an accessory in granitic igneous rocks, gneissic metamorphic rocks, and detrital sands. The most important deposits for the exploitation of xenotime are in the Bayan Obo and Maoniuping carbonatites in China and in Mount Weld in Australia, among others.

77.2 Producing Countries

The main world reserves of rare earths are in China (44 Mt), Brazil and Vietnam (22 Mt), Russia (12 Mt), India (6.9 Mt), Australia (3.4 Mt), Greenland (1.5 Mt), and the United States (1.4 Mt), and also in Myanmar, Canada, Madagascar, and Thailand (Fig. 77.2).

77.3 Applications

Manufacture of Magnets
Samarium-cobalt magnets are very powerful (after neodymium, the most powerful), have a high resistance to demagnetization, and work perfectly at temperatures of up to 350 °C. That makes them very useful in aerospace (Fig. 77.3), microwave devices, and in military helicopters.

Samarium is also used in magnets for the electromagnetic pads of electric guitars and electronic musical instruments.

Medicine
Samarium-153 is used as a palliative medicine to relieve the severe pain of bone cancer. The trade name is Quadramet (Fig. 77.4).

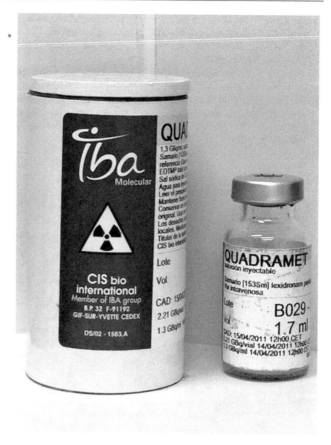

Fig. 77.4 Quadramet medicine. (*Photo* Joaquim Sanz. MGVM)

Glass Industry
Samarium is added to glass to absorb infrared radiation.

Power Generation
Samarium has a high neutron absorption capacity, which is why it is used in the manufacture of nuclear reactor control rods, whose role is to regulate the nuclear chain reaction that produces heat.

77.4 Recycling

Recycling of this element is being looked into; systems of metal separation using samarium-cobalt magnets are being sought and their profitability investigated.

References

Aagaard P (1974) Rare earth elements adsorption on clay minerals. Bull Du Groupe Français Des Argiles 26(2):193–199. https://doi.org/10.3406/argil.1974.1217

Fig. 77.3 Commercial aircraft (*Photo* Joaquim Sanz. MGVM)

Bao Z, Zhao Z (2008) Geochemistry of mineralization with exchange-able REY in the weathering crusts of granitic rocks in South China. Ore Geol Rev 33(3–4):519–535. https://doi.org/10.1016/j.oregeorev.2007.03.005

Chi R, Tian J (2008) Weathered Crust Elution-deposited Rare Earth Ores. Nova Science Publishers

Goodenough KM, Wall F, Merriman D (2018) The rare earth elements: demand, global resources, and challenges for resourcing future generations. Nat Resour Res 27(2):201–216. https://doi.org/10.1007/s11053-017-9336-5

Further Reading

Alkane Resources (2021) Rare Earths. http://www.alkane.com.au/products/rare-earths-overview/ (last accessed May 2021)

Roskill (2021) Market Reports. Rare Earths. https://roskill.com/market-report/rare-earths/ (last accessed May 2021)

USGS (2021) Commodity Statistcs and Information. Rare Earths. https://www.usgs.gov/centers/nmic/rare-earths-statistics-and-information (last accessed May 2021)

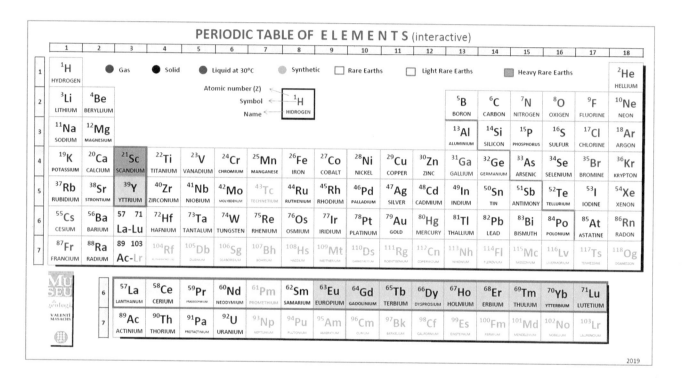

PERIODIC TABLE OF ELEMENTS (interactive)

- Classified as a rare earths element because it presents similar characteristics, scandium is not actually an REE
- Predicted theoretically by Mendeleyev in 1871
- Discovered in 1879 by Lars Fredrik Nilson, who named it after his homeland of Scandinavia
- A rare, light, and acid-resistant metal
- 40% lighter than titanium
- Assigned the status of a strategic metal by the EU in 2017
- Mainly obtained as a by-product of mining rare earth and iron in China, phosphates in Kola (Fig. 78.1), uranium in Ukraine and Kazakhstan, and nickel laterite ore in the Philippines

- Can also be extracted by carbonization of the red sludge remaining after aluminum production (red mud).

78.1 Geology

Approximately 90% of global production is from the Bayan Obo deposit in China, where scandium is a by-product of mining other REE and iron. It is hosted mostly by aegirine, although a small but significant proportion is present in bastnäsite-Ce, monazite-Ce, and fluorite.

Fig. 78.1 Apatite with scandium. (Kovdor, Kola Peninsula) Russia (*Photo* Joaquim Sanz. MGVM)

After China, Russia is the most important scandium-producing country. The carbonatite-phosphorite deposit at Kovdor, Kola Peninsula, is the most important producer of scandium in baddeleyite-magnetite-apatite, and Tomtor (Yakutia) is one of the world's largest carbonatite deposits.

Outside of China and Russia, nickel laterite deposits in the Philippines and Australia are the most important scandium resources (Vasyukova and Williams-Jones 2018).

Uranium deposits in Kazakhstan are good scandium producers as a by-product from the leaching uranium process (ROSATOM). Research is being carried out to try to produce scandium from red mud wastes (the bauxite-aluminum process) (Siegfried and Wall 2018).

78.2 Producing Countries

There are no data on world reserves of scandium. Due to its low concentration, it is obtained as a by-product of processing other metals (iron, REE, titanium, zirconium, uranium, and apatite). However, its presence is observed in low concentrations in more than 100 minerals. It has been identified in Australia, Canada, China, Kazakhstan, Madagascar, Norway, the Philippines, Russia, Ukraine, and the United States (Fig. 78.2).

78.3 Applications

Metallurgical Industry

The main application of scandium is in alloys with aluminum, such as AlMgSc, with which a very corrosion-resistant, light, and durable alloy is achieved with a very good capacity for laser-beam fusion welding, obtaining excellent properties in

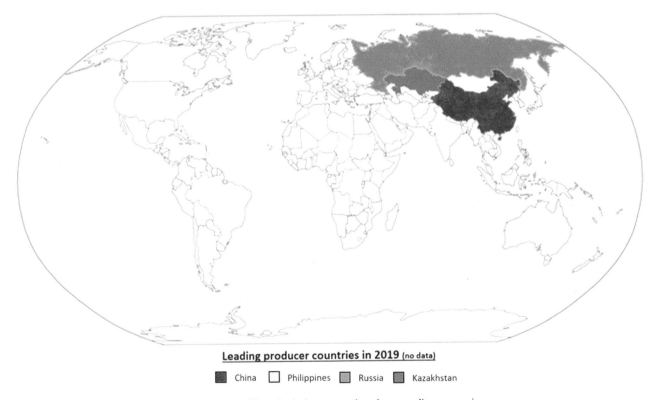

Leading producer countries in 2019 (no data)

⬛ China ☐ Philippines ◻ Russia ◼ Kazakhstan

Fig. 78.2 List of producing countries based on the US geological survey, mineral commodity summaries

welded joints and eliminating the need for rivets. The Airbus Group company has been researching AlMgSc (Scalmalloy®) alloys for some years.

APWorks, a subsidiary of Airbus, is designing a super-light motorcycle with this AlMgSc alloy, to be manufactured by 3D metal printing (Additive Manufacturing).

Also being tested is the combination of aluminum-scandium-zircon (AlScZr), an alloy that can be welded instead of riveted and thus reduce the weight of commercial aircraft by 20%, with a corresponding reduction in fuel consumption.

AlSc alloys may contribute to reducing the weight of electric vehicles (EV), with a consequent saving in the size of the Li-ion batteries.

Although the presence of scandium on the market is very limited and prices very high, it is believed that increases in its application will improve the extraction volumes and reduce the cost.

AlSc alloys are used in the manufacture of high-end sports equipment, such as baseball bats, bicycle frames, and golf clubs. These alloy is inert in the presence of salt water and are used in the construction of certain parts of boats and desalination plants.

Lamp Industry

Scandium oxide is used in the manufacture of high-intensity discharge lamps to illuminate stadiums, stages, and similar. Scandium iodide, added to mercury vapor lamps, increases the intensity of the light emitted and imparts a color temperature similar to that of the sun, making the images in nighttime TV broadcasts appear as if they were in daylight (Fig. 78.3).

Fig. 78.3 Football pitch lighting. (*Image* courtesy of Albert Prat)

Fuel Cells

Taking advantage of its high electrical conductivity and heat stabilization, scandium is used in the manufacture of solid oxide fuel cells (SOFCs) (Bloom Energy 2021), allowing a lower operating temperature and longer equipment life with less expensive materials.

Other Fields

The use of scandium together with gadolinium and gallium is being investigated for the manufacture of crystals for high-energy lasers.

78.4 Recycling

Recycling of scandium is unknown.

References

Li G et al (2018) Scandium from Bauxite Ore Residues. https://doi.org/10.1016/j.hydromet.2018.01.007

Kalashnikov A et al (2016) Scandium of the Kovdor magnetite-apatite-baddeleyite deposit. Ore Geol Rev 72(1):532–537. https://www.researchgate.net/publication/281271899

Williams-Jones AE, Vasyukova OV (2018) The economic geology of scandium, the runt of the rare earth element litter. Econ Geol 113(4):973–998. https://doi.org/10.5382/econgeo.2018.4579

Further Reading

Airbus APWorks (2017) Metal 3D printing. https://3dprintingindustry.com/news/apworks-and-additive-industries-advance-metal-3d-printing-to-series-production-in-aerospace-160166/ (last accessed May 2021)

Bloom Energy (2021) https://www.bloomenergy.com/ (last accessed May 2021)

Scandium International (2021) http://www.scandiummining.com/ (last accessed May 2021)

USGS (2019) Commodity Statistics and Information. Scandium . Available at: https://prd-wret.s3-us-west-2.amazonaws.com/assets/palladium/production/atoms/files/mcs-2019-scandi.pdf (last accessed May 2021)

Pete R. Siegfried. (2019) Scandium an important and critical metal. Available at: https://www.minersoc.org/wp-content/uploads/2019/05/3ICM-Siegfried.pdf (last accessed May 2021)

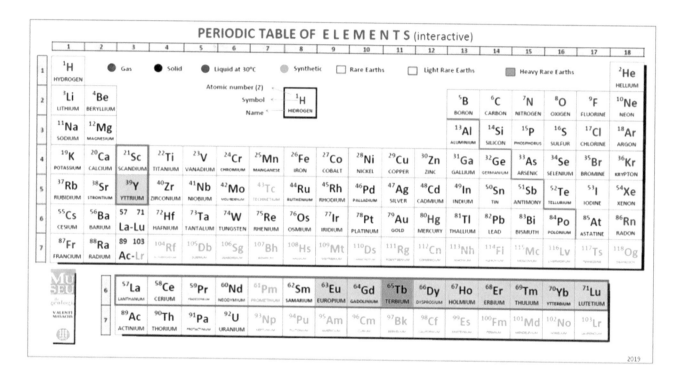

- A member of the heavy rare earths (HREE)
- Discovered in 1843 in Ytterby mine, Sweden
- Malleable and ductile
- Relatively stable in contact with air
- Oxidizes slowly
- Found in the heavy rare-earth clays of China, in xenotime (Fig. 79.1), and in monazite.

79.1 Geology

The variations are substantial within the REE class of ore deposits, and the formation of an REE ore deposit gives little information about its classification. Besides, classifying REE ore deposits on the basis of only their genetic evolution would quickly induce misinterpretations. For instance, the class of copper–gold-uranium-REE-iron (IOCG) includes

Fig. 79.1 Xenotime-Y (yttrium phosphate with cerium and terbium). Novo Horizonte (Bahia, Brazil) (*Photo* Joaquim Sanz. MGVM)

the well-known Olympic Dam deposit (South Australia), the iron deposits of Kiruna (Sweden), the iron-REE deposits of Box Bixby and Pea Ridge (Missouri, United States) and possibly the REE-rich Bayan Obo (Mongolia), the Palabora carbonatite-hosted copper, and the Vergenoug iron-fluorine deposit (South Africa). When examined in detail, these deposits are remarkably different.

There are several types of natural (primary) REE resources, including those formed by high-temperature geological processes (carbonatites, alkaline rocks, vein, and skarn deposits) and those formed by low-temperature processes (placers, laterites, bauxites, and ion-adsorption clays) (Goodenough et al. 2018).

Clay-fixing REE by adsorption in clays (Aagaard 1974) in the south of China comprises the main source of world-wide resources (ion-adsorption clays) (Bao and Zhao 2008). The various deposits produce different types of REE: weathering of two mica granites produces laterites rich in Y-HREE, whilst weathering of biotitic granites generates laterites rich in LREE (Chi and Tian 2008).

Terbium is basically obtained from ion-adsorption clays and from xenotime-(Y), YPO_4, but it can also be extracted from monazite $(Ce,La,Nd,Th)PO_4$. The most important deposits for the exploitation of xenotime are in the Bayan Obo and Maoniuping carbonatites in China and in Mount Weld in Australia, among others. Monazite is found in carbonatite deposits such as Mount Weld and Bayan Obo.

Monazite and xenotime are found in placer deposits. Both are highly resistant to transport and are common in detritic sediments, accumulating in alluvial and beach placers with

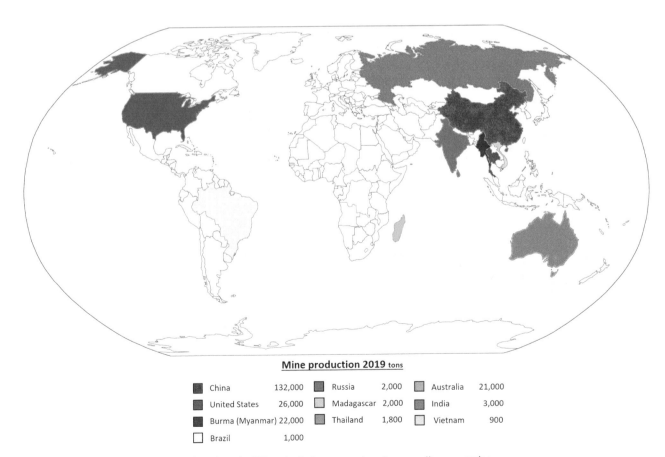

Mine production 2019 tons

China	132,000	Russia	2,000	Australia	21,000
United States	26,000	Madagascar	2,000	India	3,000
Burma (Myanmar)	22,000	Thailand	1,800	Vietnam	900
Brazil	1,000				

Fig. 79.2 List of producing countries based on the US geological survey, mineral commodity summaries

zircon, ilmenite, and rutile. Important placer deposits are found in India, Malaysia, Myanmar, and Brazil, among others. Monazite is a widely distributed mineral, appearing as an accessory in granitic igneous rocks, gneissic metamorphic rocks, and detrital sands.

79.2 Producing Countries

The main world reserves of rare earths are in China (44 Mt), Brazil and Vietnam (22 Mt), Russia (12 Mt), India (6.9 Mt), Australia (3.4 Mt), Greenland (1.5 Mt), and the United States (1.4 Mt), and also in Myanmar, Canada, Madagascar, and Thailand (Fig. 79.2).

79.3 Applications

Automotive Industry, and Power Generation
Terbium, along with neodymium and dysprosium, is part of the neodymium magnets (NdFeB) that are a fundamental part of electric motors for hybrid vehicles (Fig. 79.3) and also wind turbines, as they help to withstand high working temperatures while enhancing performance in the generation of energy from the wind.

Electronic Industry
Terbium is used as green phosphorous material (phosphorophore) in fluorescent lamps and in plasma and LCD screens (phosphorophore: compound that presents luminescence when excited by electrons) (has the properties of the element phosphorous).

Terbium (lemon-green phosphorescence) and europium (red and blue phosphorescence) produce an intense white light, brighter than the traditional fluorescent light, and achieve a better chromatic balance.

Terphenol-D is an alloy of iron, terbium, and dysprosium that is highly magnetostrictive (expands or contracts in the presence of a magnetic field). This is used in naval sonar systems and magnetomechanical sensors.

Other Fields
Terbium (III) is present in the ink used on euro banknotes that fluoresce green under ultraviolet light.

79.4 Recycling

Processes are being developed to recycle terbium from magnets containing it, recovered from electric motors, wind turbines, and electronic equipment using ISR technology (Innord's separation of REE).

At its La Rochelle factory (France), the Solvay company started recycling fluorescent tubes and energy-saving lamps in 2011 due to their content of rare earths such as cerium, lanthanum, europium, terbium, gadolinium, and yttrium. However, in 2016, this type of recycling ceased due to the fall in rare-earth prices (compared to 2011), the reduction in the supply of used fluorescent tubes (due to the progressive introduction of LED-based lighting), and the stabilization of rare earth exports from China.

References

Aagaard P (1974) Rare earth elements adsorption on clay minerals. Bull Du Groupe Français Des Argiles 26(2):193–199. https://doi.org/10.3406/argil.1974.1217

Bao Z, Zhao Z (2008) Geochemistry of mineralization with exchangeable REY in the weathering crusts of granitic rocks in South China. Ore Geol Rev 33(3–4):519–535. https://doi.org/10.1016/j.oregeorev.2007.03.005

Chi R, Tian J (2008) Weathered crust elution-deposited rare Earth ores. Nova Science Publishers

Goodenough KM, Wall F, Merriman D (2018) The rare earth elements: demand, global resources, and challenges for resourcing future generations. Nat Resour Res 27(2):201–216. https://doi.org/10.1007/s11053-017-9336-5

Further Reading

Alkane Resources (2021) Rare Earths. http://www.alkane.com.au/products/rare-earths-overview/ (last accessed May 2021)

Geomega. Innord Inc. (2020) Rare Earths Separation. https://ressourcesgeomega.ca (last accessed May 2021)

Roskill (2021) Market Reports. Rare Earths. https://roskill.com/market-report/rare-earths/ (last accessed May 2021)

USGS (2021) Commodity Statistics and Information. Rare Earths. https://www.usgs.gov/centers/nmic/rare-earths-statistics-and-information (last accessed May 2021)

Fig. 79.3 Hybrid car engine. (*Photo* Joaquim Sanz. MGVM)

PERIODIC TABLE OF E L E M E N T S (interactive)

1	2	3	4	5	6	7	8	9	10	11	12	13	14	15	16	17	18
¹H HYDROGEN																	²He HELLIUM
³Li LITHIUM	⁴Be BERYLLIUM											⁵B BORON	⁶C CARBON	⁷N NITROGEN	⁸O OXIGEN	⁹F FLUORINE	¹⁰Ne NEON
¹¹Na SODIUM	¹²Mg MAGNESIUM											¹³Al ALUMINIUM	¹⁴Si SILICON	¹⁵P PHOSPHORUS	¹⁶S SULFUR	¹⁷Cl CHLORINE	¹⁸Ar ARGON

Gas • Solid • Liquid at 30°C • Synthetic □ Rare Earths □ Light Rare Earths ▨ Heavy Rare Earths

Atomic number (Z), Symbol, Name

¹⁹K POTASSIUM	²⁰Ca CALCIUM	²¹Sc SCANDIUM	²²Ti TITANIUM	²³V VANADIUM	²⁴Cr CHROMIUM	²⁵Mn MANGANESE	²⁶Fe IRON	²⁷Co COBALT	²⁸Ni NICKEL	²⁹Cu COPPER	³⁰Zn ZINC	³¹Ga GALLIUM	³²Ge GERMANIUM	³³As ARSENIC	³⁴Se SELENIUM	³⁵Br BROMINE	³⁶Kr KRYPTON
³⁷Rb RUBIDIUM	³⁸Sr STRONTIUM	³⁹Y YTTRIUM	⁴⁰Zr ZIRCONIUM	⁴¹Nb NIOBIUM	⁴²Mo MOLYBDENUM	⁴³Tc TECHNETIUM	⁴⁴Ru RUTHENIUM	⁴⁵Rh RHODIUM	⁴⁶Pd PALLADIUM	⁴⁷Ag SILVER	⁴⁸Cd CADMIUM	⁴⁹In INDIUM	⁵⁰Sn TIN	⁵¹Sb ANTIMONY	⁵²Te TELLURIUM	⁵³I IODINE	⁵⁴Xe XENON
⁵⁵Cs CESIUM	⁵⁶Ba BARIUM	57 71 La-Lu	⁷²Hf HAFNIUM	⁷³Ta TANTALUM	⁷⁴W TUNGSTEN	⁷⁵Re RHENIUM	⁷⁶Os OSMIUM	⁷⁷Ir IRIDIUM	⁷⁸Pt PLATINUM	⁷⁹Au GOLD	⁸⁰Hg MERCURY	⁸¹Tl THALLIUM	⁸²Pb LEAD	⁸³Bi BISMUTH	⁸⁴Po POLONIUM	⁸⁵At ASTATINE	⁸⁶Rn RADON
⁸⁷Fr FRANCIUM	⁸⁸Ra RADIUM	89 103 Ac-Lr	¹⁰⁴Rf	¹⁰⁵Db	¹⁰⁶Sg	¹⁰⁷Bh	¹⁰⁸Hs	¹⁰⁹Mt	¹¹⁰Ds	¹¹¹Rg	¹¹²Cn	¹¹³Nh	¹¹⁴Fl	¹¹⁵Mc	¹¹⁶Lv	¹¹⁷Ts	¹¹⁸Og

| 6 | ⁵⁷La LANTHANUM | ⁵⁸Ce CERIUM | ⁵⁹Pr PRASEODYMIUM | ⁶⁰Nd NEODYMIUM | ⁶¹Pm PROMETHIUM | ⁶²Sm SAMARIUM | ⁶³Eu EUROPIUM | ⁶⁴Gd GADOLINIUM | ⁶⁵Tb TERBIUM | ⁶⁶Dy DYSPROSIUM | ⁶⁷Ho HOLMIUM | ⁶⁸Er ERBIUM | ⁶⁹Tm THULIUM | ⁷⁰Yb YTTERBIUM | ⁷¹Lu LUTETIUM |
| 7 | ⁸⁹Ac ACTINIUM | ⁹⁰Th THORIUM | ⁹¹Pa PROTACTINIUM | ⁹²U URANIUM | ⁹³Np NEPTUNIUM | ⁹⁴Pu PLUTONIUM | ⁹⁵Am AMERICIUM | ⁹⁶Cm CURIUM | ⁹⁷Bk BERKELIUM | ⁹⁸Cf CALIFORNIUM | ⁹⁹Es EINSTENIUM | ¹⁰⁰Fm FERMIUM | ¹⁰¹Md MENDELEVIUM | ¹⁰²No NOBELIUM | ¹⁰³Lr LAURENCIUM |

2019

- A member of the heavy rare earths (HREE)
- Discovered in 1879 by the Swedish chemist, Per Teodor Cleve
- Named by Cleve after Thule, the ancient name for Scandinavia
- The least abundant of the entire REE group
- Malleable, ductile, and very rare
- Oxidizes slowly in the presence of moisture
- Found in clays from China that are impregnated with rare earths (Fig. 80.1), also in xenotime.

80.1 Geology

The variations are substantial within the REE class of ore deposits, and the formation of an REE ore deposit gives little information about its classification. Besides, classifying REE ore deposits on the basis of only their genetic evolution would quickly induce misinterpretations. For instance, the class of copper–gold-uranium-REE-iron (IOCG) includes the well-known Olympic Dam deposit (South Australia), the iron deposits of Kiruna (Sweden), the iron-REE deposits of Box Bixby and Pea Ridge (Missouri, United States) and

Fig. 80.1 Clay with rare earths and thulium. (China) (*Photo* Joaquim Sanz. MGVM)

possibly the REE-rich Bayan Obo (Mongolia), the Palabora carbonatite-hosted copper, and the Vergenoug iron-fluorine deposit (South Africa). When examined in detail, these deposits are remarkably different.

There are several types of natural (primary) REE resources, including those formed by high-temperature geological processes (carbonatites, alkaline rocks, vein, and skarn deposits) and those formed by low-temperature processes (placers, laterites, bauxites, and ion-adsorption clays) (Goodenough et al. 2018).

Clay-fixing REE by adsorption in clays (Aagaard 1974) in the south of China comprises the main source of world-wide resources (ion-adsorption clays) (Bao and Zhao 2008). The various deposits produce different types of REE: weathering of two mica granites produces laterites rich in Y-HREE, whilst weathering of biotitic granites generates laterites rich in LREE (Chi and Tian 2008).

Thulium is obtained basically from ion-adsorption clays and from xenotime-(Y), YPO₄. The most important deposits in terms of the exploitation of xenotime are the Bayan Obo and Maoniuping carbonatites in China and at Mount Weld in Australia. Xenotime is also found in placer deposits. It is highly resistant to transport and common in detritic sediments, accumulating in alluvial and beach placers with zircon,

Mine production 2019 tons

China	132,000	Russia	2,000	Brazil	1,000
United States	26,000	Madagascar	2,000	Vietnam	900
Burma (Myanmar)	22,000	Thailand	1,800	India	3,000
Australia	21,000				

Fig. 80.2 List of producing countries based on the US geological survey, mineral commodity summaries

ilmenite, and rutile. Important placer deposits are in India, Malaysia, Myanmar, and Brazil, among other countries.

80.2 Producing Countries

The main world reserves of rare earths are in China (44 Mt), Brazil and Vietnam (22 Mt), Russia (12 Mt), India (6.9 Mt), Australia (3.4 Mt), Greenland (1.5 Mt), and the United States (1.4 Mt), with other countries such as Myanmar, Madagascar, Canada, and Thailand (Fig. 80.2).

80.3 Applications

Medicine

The laser (Ho-Cr-Tm)-YAG is used in medicine to perform surface tissue ablation (see rare earths: holmium). The thulium laser (YAG Tm 200w) is used with great efficiency in day surgery for benign prostate enlargement (HPE), up to large sizes, extending to the very limits of the capsule and thus discouraging any regrowth. After the intervention, the patient recovers within two to three days and experiences rapid relief (see: rare earths: holmium).

Other Fields

The thulium-170 isotope, with an energy of 85 keV, is used as a source of gamma rays in portable industrial radiography, archaeology, art, medicine, etc., despite its high cost and limited availability in the market.

The laser (Ho-Cr-Tm)-YAG is highly efficient and has many uses in meteorology and the military.

Thulium (III) is present in the ink for euro banknotes that fluoresce blue under ultraviolet light (Fig. 80.3).

80.4 Recycling

Thulium recycling is unknown.

Fig. 80.3 Euro banknotes under ultraviolet light. (*Photo* Joaquim Sanz. MGVM)

References

Aagaard P (1974) Rare earth elements adsorption on clay minerals. Bull Du Groupe Français Des Argiles 26(2):193–199. https://doi.org/10.3406/argil.1974.1217

Bao Z, Zhao Z (2008) Geochemistry of mineralization with exchangeable REY in the weathering crusts of granitic rocks in South China. Ore Geol Rev 33(3–4):519–535. https://doi.org/10.1016/j.oregeorev.2007.03.005

Chi R, Tian J (2008) *Weathered Crust Elution-deposited Rare Earth Ores*. Nova Science Publishers.

Goodenough KM, Wall F, Merriman D (2018) The rare earth elements: demand, global resources, and challenges for resourcing future generations. Nat Resour Res 27(2):201–216. https://doi.org/10.1007/s11053-017-9336-5

Further Reading

Alkane Resources (2021) Rare Earths. http://www.alkane.com.au/products/rare-earths-overview/ (last accessed May 2021)

Roskill (2021) Market Reports. Rare Earths. https://roskill.com/market-report/rare-earths/ (last accessed May 2021)

USGS (2021) Commodity Statistics and Information. Rare Earths. https://www.usgs.gov/centers/nmic/rare-earths-statistics-and-information (last accessed May 2021)

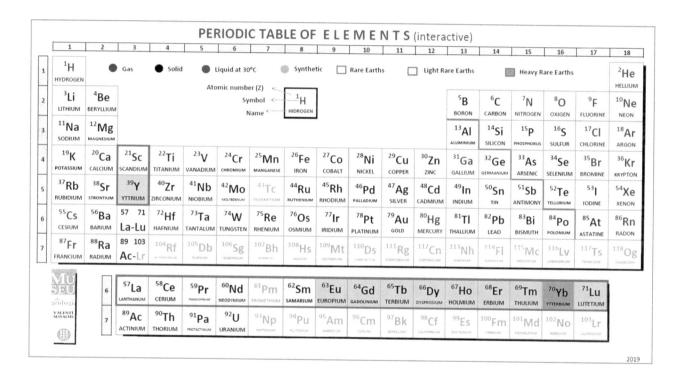

- A member of the heavy rare earths (HREE)
- Discovered in 1878 by the French chemist Jean de Marignac and named after the Swedish village of Ytterby
- Malleable and ductile
- Reacts with water and oxidizes in air
- Mainly found in the heavy rare earth-impregnated clays of China (Fig. 81.1), and in xenotime.

81.1 Geology

The variations are substantial within the REE class of ore deposits, and the formation of an REE ore deposit gives little information about its classification. Besides, classifying REE ore deposits on the basis of only their genetic evolution would quickly induce misinterpretations. For instance, the class of copper–gold-uranium-REE-iron (IOCG) includes the well-known Olympic Dam deposit (South Australia), the

Fig. 81.1 Clay with rare earths and ytterbium. (*China*) (*Photo Joaquim Sanz. MGVM*)

iron deposits of Kiruna (Sweden), the iron-REE deposits of Box Bixby and Pea Ridge (Missouri, United States) and possibly the REE-rich Bayan Obo (Mongolia), the Palabora carbonatite-hosted copper, and the Vergenoug iron-fluorine deposit (South Africa). When examined in detail, these deposits are remarkably different.

There are several types of natural (primary) REE resources, including those formed by high-temperature geological processes (carbonatites, alkaline rocks, vein, and skarn deposits) and those formed by low-temperature processes (placers, laterites, bauxites, and ion-adsorption clays) (Goodenough et al. 2018).

Clay-fixing REE by adsorption in clays (Aagaard 1974) in the south of China comprises the main source of worldwide resources (ion-adsorption clays) (Bao and Zhao 2008). The various deposits produce different types of REE: weathering of two mica granites produces laterites rich in Y-HREE, whilst weathering of biotitic granites generates laterites rich in LREE (Chi and Tian 2008).

Ytterbium is obtained basically from ion-adsorption clays and from xenotime-(Y), YPO$_4$. The most important deposits in terms of the exploitation of xenotime are the Bayan Obo and Maoniuping carbonatites in China and at Mount Weld in Australia. Xenotime is also found in placer deposits. It is highly resistant to transport and common in detritic sediments, accumulating in alluvial and beach placers with zircon, ilmenite, and rutile. Important placer deposits are in India, Malaysia, Myanmar, and Brazil, among other countries.

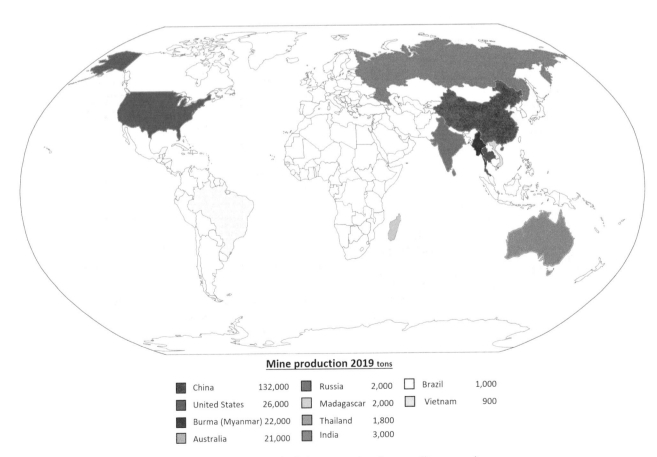

Mine production 2019 tons

China	132,000	Russia	2,000	Brazil	1,000
United States	26,000	Madagascar	2,000	Vietnam	900
Burma (Myanmar)	22,000	Thailand	1,800		
Australia	21,000	India	3,000		

Fig. 81.2 List of producing countries based on the US geological survey, mineral commodity summaries

81.2 Producing Countries

The world's main reserves of rare earths are in China (44 Mt), Brazil and Vietnam (22 Mt), Russia (12 Mt), India (6.9 Mt), Australia (3.4 Mt), Greenland (1.5 Mt), and the United States (1.4 Mt), also in Myanmar, Canada, Madagascar, and Thailand (Fig. 81.2).

81.3 Applications

Metallurgical Industry

Ytterbium is added to stainless steel to improve its refinement and strength.

The isotope yterbium-169 is an important x-ray emitter between 49 and 59 keV, as well as being a gamma emitter from 63 to 308 keV. It is used as an x-ray source in portable industrial radiography for steel thicknesses of between 4 and 15 mm (Gilligan Engineering, 2021). However, selenium-75, which has a longer half-life than ytterbium-169 (120 days versus 32 days) and is more affordable, is replacing it (Rueda 2006).

Power Generation

Ytterbium is used in solar cells, along with silicon, to convert solar energy into electricity by improving the efficiency of the absorption of infrared wavelengths.

Electronics Industry

Ytterbium can be used as an amplifier in fiber-optic cables. By interleaving segments of optical fibers doped with this element along the entire length of the cable, the signal level is increased.

Ytterbium is involved in the manufacture of industrial product-marker lasers.

Other Fields

Ytterbium, when subjected to high pressure, increases its electrical resistance, which is why it is used to measure ground deformation in nuclear explosions, blastings (Fig. 81.3), and earthquakes.

81.4 Recycling

Ytterbium recycling is unknown.

Fig. 81.3 Blasting in a limestone quarry. (*Image* courtesy of PROMSA)

References

Aagaard P (1974) Rare earth elements adsorption on clay minerals. Bull Du Groupe Français Des Argiles 26(2):193–199. https://doi.org/10.3406/argil.1974.1217

Bao Z, Zhao Z (2008) Geochemistry of mineralization with exchangeable REY in the weathering crusts of granitic rocks in South China. Ore Geol Rev 33(3–4):519–535. https://doi.org/10.1016/j.oregeorev.2007.03.005

Chi R, Tian J (2008) Weathered crust elution-deposited rare Earth ores. Nova Science Publishers

Goodenough KM, Wall F, Merriman D (2018) The rare earth elements: demand, global resources, and challenges for resourcing future generations. Nat Resour Res 27(2):201–216. https://doi.org/10.1007/s11053-017-9336-5

Further Reading

Alkane Resources (2021) Rare Earths. http://www.alkane.com.au/products/rare-earths-overview/ (last accessed May 2021)

Gilligan Engineering (2021) Ytterbium. Yb-169. https://gilligans.co.uk/products/ytterbium-169/ (last accessed May 2021)

Roskill (2021) Market Reports. Rare Earths. https://roskill.com/market-report/rare-earths/ (last accessed May 2021)

USGS (2021) Commodity Statistics and Information. Rare Earths. https://www.usgs.gov/centers/nmic/rare-earths-statistics-and-information (last accessed May 2021)

PERIODIC TABLE OF E L E M E N T S (interactive)

Periodic table of elements, 2019. Legend: Gas, Solid, Liquid at 30°C, Synthetic, Rare Earths, Light Rare Earths, Heavy Rare Earths. Atomic number (Z), Symbol, Name.

- Classified as an element of the rare earths as it presents similar characteristics, yet yttrium is not actually an REE
- Identified in 1789 by Johan Gadolin, who named after Ytterby, a Swedish village—the elements ytterbium, terbium, and erbium were also discovered here
- Has the greatest known affinity for oxygen of any element
- Relatively stable in contact with air
- Has high impact resistance and a high melting point
- Found in xenotime (Fig. 82.1), in the impregnated rare-earth clays of China, in monazite, and in bastnäsite.

82.1 Geology

The variations are substantial within the REE class of ore deposits, and the formation of an REE ore deposit gives little information about its classification. Besides, classifying REE ore deposits on the basis of only their genetic evolution would quickly induce misinterpretations. For instance, the class of copper–gold–uranium–REE–iron (IOCG) includes the well-known Olympic Dam deposit (South Australia), the iron deposits of Kiruna (Sweden), the iron-REE deposits of Box Bixby and Pea Ridge (Missouri, United States) and possibly the REE-rich Bayan Obo (Mongolia), the Palabora carbonatite-hosted copper, and the Vergenoug iron-fluorine deposit (South Africa). When examined in detail, these deposits are remarkably different.

J. Sanz et al., *Elements and Mineral Resources*, Springer Textbooks in Earth Sciences,
Geography and Environment, https://doi.org/10.1007/978-3-030-85889-6_82

Fig. 82.1 Xenotime-Y (yttrium phosphate with cerium). Novo Horizonte (Bahia, Brazil) (*Photo* Joaquim Sanz. MGVM)

There are several types of natural (primary) REE resources, including those formed by high-temperature geological processes (carbonatites, alkaline rocks, vein, and skarn deposits) and those formed by low-temperature processes (placers, laterites, bauxites, and ion-adsorption clays) (Goodenough et al. 2018).

Clay-fixing REE by adsorption in clays (Aagaard 1974) in the south of China comprises the main source of worldwide resources (ion-adsorption clays) (Bao and Zhao 2008). The various deposits produce different types of REE: weathering of two mica granites produces laterites rich in Y-HREE, whilst weathering of biotitic granites generates laterites rich in LREE (Chi and Tian 2008).

Yttrium is basically obtained from ion-adsorption clays and from xenotime-(Y), YPO_4, but it can also be extracted from monazite $(Ce,La,Nd,Th)PO_4$ and from bastnäsite, $(Ce,La,Y,)$ $(CO_3)F$. The most important deposits for the exploitation of xenotime and bastnäsite are in the Bayan Obo and Maoniuping carbonatites in China and Mount Weld in Australia, among others. Monazite, too, is found in carbonatite deposits such as Mount Weld and Bayan Obo.

Monazite and xenotime are also found in placer deposits. Both are highly resistant to transport and are common in detritic sediments, accumulating in alluvial and beach placers with zircon, ilmenite, and rutile. Important placer deposits are in India, Malaysia, Myanmar, Brazil, among other countries. Monazite is a widely distributed mineral, appearing as an accessory in granitic igneous rocks, gneissic metamorphic rocks, and detrital sands.

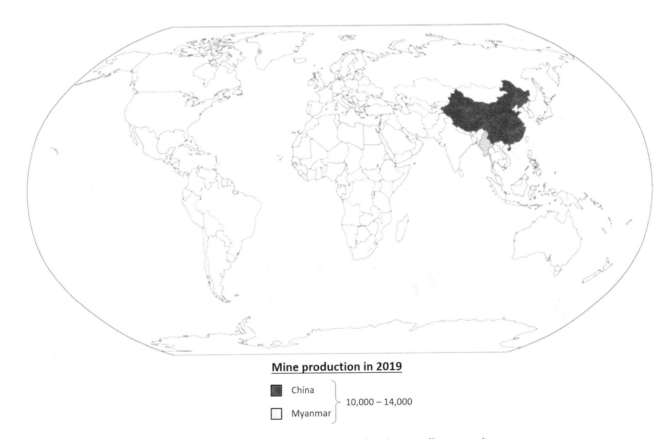

Mine production in 2019

China
Myanmar
} 10,000 – 14,000

Fig. 82.2 List of producing countries based on the US geological survey, mineral commodity summaries

82.2 Producing Countries

The main yttrium reserves are located in: Australia, Brazil, Canada, China, and India. The estimated world reserves of this element are 500,000 tons (Fig. 82.2).

82.3 Applications

Metallurgical Industry

Yttrium is added to magnesium and aluminum alloys to impart resistance to corrosion. Synthetic yttrium–aluminum garnet (YAG), doped with neodymium, is the basic component of the powerful Nd-YAG lasers used in welding and cutting of steels and other metals.

Electronics and Lighting Industry

Yttrium has a prominent use in white LED technology. It is also used in fluorescent tubes to produce an intense white light with significant energy savings (Fig. 82.3).

Yttrium, doped with europium, is used as a red phosphorescent (phosphorophore) material in the manufacture of plasma and LCD screens (phosphorophore: a compound that exhibits luminescence when excited by electrons) (has the properties of phosphorus element).

Ceramic Industry

Yttrium oxide added to zircon oxide is part of an extremely hard and heat-resistant ceramic used in the manufacture of knives, in dentistry, and in engines.

Medicine

NdYVO$_4$-YAG   (neodymium-yttrium-vanadium/yttrium–aluminum-garnet) lasers are vital in ophthalmology, dermatology, and otolaryngology.

The isotope yttrium-90 is used in the treatment of bone, ovarian, and pancreatic tumors and leukemia.

Other Fields

Yttrium oxide and cerium oxide are substitutes for thorium in gas-light casings, because they are heat resistant, give an intense light when heated, and are not radioactive.

82.4 Recycling

At its La Rochelle factory (France), the Solvay company started recycling fluorescent tubes and energy-saving lamps in 2011 due to their content of rare earths such as cerium, lanthanum, europium, terbium, gadolinium, and yttrium. However, in 2016, this type of recycling ceased due to the fall in rare-earth prices (compared to 2011), the reduction in the supply of used fluorescent tubes (due to the progressive introduction of LED-based lighting), and the stabilization of rare earth exports from China.

Research continues into the recycling of REE elements obtained from old consumer electronics equipment ('urban mining'), since current chemical separation methods are expensive and unprofitable for recycling companies.

References

Aagaard P (1974) Rare earth elements adsorption on clay minerals. Bull Du Groupe Français Des Argiles 26(2):193–199. https://doi.org/10.3406/argil.1974.1217

Bao Z, Zhao Z (2008) Geochemistry of mineralization with exchangeable REY in the weathering crusts of granitic rocks in South China. Ore Geol Rev 33(3–4):519–535. https://doi.org/10.1016/j.oregeorev.2007.03.005

Chi R, Tian J (2008) Weathered crust elution-deposited rare Earth ores. Nova Science Publishers

Goodenough KM, Wall F, Merriman D (2018) The rare earth elements: demand, global resources, and challenges for resourcing future generations. Nat Resour Res 27(2):201–216. https://doi.org/10.1007/s11053-017-9336-5

Further Reading

Alkane Resources (2021) Rare Earths. http://www.alkane.com.au/products/rare-earths-overview/ (last accessed May 2021)

Roskill (2021) Market Reports. Rare Earths. https://roskill.com/market-report/rare-earths/ (last accessed on May 2021)

USGS (2021) Commodity Statistics and Information. Rare Earths. https://www.usgs.gov/centers/nmic/rare-earths-statistics-and-cinformation (last accessed May 2021)

Fig. 82.3 Compact fluorescent tube. (*Photo* Joaquim Sanz. MGVM)

Introduction

83.1 The Minerals

In this part of the book you will find most of what are termed the industrial minerals (generally excluding metals and mineral fuels) of high economic value that, rather than because of the compounds or elements that can be extracted from them, are used directly, depending on their physical or chemical properties either in industrial processes or after appropriate preparation. A family home contains up to 150 tons of industrial minerals and a car up to 250 kg; these minerals represent 50% of the content of paint and paper, while ceramics and glass are made entirely of these minerals. They are essential to our daily lives.

About 50% of these minerals are recycled indirectly; that is, when many of the products in which they have been used are reused or transformed. Ceramics, concrete, glass, paper, plastics, etc. are being recycled and primary industrial minerals are being saved (http://www.ima-europe.eu).

83.2 Applications

The most important applications of these minerals fall under the following four headings.

Construction

Essential minerals for building houses, making roads, glass, mortar, concrete, ceramic tiles, and so on, include limestone, gypsum, dolomite, thenardite, silica sand, and kaolinite (Fig. 83.1).

Industry

Industrial minerals are involved in the manufacture of refractories, cements, insulation, abrasives, paper, plastic, tires, paint, detergents, cosmetics, medicines, fertilizers (minerals such as magnesite, thenardite, talc, silica, calcite, mica, sylvinite, and dolomite) (Fig. 83.2).

Environment

Such minerals are involved in the treatment of soils, water, and sewage, in the restoration of degraded land (minerals such as: gypsum, talc, dolomite, zeolites, calcite, and magnesite) (Fig. 83.3).

Agriculture and Livestock

Minerals are essential as fertilizers for land, to correct soil pH, and as additives in animal feed (among others, sylvinite, magnesite, phosphorite, calcite, and dolomite) (Fig. 83.4).

J. Sanz et al., *Elements and Mineral Resources*, Springer Textbooks in Earth Sciences, Geography and Environment, https://doi.org/10.1007/978-3-030-85889-6_83

Fig. 83.1 Highway (Image courtesy of Salvador Redó); A university building. (*Photos* Joaquim Sanz. MGVM)

Fig. 83.2 Plastics in a car interior; Magazines and journals. (*Photos* Joaquim Sanz. MGVM)

Fig. 83.3 Wastewater treatment plant; Field of sunflowers on reclaimed land. (*Photos* Joaquim Sanz. MGVM)

Fig. 83.4 Healthy cows; Plump cherries, thanks to agrichemicals. (*Photos* Joaquim Sanz. MGVM)

Fig. 84.1 Andalusite. (Cap de Creus) Catalonia. *Photo* Joaquim Sanz. MGVM

- An aluminum silicate
- Very hard (7.5 on the Mohs scale)
- Fragile
- Has prismatic crystals with a square section (Fig. 84.1)
- Highly refractory, and resistant to both thermal shock and chemical attack
- Transformed into mullite between 1200 °C and 1550 °C
- From metamorphic rock.

84.1 Geology

Andalusite is an aluminum silicate (Al_2SiO_5), sharing its chemical composition with kyanite and sillimanite. These three minerals are polymorphs, and are often summarized as the sillimanite minerals. Andalusite is stable at pressures below 4 kbar, and is formed especially by low-grade metamorphism, by low-grade contact, or regional metamorphism of argillaceous sediments (Okrusch and Frimmel 2020).

Andalusite is a typical constituent of metapelites such as mica schists and it can intergrow with quartz, especially in contact metamorphic aureoles (Clarke et al. 2005). Moreover, it can occur in migmatites and in relatively Al-rich magmatic rock such as rhyolite or granite, aplite, and pegmatite.

Metasomatic or hydrothermal depletion of SiO_2 can also lead to the formation of aluminosilicates. An element could be enriched or depleted by these processes, for instance acidic igneous rocks in contact with ultramafic rocks and in epithermal alteration zones (Neukirchen and Ries 2020). Moreover, andalusite can accumulate together with other heavy minerals in placer deposits and as a detrital mineral in some sandstones.

The main andalusite ore deposits correspond to contact metamorphic metapelites in the Bushveld Complex (South Africa) and in Glomel (France). Additionally, the finest andalusite gems are found in the Santa Tereza deposit (Brazil) and in Sri Lanka's gemmiferous sands.

84.2 Producing Countries

Important deposits of andalusite are known in China, France, Peru, and South Africa, but the size of their reserves is unknown (Fig. 84.2).

84.3 Applications

Electrical Industry

Porcelains prepared with andalusite give excellent results in the manufacture of electrical insulators with resistance to high temperatures (Fig. 84.3). It is used in the manufacture of certain special ceramics.

J. Sanz et al., *Elements and Mineral Resources*, Springer Textbooks in Earth Sciences, Geography and Environment, https://doi.org/10.1007/978-3-030-85889-6_84

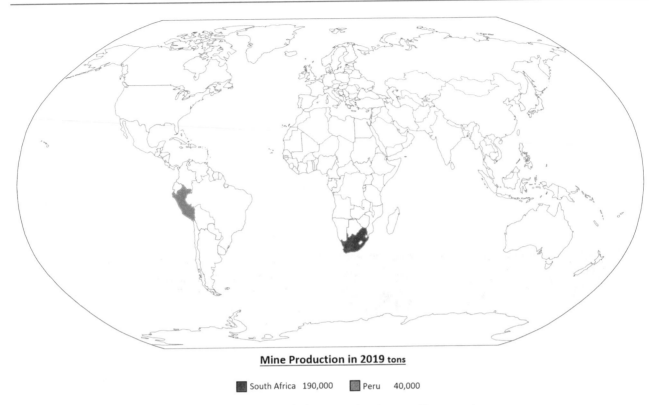

Mine Production in 2019 tons

■ South Africa 190,000   ■ Peru   40,000

Fig. 84.2 List of producing countries based on the US geological survey, mineral commodity summaries

Fig. 84.3 Electrical insulator. (*Photo* Joaquim Sanz. MGVM)

Steel Castings

Mixed with silica sand, andalusite is used to make refractory molds to receive molten steel.

Refractory Bricks

The manufacture of fireproof bricks for kilns, cement, glass, non-ferrous metals, ceramics, etc., uses this mineral as a basic element to guarantee resistance to the high working temperatures experienced on a continuous basis.

Other Fields

Due to the high hardness of andalusite, it is used as an abrasive in many industries. Its temperature resistance and hardness make it the mineral of choice in the manufacture of vehicle brake pads.

Gemology

Andalusite, when it has a good greenish brown color and transparency, is greatly valued as a carved gem, and Sri Lanka and Brazil are the countries with the best deposits of this semi-precious stone.

84.4 Recycling

In the preparation of molds with silica sand for casting, part of andalusite refractory sand is reused, but the recycling is insignificant.

References

Clarke DB, Dorais M, Barbarin B, Barker D, Cesare B, Clarke G, Woodard HH et al (2005) Occurrence and origin of andalusite in peraluminous felsic igneous rocks. J Petrol 46(3):441–472. https://doi.org/10.1093/petrology/egh083

Neukirchen F, Ries G (2020) The world of mineral deposits: a beginner's guide to economic geology. Springer Nature. https://doi.org/10.1007/978-3-030-34346-0

Okrusch M, Frimmel HE (2020) Mineralogy: an introduction to minerals, rocks, and mineral deposits. Springer Nature. https://doi.org/10.1007/978-3-662-57316-7

Further Reading

Anthony JW, Bideaux RA, Bladh KW, Nichols MC (eds) (2003) Handbook of mineralogy, reprinted. Chantilly, VA. Mineralogical Society of America, pp. 20151–1110. http://www.handbookofmineralogy.org/

IMERYS (2021) Refractory Minerals. https://imerys-refractoryminerals.com/andalusite/ (last accessed May 2021)

Mata JM, Sanz J (2007) Guia d'identificació de minerals. Manresa, Catalonia. Edicions UPC/Parcir (Catalan 2nd paper edition). p 262. ISBN: 9788483019023. http://hdl.handle.net/2117/90445

Sanz J, Tomasa O (2017) Elements i Recursos minerals: Aplicacions i reciclatge. Manresa, Catalonia. Zenobita Edicions /Iniciativa Digital Politècnica (Catalan 3rd digital edition). http://hdl.handle.net/2117/105113

USGS (2021) Commodity Statistics and Information. Kyanite and Related Minerals. https://www.usgs.gov/centers/nmic/kyanite-and-related-minerals-statistics-and-information (last accessed May 2021).

Fig. 85.1 Chalk (Calcium carbonate). L'Arbós (Catalonia) (*Photo Joaquim Sanz. MGVM*)

- The basic component of calcite
- Fragile and not very hard
- Reactive with acids, effervescing
- Forms sedimentary and metamorphic rocks, such as limestone, chalk (Fig. 85.1), travertine, and marble.

Calcium carbonate of great beauty has been used as ornamental stone since ancient times, for instance by the Romans, to the present day (Fig. 85.2).

85.1 Geology

Limestone is a sedimentary rock comprised chiefly of calcium carbonate ($CaCO_3$). Deposits are extensive around the world. Therefore, there is a high variability of limestone deposits. Typically, they are formed in two main environments.

The first setting is shallow, calm, and warm waters, nowadays found between latitudes 30 degrees north and 30 degrees south, e.g. the Caribbean Sea and around the Pacific islands. In this setting, organisms can extract the necessary ingredients from ocean water to form calcium carbonate shells and skeletons. These accumulate as sediment and are lithified into limestone. Therefore, limestones formed in this type of deposits are biogenic sediments (Geyssant 2001). After uplift, such deposits become the rock that is widespread around the world.

Alternatively, limestone can be formed through evaporation of groundwater with a high percentage of $CaCO_3$. When moisture from fractures or other pore spaces reaches the roof of a cave, they may form droplets that evaporate before falling to the cave floor. Any dissolved calcium carbonate is deposited on the cave's roof, forming stalactites. Should the droplets fall to the ground, evaporation will contribute to the construction of a stalagmite. Stalactites and stalagmites are not used as a source of calcium; they have only ornamental and aesthetic value.

85.2 Producing Countries

World reserves of calcium carbonate, in its many natural terrestrial forms such as limestone, travertine, chalk, white marble, etc. are very large (Fig. 85.3).

85.3 Applications

Chemical and Glass Industry
Calcium carbonate extracted from chalk or limestone is used as an agricultural fertilizer in slightly acidic soils and in the treatment of drinking water. Calcium oxide (lime) is part of the glass manufacturing process.

© The Author(s), under exclusive license to Springer Nature Switzerland AG 2022
J. Sanz et al., *Elements and Mineral Resources*, Springer Textbooks in Earth Sciences,
Geography and Environment, https://doi.org/10.1007/978-3-030-85889-6_85

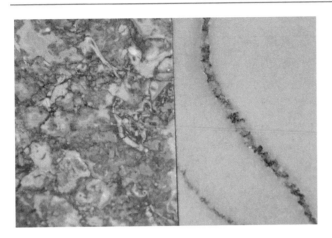

Fig. 85.2 **Fossiliferous limestone** (*brocatello*) (*Tivissa, Catalonia*) and **marble** (*cipollin*o), (*Eubea, Greece*) (*Photo* Joaquim Sanz. MGVM)

Steel Industry

Limestone and lime are essential in the steel industry to smelt iron and steel.

Construction

Also important is the consumption of limestone as an aggregate and in the manufacture of cement, concrete, and lime. Certain limestone and marble rocks of good quality are used as ornamental facades and interiors of buildings, and others as ornamental gravel for gardens.

Fillers

Calcium carbonate is used as a filler to give consistency to paper, cardboard, paint, plastics (PVC), rubber, polymers, adhesives, and gum (Fig. 85.4).

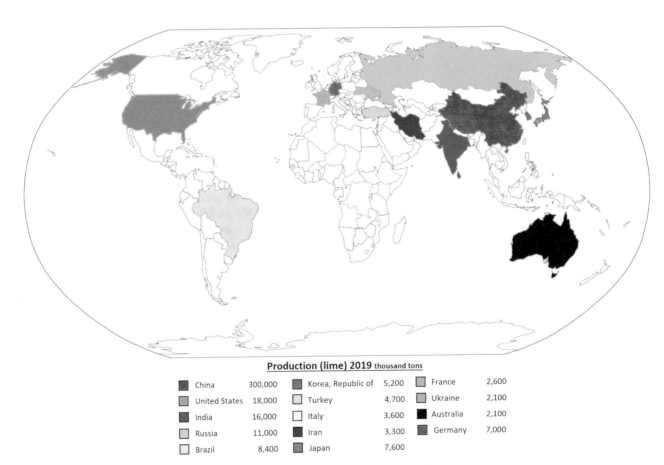

Production (lime) 2019 thousand tons

China	300,000	Korea, Republic of	5,200	France	2,600
United States	18,000	Turkey	4,700	Ukraine	2,100
India	16,000	Italy	3,600	Australia	2,100
Russia	11,000	Iran	3,300	Germany	7,000
Brazil	8,400	Japan	7,600		

Fig. 85.3 List of producing countries based on the US geological survey, mineral commodity summaries

Fig. 85.4 Paper manufacturing. (*Image* courtesy of Stora Enso)

85.4 Recycling

The calcium oxide (lime) that is used in the paper industry, in sewage treatment plants, and in car factories is recycled.

Recycling and/or reuse of glass, plastic, paper, and construction waste (after demolition) contributes to reducing the consumption of limestone and calcium carbonate (Fig. 85.5).

Reference

Geyssant J (2001) Limestone deposits. In Calcium carbonate. Basel. Birkhäuser, pp 31–51. https://doi.org/10.1007/978-3-0348-8245-3_3

Further Reading

Britannica (2021) calcite. https://www.britannica.com/science/calcite (last accessed May 2021)

Global Aggregates Information Network (2021) https://www.gain.ie/ (last accessed May 2021)

IMERYS Carbonates (2021) https://www.imerys-performance-minerals.com/our-minerals/calcium-carbonate (last accessed May 2021)

Mata JM, Sanz J (2007) Guia d'identificació de minerals. Manresa, Catalonia. Edicions UPC/Parcir (Catalan 2nd paper edition). p 262. ISBN: 9788483019023. http://hdl.handle.net/2117/90445

Middleton GV et al (eds) (2016) Encyclopedia of sediments and sedimentary rocks. Springer, Dordrecht. https://doi.org/10.1007/978-1-4020-3609-5

Mindat (2021) Limestone. https://www.mindat.org/min-49160.html (last accessed May 2021)

OMYA SPAIN (2021) Carbonates. https://www.omya.com/ES-EN (last accessed May 2021)

PROVENÇALE. (2021) Calcium Carbonate. http://www.provencale.com/en/ (last accessed May 2021)

Reverté. Calcium Carbonates (2021). https://reverteminerals.es/en/ (last accessed May 2021)

Roskill (2016) Ground & Precipitated Calcium Carbonate. https://roskill.com/market-report/ground-precipitated-calcium-carbonate/ (last accessed May 2021)

Sanz J, Tomasa O (2017) Elements i Recursos minerals: Aplicacions i reciclatge. Manresa, Catalonia. Zenobita Edicions /Iniciativa Digital Politècnica (Catalan 3rd digital edition). http://hdl.handle.net/2117/105113

Warren JK (2016) Interpreting evaporite textures. In Evaporites. Cham. Springer, pp 1–83. https://doi.org/10.1007/978-3-319-13512-0_1

USGS (2021) Commodity Statistics and Information. Lime. https://www.usgs.gov/centers/nmic/lime-statistics-and-information?qt-science_support_page_related_con=0#qt-science_support_page_related_con (last accessed May 2021)

Fig. 85.5 Recovery of aggregates from the demolition of a building. (*Image* courtesy of Regió 7)

It is also involved in the manufacture of casts in artworks, in human food, and in the production of gentle abrasives and toothpaste.

Calcium carbonate is used as an oil and gas absorbent and as a feed supplement for hens to give strength to eggshell. It is also used in the manufacture of fire extinguishers, glazes, ceramics, cosmetics, detergents, and asphalt agglomerate.

Fig. 86.1 Corundum (aluminum oxide). Madagascar (*Photo* Joaquim Sanz. MGVM)

- Aluminum oxide (Fig. 86.1)
- Very hard (9 on the Mohs scale)
- A rock-forming mineral
- Found in igneous, metamorphic, and sedimentary rock
- Transparent red corundum with Fe–Cr impurities is known as ruby, and blue (with Fe-Ti impurities) as sapphire—there are also other colors (yellow, orange).

86.1 Geology

Corundum is an aluminum oxide (Al_2O_3). It is a relatively scarce mineral that appears in aluminous rocks, usually metamorphic, such as marbles, micaceous shales, and gneisses. It may also be related to basic (silica-poor) or intermediate igneous rocks such as pegmatites, syenites, or nepheline syenites. Gemstone specimens are much rarer, especially ruby, which requires the presence of chromium, usually in ultrabasic rocks. More abundant are the alluvial deposits from weathering, erosion, and transport of materials that constituted the primary deposit. For this reason, most exploitation of rubies and sapphires is situated in depositional locations in river environments or alluvial plains.

The main value of corundum lies in its hardness. The name comes from Sanskrit and means ruby: the terms ruby and sapphire refer to red and blue, respectively. The red gemological variety is due to replacement of aluminum by chrome. The typical sapphire is blue corundum, blue because the alumina has been replaced by titanium and iron (Fe^{2+}). Padparadscha, a term derived from the Sinhalese (Sri Lankan) word for lotus flower, is an orange gemological variety. It is very scarce, and because of its beautiful color it is expensive. Others that are less economically valuable are colorless, yellow, green, pink, violet, brown, etc. All are called sapphires. The world reserves of this mineral are centered on Burma, Cambodia, Sri Lanka, India, Afghanistan, Madagascar, and Tanzania, among others.

Corundum crystals that cannot be used as gems are exploited for industrial purposes.

86.2 Producing Countries

For industrial purposes, synthetic corundum is obtained from calcined bauxite (a mixture of aluminum oxides and hydroxides), as it is a reliable and regular source of corundum (Fig. 86.2).

© The Author(s), under exclusive license to Springer Nature Switzerland AG 2022
J. Sanz et al., *Elements and Mineral Resources*, Springer Textbooks in Earth Sciences,
Geography and Environment, https://doi.org/10.1007/978-3-030-85889-6_86

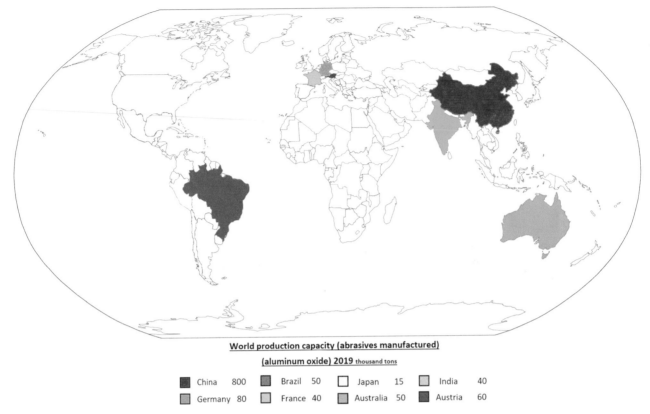

World production capacity (abrasives manufactured)
(aluminum oxide) 2019 thousand tons

| China | 800 | Brazil | 50 | Japan | 15 | India | 40 |
| Germany | 80 | France | 40 | Australia | 50 | Austria | 60 |

Fig. 86.2 List of producing countries based on the US geological survey, mineral commodity summaries

86.3   Applications

Abrasive Manufacturing

The great hardness of corundum makes it especially useful as an abrasive. It is found in nature as an impure granular variety known as emery, formed of a mixture of this mineral with iron oxides (hematite or magnetite) and spinel, of medium hardness 8 on the Mohs scale. This is used for roughening and polishing metals, dulling glass, working precious stones of a hardness equal to or less than 8, sharpening mechanical tools, etc. (Fig. 86.3). However, emery abrasive has been superseded by manufactured abrasives such as silicon carbide (carborundum) (see: silica/quartz).

These abrasives can be found on the market in the form of grinding wheels or emery cloth. Some cosmetic nail files have a layer of finely ground synthetic corundum.

Fig. 86.3 An abrasive disk. (*Photo* Joaquim Sanz. MGVM)

Jewelry

The main application of transparent or translucent corundum of a good color is as a precious stone in jewelry (Fig. 86.4).

For many years synthetic corundum has been manufactured using the Verneuil process or through hydrothermal synthesis, obtaining a quality almost identical to that of natural corundum, both ruby and sapphire, with larger crystals than those found in nature yet at a much lower price than that of the natural stone.

Fig. 86.4 Pendant with a ruby showing a star effect. (*Photo* Joaquim Sanz. MGVM)

Other Fields

Besides jewelry, synthetic corundum is used in the construction of certain mechanical parts, such as tubes, rods, bearings, and scratch-resistant optical equipment, and as a component of lasers.

86.4 Recycling

Certain metal industries that work with emery recover it to the maximum extent for reuse.

Further Reading

Mata JM, Sanz J (2007) Guia d'identificació de minerals. Manresa, Catalonia. Edicions UPC/Parcir (Catalan 2nd paper edition). p 262. ISBN: 9788483019023. http://hdl.handle.net/2117/90445.

Provençale. (2021) Synthetic Corundum. https://www.provencale.com/en/product-range/other-minerals/corindon-synthetique/ (last accessed May 2021)

Sanz J, Tomasa O (2017) Elements i Recursos minerals: Aplicacions i reciclatge. Manresa, Catalonia. Zenobita Edicions/Iniciativa Digital Politècnica (Catalan 3rd digital edition). http://hdl.handle.net/2117/105113

USGS (2021) Commodity Statistics and Information. Abrasives (manufactured). https://www.usgs.gov/centers/nmic/manufactured-abrasives-statistics-and-information (last accessed May 2021)

Dolomite

Fig. 87.1 Dolomite (calcium and magnesium carbonate). *Eugui* (Spain) (*Photo* Joaquim Sanz. MGVM)

- Dolomite **mineral** is a calcium and magnesium carbonate (Fig. 87.1)
- Calcite-like, but does not effervesce with hydrochloric acid unless heated
- More resistant to acids than calcite
- Dolomite **rock** (dolostone) is a sedimentary rock formed mainly of dolomite mineral and calcite (Fig. 87.2), with applications similar to limestone.

There are also valuable dolomitic limestones and dolomitic marbles.

87.1 Geology

Dolomite is a calcium magnesium carbonate with the chemical composition $CaMg(CO_3)_2$, constituting a rock-forming mineral found in abundance around the world. It has numerous commercial uses.

Dolomite mineral is rarely found in modern sedimentary environments; however, dolomite rocks (dolostones) are common in geological history. Furthermore, they can be geographically extensive and reach thicknesses of hundreds to thousands of meters.

Dolomite is formed by diagenesis or hydrothermal metasomatism of limestone. Therefore, rocks that are rich in dolomite were mostly deposited initially as calcium carbonate muds and then altered by magnesium-rich water to form dolomite (Anthony et al. 2011). Moreover, dolomite constitutes a major component of some contact metamorphic rocks and marbles, in carbonatites and ultramafic rocks as well as gangue in hydrothermal veins.

87.2 Producing Countries

Statistics on world dolostone production are not published, and the limited data do not distinguish between that obtained for industrial uses and that for aggregate, for construction, or quarried.

87.3 Applications

The applications described here correspond to dolostone and dolomite mineral.

J. Sanz et al., *Elements and Mineral Resources*, Springer Textbooks in Earth Sciences, Geography and Environment, https://doi.org/10.1007/978-3-030-85889-6_87

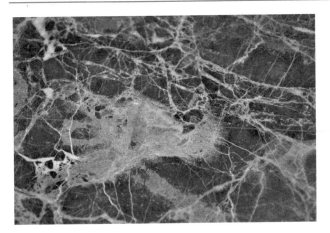

Fig. 87.2 Dolomitic recrystallized limestone. *Jumilla* (Spain) (*Photo* Joaquim Sanz. MGVM)

Agriculture

Dolomite is mixed into soil to reduce its acidity and as a source of magnesium for good plant growth. It directly influences the chlorophyll function and the assimilation of potassium (Fig. 87.3).

Powdered dolomite is an important component of many animal feeds.

Steel Industry

Sintered dolomite (calcined at 1600–1700 °C) has been used in the past as a refractory in the steel industry. Nowadays, since the presence of calcium oxide from the dolomite decreases the final quality of the cast, calcined magnesite is used (see: magnesite).

Fig. 87.3 Field of healthy maize. (*Photo* Joaquim Sanz. MGVM)

Dolomite is also useful as a flux for steel in blast furnaces.

Other Fields

Dolostone and dolomite mineral powder are used as fillers in the manufacture of soap, detergents, paint, ceramics, glazes, rubber, and paper. Dolostone, which has an attractive appearance, is used as an ornamental rock, in construction, and as an interior and exterior wall covering. It is a good thermal insulator and is fire resistant.

As an aggregate, dolostone is used in the preparation of cements, concrete, mortars, asphalt, and stucco, also in sugar refinery and as an adsorbent of heavy metals in contaminated soils, in wastewater treatment, and in the manufacture of flat glass.

87.4   Recycling

The best recycling of this mineral and rock is indirect, recovered by recycling paper and glass.

References

Anthony JW, Bideaux RA, Bladh KW, Nichols, MC (eds) (2003) Handbook of mineralogy. Chantilly, VA. Mineralogical Society of America, pp 20151–1110. http://www.handbookofmineralogy.org/

Further Reading

Calcinor (2021) Dolomite. https://www.calcinor.com/en/products/calcium-and-magnesium-carbonateca-mg-co32 (last accessed May 2021)

Mata JM, Sanz J (2007) Guia d'identificació de minerals. Manresa, catalonia. Edicions UPC/Parcir (Catalan 2nd paper edition). p 262. ISBN: 9788483019023. http://hdl.handle.net/2117/90445

Sanz J, Tomasa O (2017) Elements i Recursos minerals: Aplicacions i reciclatge. Manresa, catalonia. Zenobita Edicions/Iniciativa Digital Politècnica (Catalan 3rd digital edition). http://hdl.handle.net/2117/105113

SAMIN (2021) Dolomite. https://www.samin.fr/produits/dolomie (last accessed May 2021)

SIBELCO (2021) Dolomite. https://www.sibelco.com/materials/dolomite/ (last accessed May 2021)

Warren J (2000) Dolomite: occurrence, evolution and economically important associations. Earth Sci Rev 52(1–3):1–81. https://doi.org/10.1016/S0012-8252(00)00022-2

Fig. 88.1 Orthoclase feldspar (aluminum potassium silicate). *Montnegre (Catalonia) (Photo Joaquim Sanz. MGVM)*

- Aluminosilicates of potassium (orthoclase) (Fig. 88.1), sodium (albite), or calcium (anorthite)
- Sodium and calcium feldspars form plagioclases
- Great hardness (6 on the Mohs scale)
- High resistance to abrasion
- Low viscosity
- Very abundant in all environments.

Feldspars are rock-forming minerals (igneous (pegmatites, Figs. 88.2 and 88.3), metamorphic, and sedimentary rocks).

88.1 Geology

Feldspars form an extensive group of silicate minerals. They constitute about 50% of the Earth's crust, therefore they are found globally in igneous, metamorphic, and sedimentary rocks. The most well-known feldspars are those that include in their chemical composition K^+, Na^+, and Ca^{++} ions, thus orthoclase ($KAlSi_3O_8$), albite ($NaAlSi_3O_8$), and anorthite ($CaAl_2Si_2O_8$). The feldspar with a composition between orthoclase and albite is known as alkali feldspar owing to the presence of alkali ions such as K^+ and Na^{++}, and the version between anorthite and albite as plagioclase feldspar.

Feldspar is especially abundant in igneous rocks (e.g. granite). Granitic rocks may contain up to 60% of alkali rocks; however, commercial feldspar is mostly mined from pegmatitic or feldspathic sand deposits.

Pegmatites are intrusive igneous rocks known for their coarse texture. The most famous pegmatites have a combination of gigantic crystal size and extreme enrichment of rare elements. The term is usually used to refer to granitic pegmatite; however, pegmatitic textures can develop in any intrusive igneous rock composition, such as ultramafic or syenitic. Granitic pegmatites are composed mainly of quartz, feldspar, and accessory mica (Simmons and Webber, 2008).

Initially, feldspathic sand deposits were either igneous or metamorphic deposits. Most pegmatite granite, due to interaction with hydrothermal fluids, was eventually transformed into feldspathic sands. Besides, deposits may be affected by mechanical weathering and, due to their two directions of good cleavage reducing their particle size and increasing their surface area, feldspar crystals are exposed to chemical weathering.

The largest feldspar deposits are found in Turkey, China, and Italy; however, feldspar resources are found in more than 70 countries and more than 40 are known to exploit them (Ghalayini 2017).

88.2 Producing Countries

Feldspars are by far the most abundant group of minerals in the Earth's crust, forming about 50% of all rock.

Fig. 88.2 Pegmatite (*blau aran*) under polarized light. (*Photo* Joaquim Sanz. MGVM)

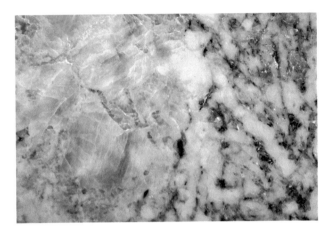

Fig. 88.3 Pegmatite (*blau aran*). *Les (Catalonia)* (*Photo* Joaquim Sanz. MGVM)

The main world reserves are in Egypt (1000 Mt), followed by Iran (630 Mt), India (320 Mt), Korea and Turkey (240 Mt), Thailand (235 Mt), (Brazil (150 Mt), and, further afield, Spain, Poland, and Czech, among others (Fig. 88.4).

88.3 Applications

Ceramics Industry

One of the main applications of feldspar is as a flow agent for the mixture formed by clay, quartz, and water in the manufacture of ceramics and as an ingredient that improves strength and hardness.

Feldspars are used as flux, together with kaolinite, in the manufacture of porcelain for tableware, sanitary ware (Fig. 88.5), insulation and electrical fuses, etc.

Glass Industry

The main use of feldspars is in the manufacture of glass, where they act as a flux, reducing the melting temperature of quartz (achieving significant energy savings in production), helping to control its viscosity, and improving the hardness and durability.

Jewelry

Some varieties of feldspars are used in jewelry. Both adularia (moonstone) and labradorite present the 'labradorescence' phenomena, with beautiful colors. An example is amazonite, an attractive bluish-green color.

Other Fields

Feldspars are used as fillers in paint, enamels, plastics, and adhesives and are added to the rubber to make tires, tapes, belts, etc.

88.4 Recycling

Recycling glass containers helps to reduce feldspar consumption, with notable energy savings.

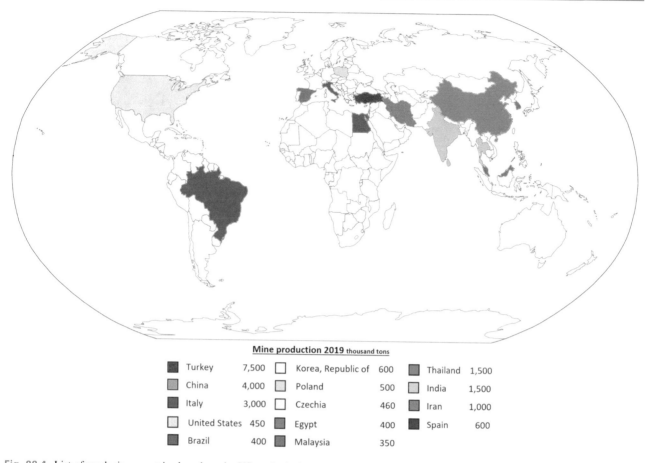

Fig. 88.4 List of producing countries based on the US geological survey, mineral commodity summaries

	Mine production 2019 thousand tons					
Turkey	7,500	Korea, Republic of	600	Thailand	1,500
China	4,000	Poland	500	India	1,500
Italy	3,000	Czechia	460	Iran	1,000
United States	450	Egypt	400	Spain	600
Brazil	400	Malaysia	350		

Fig. 88.5 Sanitary ware ceramics. (*Photo* Joaquim Sanz. MGVM)

References

Ghalayini ZT (2017) Minerals yearbook, Feldspar and Nepheline Syenite. USGS. https://prd-wret.s3.us-west-2.amazonaws.com/assets/palladium/production/atoms/files/myb1-2017-felds.pdf

Simmons WBS, Webber KL (2008) Pegmatite genesis: state of the art. Eur J Mineral 20(4):421–438. https://doi.org/10.1127/0935-1221/2008/0020-1833

Further Reading

IMA. Europe (2021) Industrial Minerals. Feldspar. http://www.ima-europe.eu/about-industrial-minerals/industrial-minerals-ima-europe/feldspar (last accessed May 2021)

IMA (2021) What is Feldspar? https://www.ima-europe.eu/sites/ima-europe.eu/files/minerals/Feldspar_An-WEB-2011.pdf (last accessed May 2021)

LLANSÀ S.A. (2021) Feldspar. https://www.llansasa.com/en/ (last accessed May 2021)

Mata JM, Sanz J (2007) Guia d'identificació de minerals. Manresa, Catalonia. Edicions UPC/Parcir (Catalan 2nd paper edition). p 262. ISBN: 9788483019023. URL: http://hdl.handle.net/2117/90445

Sanz J, Tomasa O (2017) Elements i recursos minerals: aplicacions i reciclatge. Manresa, Catalonia. Zenobita Edicions/Iniciativa Digital Politècnica (Catalan 3rd digital edition). http://hdl.handle.net/2117/105113

USGS (2021) Commodity Statistics and Information. Feldspar. https://www.usgs.gov/centers/nmic/feldspar-statistics-and-information (last accessed May 2021)

Gypsum

Fig. 89.1 Gypsum (calcium sulfate). *Vinaixa (Catalonia)* (*Photo* Joaquim Sanz. MGVM)

- Hydrated calcium sulfate (Fig. 89.1)
- Very soft mineral
- Water soluble
- A poor heat conductor, therefore a good insulator
- Obtained from evaporite deposits, but large quantities of gypsum have recently been obtained as a by-product of coal desulfurization in coal-fired power stations (FGD gypsum).

89.1 Geology

Gypsum deposits are widespread all over the world from many geological periods. Gypsum is one of the main minerals in sedimentary rocks, and it can be found in a range of mineral varieties.

Gypsum can be formed directly by precipitation due to seawater evaporation, enhanced by arid climates in hydrologically restricted marine and marginal marine settings. In the case of non-marine deposits, gypsum can be formed by the precipitation of meteoric waters under arid climate conditions. Gypsum ($CaSO_4 \cdot 2H_2O$ and $CaSO_4$) is usually present in marine and non-marine evaporite deposits. Moreover, it can be formed by the hydration of anhydrite, since both are calcium sulfates. Gypsum ore deposits are usually found layered with other sedimentary rocks and minerals, such as halite or limestone. Large, massive gypsum beds can reach thicknesses from a few centimeters to several tens of meters.

As a point of interest, the largest mineral crystals ever found are of gypsum inside the Cueva de los Cristales, 300 m deep near Naica, Mexico. Initially, the cave was full of water and above a magmatic area. For 500,000 years it was under conditions of constant pressure and temperature, therefore the gypsum molecules within the water were able to form giant crystals. A similar formation is the 8-m wide and 2-m high *geoda gigante* (giant geode) in Pulpí, Spain.

89.2 Producing Countries

The world's reserves are abundant in many countries. The most notable are China, the United States, Iran, Canada, Brazil, Spain, Turkey, and India (Fig. 89.2).

89.3 Applications

Agriculture
Gypsum is used as an agricultural fertilizer, desalinator, and soil pH corrector.

Construction
The main application of this mineral worldwide is as a thermal insulator in wall cladding, in stucco, and for partition panels (gypsum board) (Fig. 89.3).

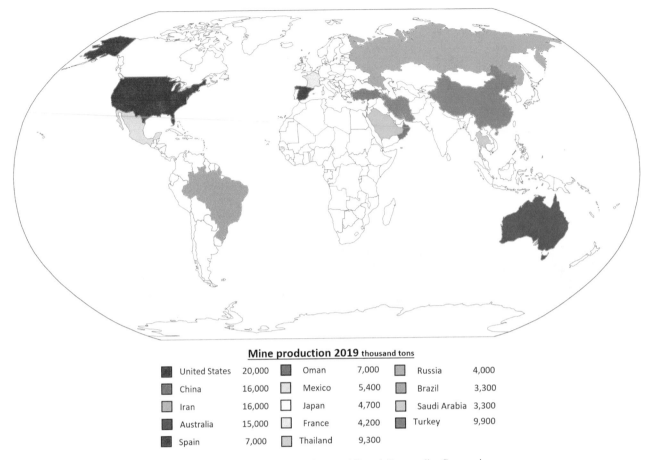

Mine production 2019 thousand tons

United States	20,000	Oman	7,000	Russia	4,000
China	16,000	Mexico	5,400	Brazil	3,300
Iran	16,000	Japan	4,700	Saudi Arabia	3,300
Australia	15,000	France	4,200	Turkey	9,900
Spain	7,000	Thailand	9,300		

Fig. 89.2 List of producing countries based on the US Geological Survey, Mineral Commodity Summaries

Fig. 89.3 Plasterboard partition. (*Image courtesy of Manel Romera*)

Medicine

In traumatology, gypsum is used in the preparation of plaster bandages to immobilize broken bones in a cast.

Chemical and Cement Industry

Gypsum is used in water treatments and sugar refineries. It is also an additive in cement manufacture as a hardening speed regulator.

Other Fields

Gypsum is used in the manufacture of molds, sanitary ware and sculpture. Plaster is used as a filler in the manufacture of paint in order to achieve better coverage.

89.4   Recycling

Gypsum is recovered from panels, boards, and offcuts generated in the manufacturing process. The recovered gypsum is used in agriculture, in the manufacture of new panels and moldings, stuccoes, drinking water treatments, and sculptures.

Further Reading

Euro Gypsum (2021) http://www.eurogypsum.org/ (last accessed May 2021).

Knauf (2021) Gypsum. http://www.knauf.com/en/ (last accessed May 2021).

Mata JM, Sanz J (2007) Guia d'identificació de minerals. Manresa, Catalonia. Edicions UPC/Parcir (Catalan 2nd paper edition). 262 p. ISBN 9788483019023. http://hdl.handle.net/2117/90445

Roskill (2018) Market Reports. Gypsum and Anhydrite. https://roskill.com/market-report/gypsum-and-anhydrite/ (last accessed May 2021).

Sanz J, Tomasa O (2017) Elements i Recursos minerals: Aplicacions i reciclatge. Manresa, Catalonia. Zenobita Edicions/Iniciativa Digital Politècnica (Catalan 3rd digital edition). http://hdl.handle.net/2117/105113

USGS (2021) Commodity Statistics and Information. Gypsum. https://www.usgs.gov/centers/nmic/gypsum-statistics-and-information (last accessed May 2021).

Fig. 90.1 Kaolinite (hydrated aluminum silicate). *Ares d'Alpont (Catalonia)* (*Photo* Joaquim Sanz. MGVM)

- Hydrated aluminum silicate (Fig. 90.1)
- Basic component of kaolin (rock) and many clays
- White and soft to the touch
- Acquires plasticity in contact with water
- Has low thermal and electrical conductivity.

90.1 Geology

Kaolin is a hydrated aluminum silicate crystalline mineral (kaolinite, $Al_2(Si_2O_5)(OH)_4$). It was termed 'China clay' from its use in China, formed commonly from weathered granite or hydrothermal activity. It is typical of three main geological environments: (1) weathering profiles; (2) hydrothermal alterations; and (3) sedimentary rocks. Kaolin is commonly a secondary mineral produced by feldspar and muscovite alteration, thus it is easily formed and is widespread in soils developed under hot, wet, intertropical climates. As a consequence, detrital kaolin minerals are important components of the sedimentary rocks deposited near these areas.

From the industrial point of view, kaolin is a clayey rock consisting mostly of kaolinite. Kaolinite is one of the most common minerals on Earth and can be easily identified by its almost pure white color, fine particle size (~ 1–2 µm), non-toxicity, very low abrasiveness, and chemical stability. Impurities, particularly Fe oxides, reduce its potential for some applications, causing color variations in the manufactured product. It is one of the most common minerals on Earth and can be easily identified by its fine particle size and plate-like structure. The world's largest reserves of premium kaolin are in Georgia in the United States and the Amazon region in Brazil.

90.2 Producing Countries

The world reserves of countries with kaolin deposits are remarkably large, but there are no known values (Fig. 90.2).

90.3 Applications

Pharmaceutical Industry
Kaolin is widely used in medicines to treat gastrointestinal problems, to relieve stomach irritation, and as an antidiarrheal.

Cosmetics Industry
Kaolin is the basis of many cosmetics. It helps to eliminate blackheads and dirt from skin, removes grease, and leaves the skin smooth and soft.

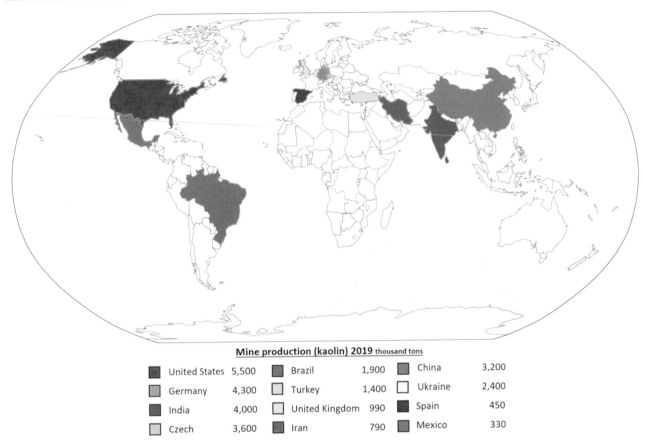

Mine production (kaolin) 2019 thousand tons

United States	5,500	Brazil	1,900	China	3,200
Germany	4,300	Turkey	1,400	Ukraine	2,400
India	4,000	United Kingdom	990	Spain	450
Czech	3,600	Iran	790	Mexico	330

Fig. 90.2 List of producing countries based on the US Geological Survey, Mineral Commodity Summaries

Paper Industry

Kaolin is used as a filler to give consistency and opacity to paper and is also applied as a surface layer to give it the glossy look often seen in magazines (Fig. 90.3).

Fig. 90.3 Paper for newspapers and magazines *(Photo Joaquim Sanz. MGVM)*

Ceramics and Glass Industry

The main use of kaolin for ceramics is in the manufacture of sanitary ware, porcelain, tiles, and tableware, as it provides strength and plasticity and reduces deformation in firing (Fig. 90.4).

In glass manufacturing, it helps to stabilize the melt at high temperatures. It is an additive when producing glass fiber.

Other Fields

Kaolin is used as a pigment in paint and plastics as it offers both shine and opacity. It is also used in the manufacture of 'chalk' for school blackboards.

Kaolin is used in the manufacture of rubber transmission belts, as it increases hardness, rigidity, and resistance to abrasion.

Kaolin is used in the manufacture of plastics, because it enhances their mechanical, electrical, and thermal properties.

Fig. 90.4   Porcelain *(Photo Joaquim Sanz. MGVM)*

90.4   Recycling

Direct recycling is negligible, but much kaolin is saved by recycling paper, glass, and plastics.

Further Reading

Garcia-Valles M, Pi T, Alfonso P, Canet C, Martínez S, Jiménez-Franco A, Tarrago M, Hernández-Cruz B (2015) Kaolin from Acoculco (Puebla, Mexico) as raw material: Mineralogical and thermal characterization. Clay Miner 50:405–416. https://doi.org/10.1180/claymin.2015.050.3.12

IMA Europe (2021) Industrial Minerals. Kaolin. http://www.ima-europe.eu/about-industrial-minerals/industrial-minerals-ima-europe/kaolin (last accessed May 2021).

IMERYS (2021) Kaolin. https://www.imerys-performance-minerals.com/our-minerals/kaolin (last accessed May 2021).

Mata JM, Sanz J (2007) Guia d'identificació de minerals, Catalan 2nd paper edn. Edicions UPC/Parcir, Manresa, Catalonia. 262 p. ISBN 9788483019023. http://hdl.handle.net/2117/90445

Ruiz MD, Nieto F, Jiménez J (1980) Genesis and evolution of the kaolin-group minerals during the diagenesis and the beginning of metamorphism. In: Brindley GW, Brown G (eds) Crystal structures of clay minerals and their X-ray identification. Mineralogical Society Monograph, Britain, pp 41–52

Roskill (2013) Market Reports. Kaolin. https://roskill.com/market-report/kaolin (last accessed May 2021).

Sanz J, Tomasa O (2017) Elements i Recursos minerals: Aplicacions i reciclatge, Catalan 3rd digital edn. Zenobita Edicions/Iniciativa Digital Politècnica, Manresa, Catalonia. http://hdl.handle.net/2117/105113

USGS (2021) Commodity Statistics and Information. Clays.https://www.usgs.gov/centers/nmic/clays-statistics-and-information (last accessed May 2021).

Magnesite

Fig. 91.1 Magnesite (magnesium carbonate). *Eugui (Spain)* (*Photo Joaquim Sanz. MGVM*)

- Magnesium carbonate (Fig. 91.1)
- The most important ore to produce magnesium oxide (MgO)
- Three types of magnesite are obtained industrially:
  - Raw (not heat treated)
  - Caustic calcification (700–1000 °C)
  - Dead-burned (sintered) (1500–2000 °C)
- Other sources of magnesium are seawater, brines, serpentines, and dolostones.

91.1 Geology

The main source of magnesium is magnesite ($MgCO_3$). Magnesite usually forms during the alteration of magnesium-rich rocks or carbonate rocks by metamorphism or chemical weathering. Magnesites can be divided into three categories based on their crystal characteristics and metallogenic environment (Zheng et al. 2015): sparry magnesite; deposits of aphanitic magnesite; and sedimentary metamorphic-hydrothermal metasomatic magnesite.

Sparry magnesite deposits, formed in sedimentary or metamorphic magnesium carbonate rocks, are mainly in layered or lenticular bodies in Precambrian dolomite marble formations. These types of magnesite mostly lay in continental platform regions, such as the Haicheng-Ashiqiao magnesite deposit of the China and North Korea paleocontinent (Hurai et al. 2011).

Aphanitic magnesite deposits are found commonly related to ultramafic rocks, mainly in serpentinized rocks such as the Serbian deposits in Sumadija district. These deposits are mostly formed in shallow lacustrine sediments of the Tertiary period. Moreover, they are usually large-sized, occurring in secondary sediments in superficial strata, such as those in Australia (Zheng et al. 2015).

Sedimentary metamorphic-hydrothermal metasomatic magnesite deposits have generally layered or lenticular orebodies mostly found in rocks formed during the Pre-Ediacaran and Ediacaran periods, such as dolomite or marble. These represent major types of large and supersized deposits. Typical examples include the deposits in the Liaodong region and the Tianshan Mountain area of Xinjiang (Dong et al. 2014).

China has dominated the world magnesia supply for decades. Over the years, it has discovered new magnesium areas all over the country, such as Haicheng in Liaoning province, Dahe in Hebei province, and Basha in Tibet (Zheng et al. 2015). Nowadays, the main magnesia-producing area is Liaoning province, focused on two areas: Dashiqiao in Yingkou; and Haicheng in Anshan (Roskill 2021). However, China has other magnesium-enriched mineral sources, such as dolomite resources, with total proven reserves of more than 3 billion tonnes, as well as Qinghai Salt Lake, which contains 3.2 billion tonnes of magnesium chloride and 1.6 billion tonnes of magnesium sulfate (Zheng et al. 2015).

91.2 Producing Countries

The world's reserves of magnesite are highly abundant: Russia and North Korea (2300 Mt), China (1000 Mt), Brazil (390 Mt), Australia (320 Mt), and Turkey, Spain, and Slovakia, among other countries. The reserves would be even greater if it were possible to extract magnesium compounds from seawater, dolostones, brines, or serpentines (Fig. 91.2).

91.3 Applications

Steel and Cement Industry
Magnesite is mainly consumed in the manufacture of refractories for blast furnaces for steel casting, rotary kilns for cement plants (Fig. 91.3), other types of furnaces, and for making foundry molds.

Pharmaceutical Industry
Magnesium carbonate is used to treat excess stomach acid and is present in several laxatives.

Magnesium is an essential nutrient in the metabolism of proteins and carbohydrates, in cell division, and the functionality of smooth muscle, especially that of the heart. Also, it is used in treatments for diseases such as rheumatism and gout.

Animal Feed
Magnesium oxide is added to feed to strengthen animals' defenses, promoting fertility and reducing stress.

Construction
Magnesite is one of the components of fireproof boards (e.g. Pladur); it acts as a flame retardant together with plaster, silicates, and fiberglass.

Agriculture
Magnesite is mixed into soil to reduce its acidity and introduce magnesium. This promotes two essential functions for good plant growth: the chlorophyll function; and the assimilation of potassium (Fig. 91.4).

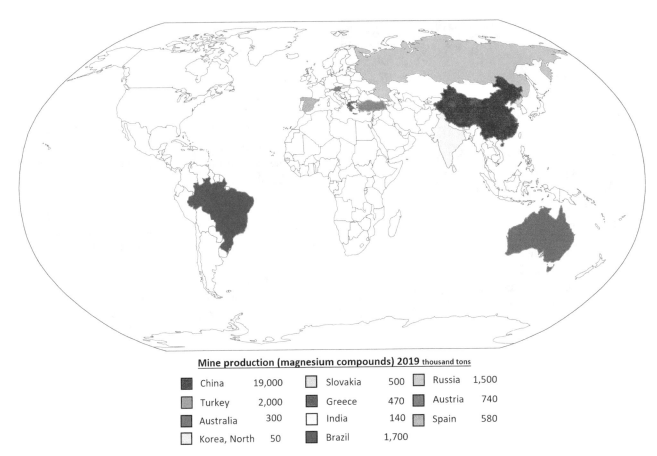

Mine production (magnesium compounds) 2019 thousand tons

China	19,000	Slovakia	500	Russia	1,500
Turkey	2,000	Greece	470	Austria	740
Australia	300	India	140	Spain	580
Korea, North	50	Brazil	1,700		

Fig. 91.2 List of producing countries based on the US Geological Survey, Mineral Commodity Summaries

Fig. 91.3 Rotary kiln (*Photo* Joaquim Sanz. MGVM)

Fig. 91.4 A healthy peach tree (*Photo* Joaquim Sanz. MGVM)

Other Fields

Obtaining metallic magnesium, decontaminants, and cements.

Magnesium oxide and hydroxide are used in wastewater treatment, in the manufacture of rubber, in gas desulfurization, in the oil industry, and in the chemical industry.

91.4 Recycling

Magnesite recycling is unknown. However, certain refractories with magnesium are reused as aggregates for construction.

References

Dong A,; Zhu X, Li S, Wang Y, Gao Z (2014) Genesis of Precambrian strata-bound magnesite deposit in NE China. Acta Geologica Sinica (English edn) 88(s2):1559–1560. https://doi.org/10.1111/1755-6724.12384_4

Hurai V, Huraiová M, Koděra P, Prochaska W, Vozárová A, Dianiška I (2011) Fluid inclusion and stable CO isotope constraints on the origin of metasomatic magnesite deposits of the Western Carpathians, Slovakia. Russian Geol Geophys 52(11): 1474–1490. https://doi.org/10.1016/j.rgg.2011.10.015

Zheng, Z, Xiaolin, C, Denghong W, Yuchuan C, Ge B, Jiankang L, Xinxing L (2015) Review of the metallogenic regularity of magnesite deposits in China. Acta Geologica Sinica (English edn) 89(5):1747–1761. https://doi.org/10.1111/1755-6724.12576

Further Reading

Luxfer Mel Technologies (2021) Extruded Magnesium. https://www.luxfermeltechnologies.com/markets/extruded-and-forging-billet/ (last accessed May 2021).

Mata JM, Sanz J (2007) Guia d'identificació de minerals, Catalan 2nd paper edn. Edicions UPC/Parcir, Manresa, Catalonia. 262 p. ISBN 9,788,483,019,023. http://hdl.handle.net/2117/90445

Roskill (2020) Market Reports. Magnesium Metal. https://roskill.com/market-report/magnesium-metal/ (last accessed May 2021).

Roullier. Groupe Roullier (2021) Magnesite. https://www.roullier.com/en/activities/magnesite (last accessed May 2021).

Sanz, J, Tomasa O (2017) Elements i Recursos minerals: Aplicacions i reciclatge, Catalan 3rd digital edn. Zenobita Edicions /Iniciativa Digital Politècnica, Manresa, Catalonia. http://hdl.handle.net/2117/105113

USGS (2021) Commodity Statistics and Information. Magnesium Metal. https://www.usgs.gov/centers/nmic/magnesium-statistics-and-information (last accessed May 2021).

Fig. 92.1 Muscovite mica (potassium aluminosilicate hydroxide). *Tamariu (Catalonia)* (*Photo* Joaquim Sanz. MGVM)

- Potassium aluminosilicate (Fig. 92.1)
- Laminar structure
- Excellent thermal and electrical insulation
- Greatest dielectric strength of known materials
- Several kinds: muscovite, biotite, and lepidolite, among others
- Abundant in acidic igneous rocks and some metamorphic rocks.

92.1 Geology

Mica constitutes a group of hydrous aluminum silicate minerals. Their characteristic crystalline structure forms layers that can be cleaved into very thin sheets. Mica crystals range in size from a few centimeters to several meters. They have a wide range of colors, from green or purple to almost colorless. Muscovite and phlogopite are the most economically significant micas, and their main ore deposit is pegmatites.

Pegmatites are intrusive igneous rocks known for their very coarse texture. The most famous pegmatites have a combination of gigantic crystal size and extreme enrichment of rare elements. The term usually refers to granitic pegmatite; however, pegmatitic textures can be developed in any intrusive igneous rock composition type, such as ultramafic or syenitic. Granitic pegmatites are composed mainly of quartz, feldspar, and accessory mica (Simmons and Webber, 2008).

The ore bodies where micas are found are usually intrusions such as sills and dikes in the surrounding rock, including granite, gneiss, and schist (Tanner 2012). Moreover, mica can be found in pegmatitic granite deposits altered by hydrothermal fluids and as a by-product of feldspar and kaolin processing. It can be found related to medium-grade metamorphic schist. The mineralogical composition of schists is highly variable, so it is usually indicated by a prefix; for instance, when the predominant mineral is mica it is known as mica-schist.

The largest known deposits are in China (e.g. Nanling district), Brazil (e.g. in Mina Gerais State), Finland (e.g. in the Eräjärvi pegmatite area), Madagascar (e.g. the Saharakara and Ampandrandava deposits), and South Korea (e.g. South Chungcheong Province), among others.

92.2 Producing Countries

The world's reserves of mica are extensive, since many rocks contain it (granites, pegmatites, schists, clay deposits), although not all are currently economically viable to extract (Fig. 92.2).

92.3 Applications

Automotive Industry

Mica is used in the manufacture of vehicle engine brakes and clutch pads because it has low thermal conductivity. It is also

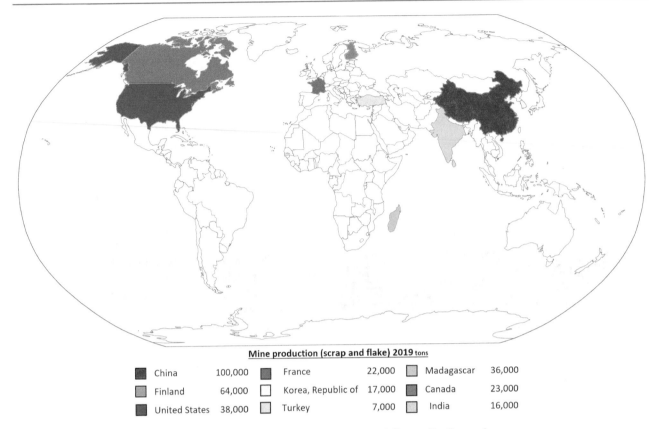

Mine production (scrap and flake) 2019 tons

China	100,000	France	22,000	Madagascar	36,000
Finland	64,000	Korea, Republic of	17,000	Canada	23,000
United States	38,000	Turkey	7,000	India	16,000

Fig. 92.2 List of producing countries based on the US Geological Survey, Mineral Commodity Summaries

used as a filler in the manufacture of vehicle tires and plastic materials to enhance their quality (Fig. 92.3).

Electrical Industry

Mica is used as a filler in the manufacture of plastics for electrical cables, as it increases the electrical resistance of the insulation.

Fig. 92.3 Plastics used in a car interior (*Photo* Joaquim Sanz. MGVM)

Construction

An excellent thermal insulator, mica imparts a good consistency, which is why it is used in the manufacture of plasterboard.

Ceramics Industry

Muscovite and lepidolite mica are used in the manufacture of special ceramics and glazes.

Other Fields

Mica is a component of drilling muds for oil wells, as it is laminated and seals the walls, preventing leaks and pressure losses when the drill head encounters fractured areas. Mica is a component of dry powder fire extinguishers because of its laminar structure and great heat resistance.

Mica is used in the manufacture of glossy paper (for wallcoverings), decorative paint and outdoor paint, where it improves corrosion resistance, and in cosmetics to make eyeshadow and nail polish.

92.4   Recycling

Recycling plastics helps to save mica.

References

Simmons WBS, Webber KL (2008) Pegmatite genesis: state of the art. Eur J Mineral 20(4):421–438. https://doi.org/10.1127/0935-1221/2008/0020-1833

Tanner JT (2012) Mica. In: Kirk-Othmer encyclopedia of chemical technology (online). https://doi.org/10.1002/0471238961.1309030120011414.a01.pub2

Further Reading

Anthony JW, Bideaux RA, Bladh KW, Nichols MC (eds) (2003) Handbook of mineralogy. Mineralogical Society of America, Chantilly, VA, pp 20151–1110. http://www.handbookofmineralogy.org/

Curry KA (2017) Minerals yearbook. USGS, Mica. https://prd-wret.s3.us-west-2.amazonaws.com/assets/palladium/production/atoms/files/myb1–2017-mica.pdf

IMA Europe (2021) Industrial Minerals. Mica. http://www.ima-europe.eu/about-industrial-minerals/industrial-minerals-ima-europe/mica (last accessed May 2021).

IMERYS (2021) Mica. https://www.imerys-performance-additives.com/our-minerals/mica (last accessed May 2021).

Mata JM, Sanz, J (2007) Guia d'identificació de minerals, Catalan 2nd paper edn. Edicions UPC/Parcir, Manresa, Catalonia. 262 p. ISBN 9788483019023. http://hdl.handle.net/2117/90445

Peng J, Zhou MF, Hu R, Shen N, Yuan S, Bi X, Du A, Qu W (2006) Precise molybdenite Re–Os and mica Ar–Ar dating of the Mesozoic Yaogangxian tungsten deposit, central Nanling district, South China. Miner Deposita 41(7):661–669. https://doi.org/10.1007/s00126-006-0084-4

Sanz J, Tomasa O (2017) Elements i Recursos minerals: Aplicacions i reciclatge, Catalan 3rd digital edn. Zenobita Edicions /Iniciativa Digital Politècnica, Manresa, Catalonia. http://hdl.handle.net/2117/105113

USGS (2021) Commodity Statistics and Information. Mica. https://www.usgs.gov/centers/nmic/mica-statistics-and-information (last accessed May 2021).

Fig. 93.1 Phosphorite (calcium phosphate). *Logrosán (Spain)* (*Photo* Joaquim Sanz. MGVM)

- A sedimentary rock with high quantities of phosphate minerals (Fig. 93.1)
- Considered as a cryptocrystalline variety of apatite
- Main source of phosphorus
- An essential nutrient for plants and animals (see: phosphorus)
- Assigned the status of a strategic mineral by the EU in 2017.

93.1 Geology

Phosphate rock resources originate predominantly in sedimentary marine phosphorites. Sedimentary phosphates (mostly phosphorites) are the main source of phosphate rock. Modern phosphorites are characterized by grains of cryptocrystalline or amorphous carbonate (CO_3)–fluorapatite $(Ca_5(PO_4)_3F)$, variously referred as collophane or francolite, occurring as beds varying in thickness from a few centimeters up to tens of meters.

Phosphorus is derived from organic material, such as fecal matter, bone material, and decaying marine organisms, either accumulated in situ or carried into shallow coastal regions by upwellings. These nutrients allow a diverse biota to flourish and eventually produce organic-rich sediments. A changeable depositional environment with alternating reduced deposition and rebuilding of sediments in shallow seas helps this phosphogenesis. Nevertheless, settings and methods of deposition vary widely and are subject to debate (Dar et al. 2017).

Older phosphorites may have undergone diagenesis, deformation, and metamorphism. During the initial stages of diagenesis, collophane precipitation occurs within the topmost layers of these sediments from pore waters rich in phosphorus leaching from the organic remains; precipitation improves where phosphatic nuclei are already present (Dar et al. 2017).

The largest sedimentary deposits are found in northern Africa and China. One of the world's greatest phosphate deposits is in Morocco, where Late Cretaceous marine sediments occur on the flatlands in front of the Atlas Mountains.

93.2 Producing Countries

The world's main reserves are in Morocco and in the Western Sahara (50,000 Mt), followed at some distance by China (3200 Mt), Algeria (2200 Mt), Syria (1800 Mt), Brazil (1700 Mt), South Africa (1500 Mt), and Saudi Arabia (1400 Mt); Australia, the United States, Finland, and Senegal have lesser volumes (Fig. 93.2).

It should be borne in mind that there are igneous rocks with a phosphate content from which phosphates could be extracted, were it economically viable.

© The Author(s), under exclusive license to Springer Nature Switzerland AG 2022
J. Sanz et al., *Elements and Mineral Resources*, Springer Textbooks in Earth Sciences,
Geography and Environment, https://doi.org/10.1007/978-3-030-85889-6_93

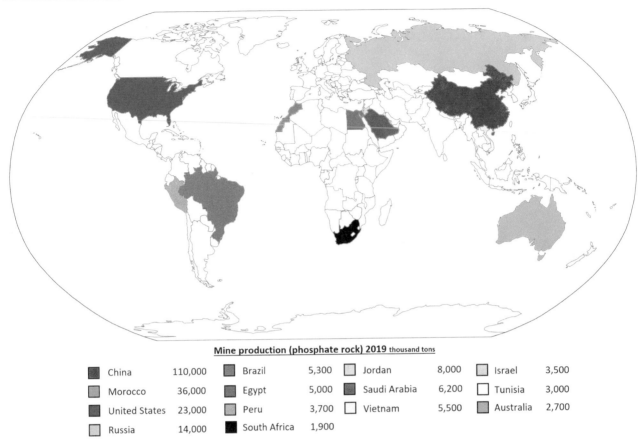

Mine production (phosphate rock) 2019 thousand tons

China	110,000	Brazil	5,300	Jordan	8,000	Israel	3,500
Morocco	36,000	Egypt	5,000	Saudi Arabia	6,200	Tunisia	3,000
United States	23,000	Peru	3,700	Vietnam	5,500	Australia	2,700
Russia	14,000	South Africa	1,900				

Fig. 93.2 List of producing countries based on the US Geological Survey, Mineral Commodity Summaries

93.3 Applications

Chemical Industry

Phosphorite is the main source of phosphoric acid and superphosphoric acid, which are used in the manufacture of agricultural fertilizers (superphosphates) to provide soil with the optimum amount of phosphorus necessary for good plant and tree growth (Fig. 93.3).

Superphosphates are used in the manufacture of animal feed supplements (for poultry, cattle, pigs, shellfish, and farmed fish).

Although the European Union banned their use in washing machine and dishwasher detergents from January 2017, phosphates form part of the composition of many detergents.

Fig. 93.3 Field fertilized with superphosphates (*Photo* Joaquim Sanz. MGVM)

93.4 Recycling

Recycling of phosphorite is unknown.

Reference

Dar SA, Khan KF, Birch WD (2017). Sedimentary: Phosphates. Reference Module in Earth Systems and Environmental Sciences; Elsevier: Amsterdam, The Netherlands, 10. https://doi.org/10.1016/B978-0-12-409548-9.10509-3

Further Reading

Mata JM, Sanz J (2007) Guia d'identificació de minerals, Catalan 2nd paper edn. Edicions UPC/Parcir, Manresa, Catalonia, 262 p. ISBN 9788483019023. http://hdl.handle.net/2117/90445

Phosphea (2021) Phosphates. http://phosphea.com/timab-phosphates-now-phosphea-detail/ (last accessed May 2021).

Roullier (2021) Phosphates. https://www.roullier.com/en/activities/phosphates (last accessed May 2021).

Sanz J, Tomasa O (2017) Elements i Recursos minerals: Aplicacions i reciclatge, Catalan 3rd digital edn. Zenobita Edicions/Iniciativa Digital Politècnica, Manresa, Catalonia. http://hdl.handle.net/2117/105113

USGS (2021) Commodity Statistics and Information. Phosphate rock. https://www.usgs.gov/centers/nmic/phosphate-rock-statistics-and-information (last accessed May 2021).

Salt (Halite)

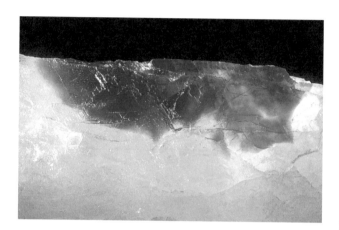

Fig. 94.1 Halite (sodium chloride). *Sallent (Catalonia)* (*Photo* Joaquim Sanz. MGVM)

- Sodium chloride (Fig. 94.1)
- Known as common salt
- Has a characteristic taste
- Fragile
- Highly soluble in water
- Obtained from evaporite mining deposits and natural brines (marine and land-based).

94.1 Geology

Halite (NaCl) is the usual mineral source of sodium. The main halite deposits are the saline flats typical of arid basins. A flat is a shallow depression with layered salts, and it remains dry until storm flooding turns it and the surrounding area into a temporary lake. Continental saline flats occupy the lowest parts of closed arid basins. They are surrounded by brine-soaked mudflats permeated with evaporite minerals that grow within the sediments. Saline flats occur in both continental and marginal marine areas, called sabkha. They vary in size, from smaller than a square kilometer to the largest, thousands of square kilometers in area, such as Lake Uyuni in Bolivia.

Another kind of halite deposit is a salt dome. This is formed when salt deposits are buried by layers of sediment throughout the ages. Confined salt has the ability to deform plastically under pressure and temperature, becoming mobile, and it is forced by unequal pressures to flow upward through weaker overlying strata. The resultant formation takes on a characteristic elongated mushroom shape and is made of relatively pure halite.

94.2 Producing Countries

The world's land reserves of salt are vast in those countries with deposits. Moreover, the seas and oceans contain inexhaustible reserves (Fig. 94.2).

94.3 Applications

Chemical Industry

A mixture of salt and water has a lower freezing point than water alone, and this property is exploited to retard the formation of ice on streets and roads (Fig. 94.3).

Salt is used in tanneries in the treatment of animal hides. It is also used in the manufacture of detergents and soaps.

Halite is used as a descaler. It softens water and prevents encrustations of dissolved salts.

The chlorine extracted from halite is used in the manufacture of polyvinyl chloride (PVC) (see: chlorine).

Halite is the source of caustic soda (NaOH), hydrochloric acid (HCl), and sodium carbonate (Na_2CO_3).

Sodium oxide is involved in the manufacture of glass. Metallic sodium is also extracted from halite.

© The Author(s), under exclusive license to Springer Nature Switzerland AG 2022
J. Sanz et al., *Elements and Mineral Resources*, Springer Textbooks in Earth Sciences, Geography and Environment, https://doi.org/10.1007/978-3-030-85889-6_94

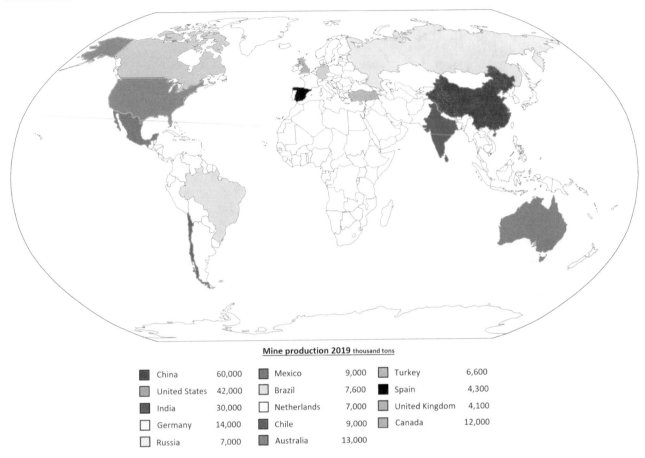

Mine production 2019 thousand tons

■ China	60,000	■ Mexico	9,000	■ Turkey	6,600
■ United States	42,000	■ Brazil	7,600	■ Spain	4,300
■ India	30,000	□ Netherlands	7,000	■ United Kingdom	4,100
□ Germany	14,000	■ Chile	9,000	■ Canada	12,000
□ Russia	7,000	■ Australia	13,000		

Fig. 94.2 List of producing countries based on the US Geological Survey, Mineral Commodity Summaries

Fig. 94.3 Salt-spreading on an icy road (*Image courtesy* Town Hall, Manresa, Catalonia)

Fig. 94.4 Smoked salmon, preserved with salt (*Photo* Joaquim Sanz. MGVM)

Food Industry

Halite, or common salt, is used in the preservation of meat, fish, vegetables, cheese, etc. It plays a fundamental role in human and animal nutrition, and in the preparation of food it enhances the taste (Fig. 94.4).

Grazing animals need salt to digest their grass and feedstuff.

Medicine

Sodium chloride is used in isotonic glucosaline solutions.

Other Fields

Salt is used in skin treatments (peels) as a bactericide and fungicide.

94.4 Recycling

Salt recycling is negligible.

Further Reading

ICL Group (2021) Salt. https://www.icl-ip.com/specialty-minerals/sm-product/salt/ (last accessed May 2021).

Mata JM, Sanz J (2007) Guia d'identificació de minerals, Catalan 2nd paper edn. Edicions UPC/Parcir, Manresa, Catalonia. 262 p. ISBN 9788483019023. http://hdl.handle.net/2117/90445

Roskill (2021) Market Reports. Salt. https://roskill.com/market-report/salt/ (last accessed May 2021).

Sanz J, Tomasa O (2017) Elements i Recursos minerals: Aplicacions i reciclatge, Catalan 3rd digital edn. Zenobita Edicions/Iniciativa Digital Politècnica, Manresa, Catalonia. http://hdl.handle.net/2117/105113

USGS (2021) Commodity Statistics and Information. Salt. https://www.usgs.gov/centers/nmic/salt-statistics-and-information (last accessed May 2021).

Sepiolite

Fig. 95.1 Sepiolite. *Vallecas (Spain)* (*Photo* Joaquim Sanz. MGVM)

- Hydrated magnesium silicate (Fig. 95.1)
- A special clay
- Soft, light, and very porous, like a sponge
- A good thermal insulator, with a high absorption and adsorption capacity.

95.1 Geology

Sepiolite is a hydrated magnesium silicate with a microfibrous morphology and a particular texture that provides a high specific surface area. Together with palygorskite, it forms the palygorskite–sepiolite series. The usual settings for sepiolite occurrences range from soils to marine and lacustrine deposits, hydrothermal veins in serpentinite, diorite, and dolostone, and the weathering of volcanic rocks (Galán and Pozo 2015). There are four essential factors influencing its distribution (Jones and Conko 2011): source

material; climate; physical parameters; and associated phase relations. The main process governing the occurrence of sepiolite is direct precipitation from solution; however, it can also be derived from the dissolution–precipitation transformation of precursor phases.

The largest ore deposits are in Spain, the United States, and Turkey. The Spanish sepiolite deposits of the Madrid Basin are the most important, economically. These were mainly formed by direct precipitation in the saline lacustrine-palustrine environments of the Miocene, with other Mg^{2+}-rich trioctahedral clay minerals. The sepiolite beds are composed mainly of sepiolite (>80%), smectites (15%), calcite and dolomite (2%), and quartz (<1%) (Galán and Pozo 2015).

The US sepiolite deposit is in the southwestern Basin and Range physiographic province, in the sedimentary filling of grabens. One of the best-known ore deposits is in the Amargosa Desert on the California–Nevada border. Based on textural data, Eberl et al. (1982) indicate that the sepiolite could have been formed by kerolite/stevensite dissolution, and not just as the result of direct precipitation (Miles 2011).

Turkish sepiolite is structured in layers and nodules in the Eskişehir Basin. Bedded sepiolite may present a variety of colors, from white to black, ranging from pure sepiolite to sepiolitic dolomite. This lacustrine deposit dates from the Miocene. It was formed by direct precipitation in saline-alkaline lake waters with a supersaturation of silica. The environment was alkaline-saline, enhanced by the arid to semi-arid climate with probable wet intervals due to seasonal fluctuations (Ece and Çoban 1994).

95.2 Producing Countries

The world's reserves of sepiolite are little known. The largest mine is in Spain, with total reserves of over 3800 Mt. The reserves of countries such as Turkey, the United States, and China are unknown (Fig. 95.2).

J. Sanz et al., *Elements and Mineral Resources*, Springer Textbooks in Earth Sciences, Geography and Environment, https://doi.org/10.1007/978-3-030-85889-6_95

Leading producer countries in 2017 thousand tons

■ Spain   483   ☐ United States (No data)   ■ Turkey   16   ■ China   (No data)

Fig. 95.2 List of producing countries based on the British Geological Survey, World Mineral Production 2013–2017)

95.3   Applications

Manufacture of Absorbents

Sepiolite is used in the manufacture of cat litter (Fig. 95.3), as it is a lightweight product, an excellent absorbent of urine, and has a dehydrating effect on solid waste that minimizes odor. It is used in the manufacture of diaper pads, smoke filters, and cigarette filters.

Fig. 95.3 Absorbent cat litter in use (*Photo* Joaquim Sanz. MGVM)

Sepiolite is an industrial absorbent used in waste treatments to absorb toxic products.

Construction

As it is porous, sepiolite is a good thermal insulator and is used to make fireproof boards. It is incorporated into cement and mortar to give a better-quality finish.

Sepiolite is used in the manufacture of asphalt emulsions, which are used as stabilizers in waterproofing coatings.

Other Fields

Sepiolite is used in drilling muds in areas with salt water or at high temperatures, replacing bentonite.

Sepiolite is an animal feed additive, as it is an excellent binding agent. It is used in the manufacture of feed because it improves its profitability, as it involves lower temperatures and compression than other binders.

Sepiolite is an excellent material for removing pollutants from the environment. It is also used as a filler material in the manufacture of rubber, as a thickener for grease, paint, pesticide, and adhesive, and as a bleach, an excipient in tablets, and in suspensions of pharmaceutical products.

95.4   Recycling

Sepiolite recycling is unknown.

References

Eberl DD, Jones BF, Khoury HN (1982) Mixed-layer kerolite/stevensite from the Amargosa Desert. Nevada. Clays Clay Miner 30(5):321–326. https://doi.org/10.1346/CCMN.1982.0300501

Ece ÖI, Çoban F (1994) Geology, occurrence, and genesis of Eskişehir sepiolites, Turkey. Clays Clay Miner 42:81–92. https://doi.org/10.1346/CCMN.1994.0420111

Galán E, Pozo M (2015) The mineralogy, geology, and main occurrences of sepiolite and palygorskite clays. In: Pasbakhsh P, Churchman GJ (eds) Natural mineral nanotubes. CRC Press, pp 117–130

Jones BF, Conko KM (2011) Environmental influences on the occurrences of sepiolite and palygorskite: a brief review. Dev Clay Sci 3:69–83. https://doi.org/10.1016/B978-0-444-53607-5.00003-7

Miles WJ (2011) Amargosa sepiolite and saponite: Geology, mineralogy and markets. Dev Clay Sci 3:265–277. https://doi.org/10.1016/B978-0-444-53607-5.00011-6

Further Reading

BGS (2021) British Geological Survey. World mineral production 2013–2017. https://www2.bgs.ac.uk/mineralsUK/statistics/worldStatistics.html (last accessed May 2021).

Galán E, Castillo A (1984) Sepiolite-palygorskite in Spanish Tertiary basins: genetical patterns in continental environments. Dev Sedimentol 37:87–124. https://doi.org/10.1016/S0070-4571(08)70031-1

Galán E, Pozo M (2011) Palygorskite and sepiolite deposits in continental environments. Description, genetic patterns and sedimentary settings. Dev Clay Sci 3:125–173. https://doi.org/10.1016/B978-0-444-53607-5.00006-2

IMA (2021) Industrial Minerals. Sepiolite. https://www.ima-europe.eu/about-industrial-minerals/industrial-minerals-ima-europe/sepiolite (last accessed May 2021).

Murray HH, Pozo M, Galán E (2011) An introduction to palygorskite and sepiolite deposits—Location, geology and uses. Dev Clay Sci 3:85–99. https://doi.org/10.1016/B978-0-444-53607-5.00004-9

MYTA (2021) Mining and Clay Technology. https://myta.es/ (last accessed May 2021).

Sanz J, Tomasa O (2017) Elements i Recursos minerals: Aplicacions i reciclatge, Catalan 3rd digital edn. Zenobita Edicions/Iniciativa Digital Politècnica. Manresa, Catalonia. http://hdl.handle.net/2117/105113

Suárez M, García-Romero E (2013) Sepiolite-palygorskite: A continuous polysomatic series. Clays Clay Miner 61:461–472. https://doi.org/10.1346/CCMN.2013.0610505

Sepiolsa. Minersa Group (2021) Sepiolite. https://www.sepiolsa.com/ (last accessed May 2021).

Tian G, Han G, Wang F, Liang J (2019) Sepiolite nanomaterials: structure, properties and functional applications. In Wang A, Wang W (eds) Nanomaterials from clay minerals—A new approach to green functional materials: micro and nano technologies. Elsevier, pp 135–201. https://doi.org/10.1016/B978-0-12-814533-3.00003-X

Fig. 96.1 Amethyst quartz (silicon oxide). *Minas Gerais (Brazil)* (*Photo* Joaquim Sanz. MGVM)

- A silicate (silicon oxide) (Fig. 96.1)
- Hard (7 on the Mohs scale) and fragile
- When an electric current is applied to the ends of a quartz crystal, it vibrates at a precise frequency (resonant behavior)
- Various forms have different applications: macrocrystalline quartz; cryptocrystalline quartz; and silica sand.

Quartz is a rock-forming mineral (igneous, metamorphic, and sedimentary), comprising granites, quartzites, and quartz sandstones (Fig. 96.2), among others.

96.1 Geology

Quartz (SiO_2) is a tectosilicate belonging to the silica group. It is widely used in industry for its hardness, resistance to chemical attacks, and so on, especially synthetic quartz, which is not twinned thus has perfect crystals that exploit its piezo-electric properties to the full.

The most common gemstone variety of quartz is rock crystal, which is colorless and transparent. The primary deposits are in igneous rocks. Rock crystal is often found in pegmatites. Some of the best-developed rock crystal is found in Brazil, in the Alps, and in the United States (Arkansas). Another important variety is amethyst, the violet variety, whose color is due to the presence of iron. It is found in Uruguay, Brazil, and other countries. Citrine is the yellow variety due to the presence of Fe^{3+}. Primary citrine deposits include pegmatites. Citrine can be found in Brazil and Madagascar. An interesting exception is ametrine quartz; half amethyst (violet, Fe^{2+}) and half citrine (yellow, Fe^{3+}), it has been found only in Bolivia.

Silica sand is made up of small grains of quartz, the result of the weathering of igneous, metamorphic, and sedimentary rocks and subsequent sedimentation.

Flint is a hard, tough chemical or biochemical sedimentary rock. It is a form of microcrystalline quartz (chert). It often forms nodules in sedimentary rocks such as chalk and marine limestones. The nodules can be dispersed randomly throughout the unit, but are often concentrated into distinct layers.

96.2 Producing Countries

Reserves of good-quality natural quartz crystals are scarce, and are limited to certain countries. By contrast, the world's reserves of silica sand are abundant, but unfortunately little information is available on the volumes of such deposits.

96.3 Applications

Macrocrystalline Quartz

Natural quartz is used in decoration and jewelry. Crystalline quartz can appear in various colors due to the presence of

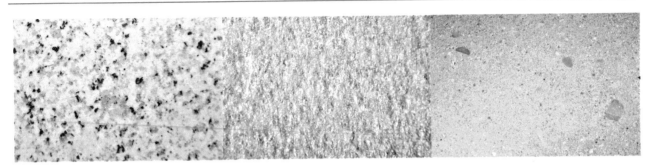

Fig. 96.2 **Granite** (crystal white) (*Cadalso de los Vidrios, Spain*); **quarzite** *(Alta, Norway)*; and **quartz sandstone** *(Barcelona, Catalonia)* (*Photo* Joaquim Sanz. MGVM)

natural impurities, turning it into semi-precious stones such as amethyst (violet), citrine (light yellow), and smoked (light brown), among others.

In general, synthetic quartz is used. Due to quartz crystals' resonant behavior, they are involved in the manufacture of digital watches, radio and TV transmitters, timers, etc.

Solid quartz is used in the manufacture of ferrosilicon, an alloy used in deoxidation and the manufacture of special steels.

Cryptocrystalline Quartz

Cryptocrystalline quartz (formed by microscopic crystals) can appear in beautiful colored forms, such as agates or jasper, for decoration and jewelry.

Cryptocrystalline quartz (such as flint) is used to make an artificial rock with resins and dyes, Silestone, much used in the construction of buildings and kitchen units.

Silica Sand

Silica sand is vital to the manufacture of glass (Fig. 96.3), fiberglass, fiber optics, and abrasives.

It is the main industrial medium in water filtration to extract solids from wastewater and to filter acidic and aggressive liquids, as it is unaffected by them, and it is also used in the manufacture of filters for diesel vehicles. Its high resistance to wear and chemical attack makes silica sand a basic component in the manufacture of tires, paint, ceramics, and refractories. It is also useful in gardening.

Silica sand is made into molds for iron, copper, and aluminum castings because it has a higher melting point than them. High-purity silica sand combined with carbon (petroleum coke) at high temperatures (over 2000 °C) produces carborundum (silicon carbide) (Fig. 96.4), a high-hardness material (9 on the Mohs scale) for use as an abrasive, in anti-slip surfaces, and in electrical lightning rods since it is insulating, resistant, and refractory.

Diatomaceous earth, formed of the silicic skeletons of unicellular algae, is used as a very fine abrasive and as a filter for wines and oils.

96.4   Recycling

By recycling glass, we save a great deal of silica sand. Glass is 100% recyclable without losing any quality (provided it does not contain heavy metals, such as lead). A glass bottle

Fig. 96.3 Glass drinking vessel, with red wine (*Photo* Joaquim Sanz. MGVM)

Fig. 96.4 Anti-slip tape with carborundum (*Photo* Joaquim Sanz. MGVM)

can be melted for reuse between 40 and 60 times with only 5% of the energy consumption that would be required for melting silica sand.

Further Reading

IMA Europe (2021) Industrial Minerals. Silica. http://www.ima-europe. eu/about-industrial-minerals/industrial-minerals-ima-europe/silica (last accessed May 2021).

Mata JM, Sanz J (2007) Guia d'identificació de minerals, Catalan 2nd paper edn. Edicions UPC/Parcir, Manresa, Catalonia 262 p. ISBN 9788483019023. http://hdl.handle.net/2117/90445

Sanz J, Tomasa O (2017) Elements i Recursos minerals: Aplicacions i reciclatge, Catalan 3rd digital edn. Zenobita Edicions/Iniciativa Digital Politècnica, Manresa, Catalonia. http://hdl.handle.net/2117/ 105113

Sibelco (2021) Quartz. https://www.sibelco.com/materials/quartz/ (last accessed May 2021).

Sibelco (2021) Silica. https://www.sibelco.com/materials/silica/ (last accessed May 2021).

USGS. (2020) Commodity Statistics and Information. Quartz Crystal (industrial) https://pubs.usgs.gov/periodicals/mcs2020/mcs2020-quartz.pdf (last accessed May 2021).

Fig. 97.1 Sylvinite (potassium and sodium chloride). *Sallent (Catalonia)* (*Photo* Joaquim Sanz. MGVM)

- Potassium chloride (KCl)
- From sylvinite (Fig. 97.1), a sedimentary rock composed of sylvite (KCl) and halite (NaCl)
- Tastes spicy and salty
- Fragile
- Highly soluble in water
- Found in abundance in evaporite deposits, in brine, and in certain seas (Dead Sea)
- In a group with potash (potassium chloride, KCl, popularly known as potash), potassium sulfate (SOP), potassium-magnesium sulfate (SOPM), and potassium muriate (MOP), with 95% KCl and NaCl, and refers to potassic fertilizers for agriculture.

97.1 Geology

The most characteristic mineral source of potassium is sylvite (KCl). Nevertheless, it is frequently observed associated with halite (NaCl), forming the mix known as sylvinite. In most cases, relatively pure sylvinite essentially has no soluble sulfate or other salts. It can occasionally be associated with carnallite (KMgCl$_3$·6H$_2$0), with a similar crystalline structure and nearly free from other soluble salts.

Sylvite deposits originate in the evaporation of epicontinental sea basins without significant drainage, either in or out. Indeed, optimal evaporation occurs in arid conditions. A frequent mineral sequence within depositional basins as evaporation of seawater increases is calcite (or dolomite), gypsum (or anhydrite), halite, and potash and, in some locations, soluble magnesium and sulfate salts.

97.2 Producing Countries

The world's potash reserves (K$_2$O equivalent) are in Canada (1,000 Mt), Belarus (750 Mt), Russia (600 Mt), China (350 Mt), Israel and Jordan (270 Mt), the United States (220 Mt), Germany (150 Mt), Chile (100 Mt), Spain (68 Mt), and Brazil (24 Mt), among others (Fig. 97.2).

97.3 Applications

Chemical Industry
Sylvite is mainly used in the production of agricultural fertilizers to improve plant growth and the quality of fruits,

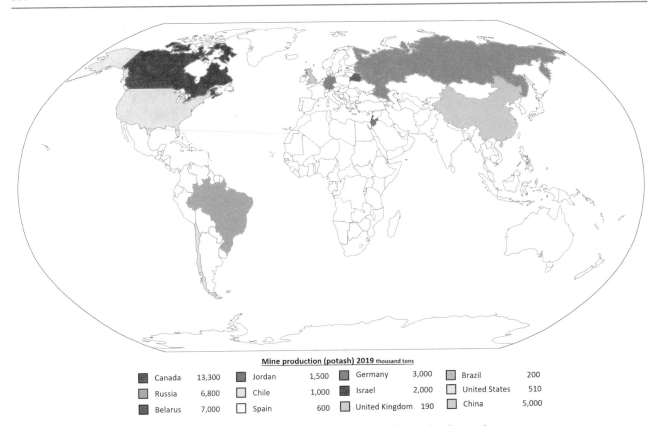

Mine production (potash) 2019 thousand tons

Canada	13,300	Jordan	1,500	Germany	3,000	Brazil	200
Russia	6,800	Chile	1,000	Israel	2,000	United States	510
Belarus	7,000	Spain	600	United Kingdom	190	China	5,000

Fig. 97.2 List of producing countries based on the US Geological Survey, Mineral Commodity Summaries

vegetables, trees, cereals, flowers, and so on, since potassium is one of the three most important macronutrients for vegetables (Fig. 97.3).

There is currently increased demand for potassium for agricultural fertilizers for crops of sunflowers, soybeans, corn, sugar cane, and so on, due to the expanding production of biofuels and the cultivation of cereals as a fundamental aspect of subsistence in emerging countries.

Glass Industry

Potassium carbonate is used as a fluxing agent in the manufacture of glass.

Other Fields

Sylvite is used in pharmaceuticals, medicines, explosives, and detergents and in the electrochemical industry and metallurgy.

Fig. 97.3 Sunflower field benefiting from potassium fertilizer (*Photo* Joaquim Sanz. MGVM)

97.4 Recycling

Sylvite recycling is unknown.

Further Reading

ICL Iberia (2021). https://www.iclfertilizers.com/browse/fertilizers/potash (last accessed May 2021).

Mata JM, Sanz J (2007) Guia d'identificació de minerals, Catalan 2nd paper edn. Edicions UPC/Parcir, Manresa, Catalonia 262 p. ISBN 9788483019023. http://hdl.handle.net/2117/90445

Sanz J, Tomasa O (2017) Elements i Recursos minerals: Aplicacions i reciclatge, Catalan 3rd digital edn. Zenobita Edicions/Iniciativa Digital Politècnica, Manresa, Catalonia. http://hdl.handle.net/2117/105113

USGS (2021) Commodity Statistics and Information. Potash. https://www.usgs.gov/centers/nmic/potash-statistics-and-information (last accessed May 2021).

Fig. 98.1 Talc (hydrated magnesium silicate) *Maçanet de Cabrenys (Catalonia)* (*Photo* Joaquim Sanz. MGVM)

- Hydrated magnesium silicate (Fig. 98.1)
- An excellent filler
- White or greenish in color
- Soft, smooth, light, and hydrophobic
- Resists temperatures up to 1300 °C
- Low thermal and electrical conductivity
- Mainly found in metamorphic rocks.

98.1 Geology

Talc is hydrated magnesium silicate ($Mg_3Si_4O_{10}(OH)_2$); however, it may contain mineral impurities such as chlorites and fibrous amphiboles, calcite, magnesite, and dolomite. Talc has a characteristic perfect-basal exfoliation known as lamellarity or platiness due to low- to moderate-grade metamorphism during its formation.

Talc has a wide range of colors, such as green, gray, or colorless, which determine its purity and commercial grade. For instance, green is related to chlorite minerals and a dark tone is usually due to organic matter.

Talc deposits were formed by metamorphism, hydrothermal activity, or metasomatic processes in Mg-rich rocks (e.g. dolomites) and mafic or ultramafic rocks, therefore the hydrothermal solutions must have been enriched with SiO_2. Moreover, deposit structures were essential to allow the hydrothermal fluids to permeate the rock massif and so generate settings favorable for these processes (McCarthy et al. 2006).

Talc deposits are classified by the initial rock from which they were derived, giving five main types.

Firstly, talc deposits derived from magnesium carbonates constitute more than half of the world's production and are generally found in ancient, metamorphosed carbonate sequences. Their whitish talc mineralization is typically found without impurities.

The second main economic type is derived from serpentinite, which is a metamorphic rock derived from mafic or ultramafic rocks (e.g. peridotite or dunite). The initial rocks suffered metasomatic processes, being heated and chemically altered by hydrothermal fluids. Their mineralization is found associated with amounts of iron, nickel, and chromium (Pi-Puig et al. 2020). For this reason, flotation techniques are generally used to increase the grade. They provide about 20% of the world's production (Eurotalc 2021).

The third type is derived from alumino-silicate rock and forms 10% of world production (Eurotalc 2021). It is usually found in combination with magnesium carbonate deposits. The mineralization is generally found with high amounts of chlorite impurities ($(Mg,Fe)_3(Si,Al)_4O_{10}(OH)_2 \cdot (Mg, Fe)_3(OH)_6$). Since the chlorite imparts some useful industrial properties, it is unnecessary to upgrade the material further (Eurotalc 2021).

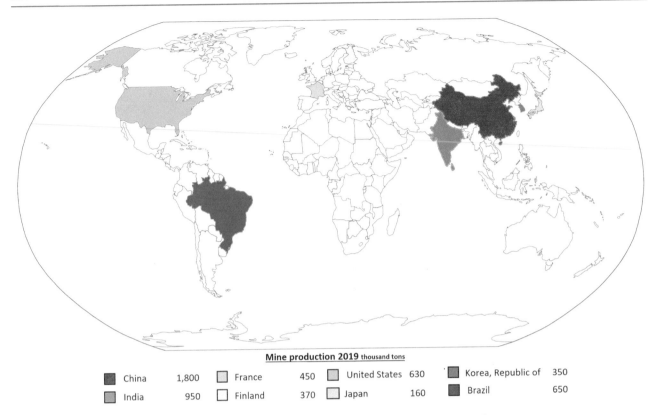

Mine production 2019 thousand tons					
China	1,800	France	450	United States	630
India	950	Finland	370	Japan	160

Korea, Republic of 350
Brazil 650

Fig. 98.2 List of producing countries based on the US Geological Survey, Mineral Commodity Summaries

The fourth type are those that were initially Mg-clays then experienced metamorphic transformation, eventually making up talc deposits. Nowadays, this type is economically insignificant.

Finally, the fifth type is the unusual black talc. Its dark color is due to the content of organic carbon and it presents a characteristic oolitic texture. It is known to be associated with dolomite, quartz, and occasional magnesite, apatite, and pyrite (Li et al. 2013). Deposits are mainly in Guangfeng County, China, estimated at more than half a billion tons (Li et al. 2013).

98.2 Producing Countries

In the world's talc reserves, the countries that stand out are the United States (140 Mt), India (110 Mt), Japan (100 Mt), China (82 Mt), and Brazil (44 Mt), and there are large reserves of no specific amounts in Finland and France (Fig. 98.2).

98.3 Applications

Plastic, Paint, and Paper Industries
Talc is used as a filler to give a good consistency to paper, paint, and all types of plastics, especially those that act as insulators for electrical conductors.

Between 20 and 30 kg of talc are used in the manufacture of a single car, enhancing the plastic that makes up the fittings, improving its resistance, and reducing its total weight (Fig. 98.3).

Talc is used in the manufacture of rubber, preventing it from sticking to the mold and making it easily removed.

Textile Industry
Talc is used as a bleach and to release the electrical charge in cotton products.

Fig. 98.3 Plastics within a car interior (*Photo* Joaquim Sanz. MGVM)

Fig. 98.4 Cosmetics (*Photo* Joaquim Sanz. MGVM)

Cosmetics Industry

Powdered talc is the basis for many cosmetic products; it provides a good texture, stability, water resistance, and adhesion to the skin. Pigment is added to obtain the desired color (Fig. 98.4).

Pharmaceutical Industry

Talc is the main ingredient in talcum powder. When applied to skin, especially babies' skin, it soothes minor irritations. It is also used as a surface lubricant for tablets to make them easier to swallow.

Ceramic Industry

A significant amount of talc is used in the manufacture of ceramics and porcelain, being added to prevent cracks in crockery, sanitary ware, tiles, etc.

Other Fields

Talc is also in agriculture, and in the manufacture of toothpaste and soap.

98.4 Recycling

Talc recycling is unknown, but by recycling paper and plastics we save a great deal of talc extraction.

References

Eurotalc (2021). Talc geology. https://www.eurotalc.eu/what-talc (last accessed May 2021).

McCarthy EF, Genco NA, Reade EH Jr (2006) Talc. In: Kogel JE, Trivedi N, Barker JM, Krukowski ST (eds) Industrial minerals and rocks, Society for Mining, Metallurgy & Exploration, Englewood, CO, pp 971–986

Pi-Puig T, Animas-Torices DY, Solé J (2020) Mineralogical and geochemical characterization of talc from two Mexican ore deposits (Oaxaca and Puebla) and nine talcs marketed in Mexico: evaluation of its cosmetic uses. Minerals 10(5):388. https://doi.org/10.3390/min10050388

Further Reading

Li C, Wang R, Lu X, Zhang M (2013) Mineralogical characteristics of unusual black talc ores in Guangfeng County, Jiangxi Province, China. Appl Clay Sci 74:37–46. https://doi.org/10.1016/j.clay.2012.12.004

IMA Europe (2021) Industrial Minerals. Talc. http://www.ima-europe.eu/about-industrial-minerals/industrial-minerals-ima-europe/talc (last accessed May 2021)

IMERYS (2021) Performance Additives. Talc. https://www.imerys-performance-additives.com/our-minerals/talc (last accessed May 2021).

Mata JM, Sanz J (2007) Guia d'identificació de minerals, Catalan 2nd paper edition. Edicions UPC/Parcir, Manresa, Catalonia, 262 p. ISBN 9788483019023. http://hdl.handle.net/2117/90445

Roskill (2015) Market Reports. Talc. https://roskill.com/market-report/talc/(last accessed May 2021).

Sanz J, Tomasa O (2017) Elements i Recursos minerals: Aplicacions i reciclatge, Catalan 3rd digital edn. Zenobita Edicions/Iniciativa Digital Politècnica, Manresa, Catalonia. http://hdl.handle.net/2117/105113

USGS (2021) Commodity Statistics and Information. Talc. https://www.usgs.gov/centers/nmic/talc-and-pyrophyllite-statistics-and-information (last accessed May 2021).

Fig. 99.1 Thenardite (sodium sulfate). *Villarrubio de Santiago (Spain)* (*Photo* Joaquim Sanz. MGVM)

- Anhydrous sodium sulfate (Fig. 99.1)
- With mirabilite (hydrous sodium sulfate) and glauberite (sodium calcium sulfate), forms the sodium sulfate group
- Tastes a little salty
- Hygroscopic
- Similar properties and uses to glauberite and mirabilite
- Found in evaporite deposits and certain brines.

99.1 Geology

Thenardite is a sodium sulfate salt (Na_2SO_4) usually found together with other sulfate salts such as mirabilite or Glauber's salt ($Na_2SO_4 \cdot 10H_2O$), glauberite ($Na_2Ca(SO_4)_2$), bloedite or astrakanite ($Na_2Mg(SO_4)_2 \cdot 4H_2O$), or burkeite ($Na_6(CO_3)(SO_4)_2$).

Sodium sulfate deposits can be classified into three main groups: lacustrine; podogenic; and those formed during the Tertiary period. Deposits related to lacustrine environments and continental-evaporitic deposits were typically in arid regions, either cold or warm, since sodium sulfate minerals can be crystallized in a wide range of climates. Examples include the largest and purest brine sources of sodium sulfate, Laguna del Rey (Mexico) and Kara-Bogaz Gol (Turkmenistan).

The Laguna del Rey evaporites precipitation was enhanced by a warm climate, while the Kara-Bogaz Gol deposits were formed in a cold climate. These deposits were brines during the Quaternary, usually hosted between mirabilite or glauberite beds together with other sodium sulfates and evaporitic salts (Warren 2016).

Thenardite found in podogenic settings is of only minor economic importance. The most relevant podogenic deposits are in the Atacama Desert, Chile.

Currently, the most economically significant sodium sulfate deposits were formed during the Tertiary period. They are layered deposits of thenardite, glauberite, and related minerals. Some well-known examples are in Spain and Turkey.

In addition, sodium sulfate can be recovered as a by-product of various chemical and waste recovery processes, including the manufacture of potash, dichromate, and phenol, among other compounds (Warren 2016).

99.2 Producing Countries

The world's reserves of sodium sulfate (USGS 2013) are in the United States (860 Mt), Spain (180 Mt), Mexico (170 Mt), Turkey (100 Mt), and Canada (84 Mt), among other countries (Fig. 99.2).

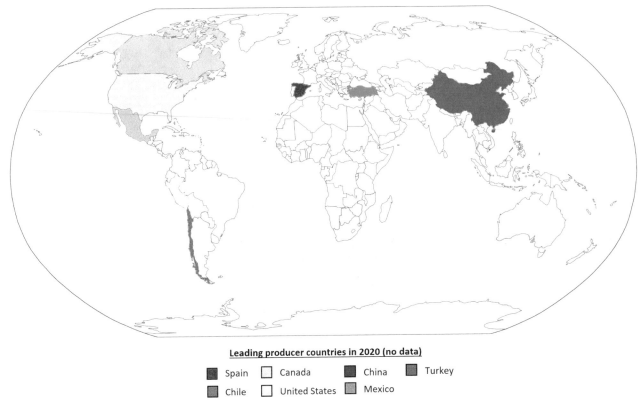

Leading producer countries in 2020 (no data)

Spain   Canada   China   Turkey

Chile   United States   Mexico

Fig. 99.2 List of producing countries based on the US Geological Survey, Mineral Commodity Summaries

According to IHS Markit, the world's main reserves and production of mirabilite are in China (no data). Large quantities of sodium sulfate are currently produced as a by-product of nitrate/iodine processing, hydrochloric acid production, and the manufacture of sodium dichromate, formic acid, and lithium carbonate, among other processes.

99.3 Applications

Chemical Industry

Thenardite's main use is as a filler in the manufacture of soap and detergent powders, whose mechanical behavior is enhanced (Fig. 99.3).

As it is hygroscopic, it is used as a moisture absorber in laboratories and in the chemical industry.

Sodium sulfate is used in the manufacture of fabrics to reduce the negative electrical charge in the fibers and thus improve the penetration of dyes.

Glass Industry

Added to glass as a clarifying agent, sodium sulfate helps to remove small air bubbles that may remain in the manufacturing process. Together with sodium carbonate (soda ash) they are the two most important sodium salts in the common and flat glass manufacturing process.

Fig. 99.3 Dishwasher detergent (*Photo* Joaquim Sanz. MGVM)

Other Fields

Sodium sulfate is also used in the manufacture of pulp paper, ceramics, and textiles and in the steel industry, and has applications in human food, animal feedstuffs, and pharmaceutical products.

99.4 Recycling

Recycling of thenardite is unknown, but by recycling glass and paper our consumption is reduced.

Reference

Warren JK (2016) Non-potash salts: Borates, Na-sulphates, Na-carbonate, lithium salts, gypsum, halite and zeolites. In: Evaporites. Springer, Cham, pp 1187–1302. https://doi.org/10.1007/978-3-319-13512-0_12

Further Reading

Criminose (2021) Crimidesa Industrial Group. Sodium sulphate. http://www.crimidesa.es/?lang=en (last accessed May 2021).

IHS Markit (2021) Sodium sulphate. https://ihsmarkit.com/products/sodium-sulfate-chemical-economics-handbook.html (last accessed May 2021).

Mata JM, Sanz J (2007) Guia d'identificació de minerals, Catalan 2nd paper edn. Edicions UPC/Parcir, Manresa, Catalonia, 262 p. ISBN 9788483019023. http://hdl.handle.net/2117/90445

Pueyo JJ, Chong G, Jensen A (2001) Neogene evaporites in desert volcanic environments: Atacama Desert, northern Chile. Sedimentology 48(6):1411–1431. https://doi.org/10.1046/j.1365-3091.2001.00428.x

Sanz J, Tomasa O (2017) Elements i Recursos minerals: Aplicacions i reciclatge, Catalan 3rd digital edn. Zenobita Edicions/Iniciativa Digital Politècnica, Manresa, Catalonia. http://hdl.handle.net/2117/105113

Fig. 100.1 Wollastonite. *Gualba (Catalonia)* (*Photo* Joaquim Sanz. MGVM)

- A white calcium silicate (Fig. 100.1)
- Often forms acicular crystals
- Has great brightness and little moisture absorption
- Valued by industry worldwide
- Usually of metamorphic origin.

100.1 Geology

Wollastonite ($CaSiO_3$) is a vital mineral because of its utility in industry. It is a typical calc-silicate formed in skarn environments. Skarn is defined by its mineralogy, usually characterized by a calc-silicate assemblage of mainly garnet, pyroxene, and amphibole, besides wollastonite. In general, skarn is formed during regional or contact metamorphism by a variety of metasomatic processes. Typical skarns are the product of a plutonic intrusion of acid composition into limestone. They are also formed when limestones are metamorphosed or when silica-bearing fluids are introduced into calcareous sediments during metamorphism, in both cases through the reaction of calcium carbonate with silicon dioxide yielding calcium metasilicate and carbon dioxide. Less commonly, wollastonite can crystallize directly from a magma. China, India, Mexico, and Finland produce more than 97% of the world's wollastonite.

100.2 Producing Countries

World reserves exceed 100 million metric tonnes. There are no data, but large deposits of wollastonite have been estimated in China, Finland, India, Mexico, the United States, and other countries (Fig. 100.2).

100.3 Applications

Ceramic Industry
Wollastonite is used in the manufacture of ceramics as it reduces the firing temperature and increases brightness and strength. Its use incorporates calcium into the ceramic paste and avoids the use of carbonates, which would otherwise lead to CO_2 emissions.

Paint and Plastics Manufacturing
Wollastonite is used as a filler in the manufacture of paint, plastics, and rubber, enhancing their strength, rigidity, and durability (Fig. 100.3).

Construction
Wollastonite is an additive in the manufacture of mortar and concrete because it increases their wear resistance, high-temperature stability, and durability.

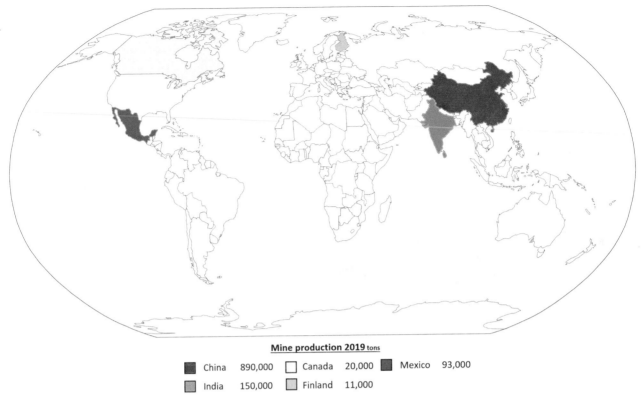

Mine production 2019 tons

■	China	890,000	□	Canada	20,000	■ Mexico 93,000
■	India	150,000	■	Finland	11,000	

Fig. 100.2 List of producing countries based on the US Geological Survey, Mineral Commodity Summaries

Fig. 100.3 Paint (*Photo* Joaquim Sanz. MGVM)

Fig. 100.4 Brake pads (*Photo* Joaquim Sanz. MGVM)

Metallurgical Industry

The addition of wollastonite to metallurgical fluxes gives steel a good fusibility and low viscosity. It also lubricates the mold walls, improving the circulation of the molten steel.

Other Fields

Wollastonite is a substitute for asbestos in the manufacture of brakes and clutches for vehicles and machinery, since it acts as a friction element and resists very well the high temperatures involved in braking (Fig. 100.4).

Wollastonite is also used in the manufacture of adhesives, elastomers, and glass.

100.4 Recycling

Recycling plastics helps to save wollastonite.

Further Reading

IMERYS (2021) Wollastonite. https://www.imerys.com/minerals/wollastonite (last accessed May 2021).

Mata JM, Sanz J (2007) Guia d'identificació de minerals, Catalan 2nd paper edn. Edicions UPC/Parcir, Manresa, Catalonia. 262 p. ISBN 9788483019023. http://hdl.handle.net/2117/90445

Meinert L, Dipple GM, Nicolescu, S (2005) World skarn deposits. In: Hedenquist JW, Thompson JFH, Goldfarb RJ, Richards JP (eds) Economic geology, 100th anniversary volume. Society of Economic Geologists

Sanz J, Tomasa O (2017) Elements i Recursos minerals: Aplicacions i reciclatge, Catalan 3rd digital edn. Zenobita Edicions/Iniciativa Digital Politècnica, Manresa, Catalonia. http://hdl.handle.net/2117/105113

USGS (2021) Commodity Statistics and Information. Wollastonite. https://www.usgs.gov/centers/nmic/wollastonite-statistics-and-information (last accessed May 2021).

Fig. 101.1 Chabazite. *Iceland* (*Photo* Joaquim Sanz. MGVM)

- A group of aluminosilicates with hydrated magnesium, potassium, calcium, and sodium
- Soft, light, and porous
- Good thermal insulators, and great absorbents and adsorbents
- High ion-exchange capabilities
- Strong affinities for ammonia (NH_3)
- Clinoptilolite and chabazite (Fig. 101.1) are the main zeolites.

101.1 Geology

Zeolites form a group of minerals found in low-temperature hydrothermal systems, especially in volcanic and volcanically derived rocks but also in a wide range of other types, typically feldspathic and also metamorphic.

Zeolites are the main mineral components in altered volcaniclastic rock of a range of ages and compositions. Zeolites form by the alteration of mainly volcanic glass in various geological environments under variable geochemical and temperature conditions. Proposed genetic models of zeolite deposits include weathering, diagenesis in open or closed hydrologic systems, low-temperature hydrothermal systems, primary magmatic environments, and impact craters. The most common zeolites, which may occur in mineable deposits, are clinoptilolite-heulandite, mordenite, chabazite, analcime, and phillipsite. Mining of zeolite deposits is widespread in many countries worldwide.

Other minerals often found associated with zeolites include apophyllite, calcite, cavansite, prehnite, epidote, quartz, pyrite, and clay minerals.

Some metamorphic rocks contain a sequence of zeolite minerals useful for assigning the relative metamorphic grade, defining the zeolite facies of metamorphism.

101.2 Producing Countries

The world reserves of natural zeolites cannot be evaluated, but they are extensive. Today, synthetic zeolites are increasingly replacing natural ones (Fig. 101.2).

101.3 Applications

Chemical Industry

The main application of zeolites is in the manufacture of detergents. They are used in the chemical industry and in swimming pools for water purification, as their porosity and specific surface area make them an excellent filtering medium for metals, odors, and some organic pollutants.

J. Sanz et al., *Elements and Mineral Resources*, Springer Textbooks in Earth Sciences, Geography and Environment, https://doi.org/10.1007/978-3-030-85889-6_101

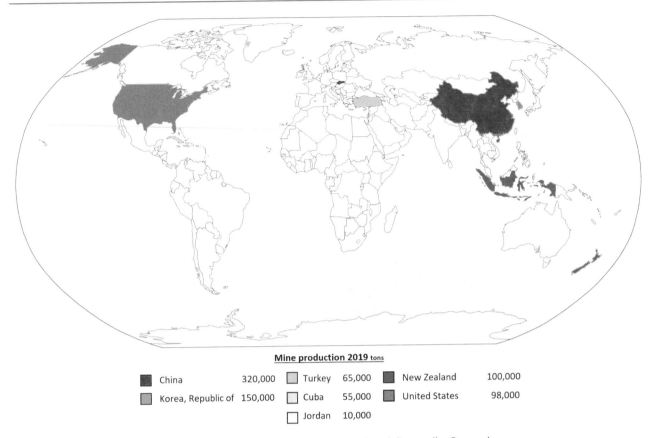

Mine production 2019 tons

■	China	320,000	■	Turkey	65,000	■	New Zealand	100,000
■	Korea, Republic of	150,000	□	Cuba	55,000	■	United States	98,000
			□	Jordan	10,000			

Fig. 101.2 List of producing countries based on the US Geological Survey, Mineral Commodity Summaries

In the petrochemical industry, they are used as catalysts in the cracking of crude oil and as a material to retain hazardous heavy metals from mining and metallurgy.

Agriculture

Zeolites are natural slow-release fertilizers for nutrients; they are used in organic farming. They retain nutrients such as potassium, magnesium, and nitrogen, as well as the water needed by the crop, so they reduce the consumption of both fertilizers and water.

Animal Breeding

Zeolites are an additive for animal feed, as they reduce the ammonia level in livestock's intestines thus help to prevent health problems and reduce bad-smelling excrement. They can also be used as an ammonia absorber for excrement on pig and chicken farms to reduce bad odors (Fig. 101.3).

Construction

Zeolites, which are natural pozzolans, are used in the manufacture of high-quality lightweight concrete.

Fig. 101.3 Pig farming (*Photo* Joaquim Sanz. MGVM)

101.4   Recycling

Zeolites from drying, gas absorption, and water purification can be reprocessed and reused.

Further Reading

International Zeolite Association (2020) http://www.iza-online.org/ (last accessed May 2021).

Marantos I, Christidis GE, Ulmanu M (2011) Zeolite formation and deposits. In: Handbook of natural zeolites. https://doi.org/10.2174/97816080526151120101

Mata JM, Sanz J (2007) Guia d'identificació de minerals, .Catalan 2nd paper en. Edicions UPC/Parcir, Manresa, Catalonia. 262 p. ISBN 9788483019023. http://hdl.handle.net/2117/90445

Mindat. Zeolite group (2021) https://www.mindat.org/min-4395.html (last accessed May 2021).

Sanz J, Tomasa O (2017) Elements i Recursos minerals: Aplicacions i reciclatge, Catalan 3rd digital edn. Zenobita Edicions/Iniciativa Digital Politècnica, Manresa, Catalonia. http://hdl.handle.net/2117/105113

USGS (2021) Commodity Statistics and Information. Zeolites (natural) https://www.usgs.gov/centers/nmic/zeolites-statistics-and-information (last accessed May 2021).

Zeolite Australia (2021) https://zeolite.com.au/ (last accessed May 2021).

Zeotech Corporation (2017) http://zeotechcorp.com (last accessed May 2021).

Correction to: Elements and Mineral Resources

Correction to:
J. Sanz et al., *Elements and Mineral Resources*, Springer Textbooks
in Earth Sciences,Geography and Environment,
https://doi.org/10.1007/978-3-030-85889-6

The original version of the book was inadvertently published with incorrect author names in the online chapter metadata. This has been updated in Chaps. 25–36 and 89–101 as "Joaquim Sanz • Oriol Tomasa • Abigail Jimenez-Franco • Nor Sidki-Rius".

The updated version of this book can be found at
https://doi.org/10.1007/978-3-030-85889-6